# VOYAGE TO THE GREAT ATTRACTOR

# VOYAGE TO THE GREAT ATTRACTOR

## EXPLORING INTERGALACTIC SPACE

## ALAN DRESSLER

 ALFRED A. KNOPF   NEW YORK   1995

THIS IS A BORZOI BOOK
PUBLISHED BY ALFRED A. KNOPF, INC.

Copyright © 1994 by Alan Dressler

Library of Congress Cataloging-in-Publication Data

Dressler, Alan Michael.
Voyage to the Great Attractor: exploring intergalactic
space / by Alan Dressler.
p.   cm.
Includes index.
ISBN 0-394-58899-1
1. Great Attractor (Astronomy)   I. Title.
QB991.G73D74   1994
523.1—dc20      93-48580   CIP

Manufactured in the United States of America
Published October 16, 1994
Reprinted Once
Third Printing, January 1995

to my parents, Charles and Gay, for starting me on my journey
to my teachers, for pointing the way
to Jacob Bronowski, for showing me the goal
and to Stephen Sondheim, for keeping my spirit company along the way

# CONTENTS

# PREFACE

This is a story of science. Unlike many books about science, this one goes beyond just listing results and explaining them: it seeks also to tell a story of how people did the actual science. It is a story in which the hurdles to understanding new and startling discoveries about our universe were as much conceptual as they were technical, so it requires no special mathematical skills or knowledge, just the desire to think hard about something far removed from everyday experience.

True revolutions in human thought have been rare, but one occurred when astronomers discovered that we live not just in a galaxy, but in a vast universe of galaxies in flight from a remote primal cataclysm, a universe that was almost unimaginably different in the past, a universe that held the seeds for our own existence long before there were stars or galaxies. Scientists like me believe that this description is of a fundamentally different nature than the notions of our universe that we humans had evolved previously, as inspired and intricate as these pictures may have been, so that, in this sense, we have in this century discovered our universe. The word "discover" has fallen into disfavor with those who think it displays a human arrogance for giving meaning or significance to something that was already there. But discover simply means to make something known for the first time, a *human* activity, so it is perfectly sensible, and not pretentious, to say that the universe was discovered in this century.

Along with our new appreciation for its size, age, and contents has come an even more important understanding that our universe is evolving, and that we are a key part of its process. This knowledge will, I think, launch our species on a new adventure that is destined to change the lives of our descendants. After centuries of displacement and alienation as we have moved from the center of our universe to a remote galactic outpost, we will reclaim our place as a most astounding offspring of creation: a creature that can gaze out and *understand*. Our minds are returning us to the center of our cosmos.

A half-century has passed since the beginning of this new journey, yet we are not far from the trailhead, still groping our way in this world of new wonders. We are scouting new paths. A few will take us further along; from most we will turn back with little more than a better feel for the lay of the land.

This book tells of the search for the nature of the universe from the perspective of travelers along one such path. The research on which my colleagues and I embarked took us into the depths of intergalactic space, where we discovered that our Milky Way galaxy and its neighbors are moving toward a distant continent of matter, a mostly invisible mass decorated with thousands of galaxies. This result blended with work by other astronomers who were finding that galaxies are not smoothly distributed in the universe, but collected into huge *superclusters* of galaxies separated by vast, empty *voids*. The strength and texture of this pattern—the contrast between superclusters and voids—may well reveal the nature of matter and energy in the first moments of the big bang, events that predestined the universe in which we came to be.

The story begins with the present and the past—the organization of our expedition and its roots in the momentous work of our predecessors. I speak of galaxies, their discovery and the discovery of the universe they outline, and why and how seven scientists, I one of them, prepared to learn more about one particular type—*elliptical* galaxies. I describe a 1981 trip to the Las Campanas Observatory, where I made my first telescopic observations for the project, and tell you more about what astronomers do. The diversion that follows into the world of curved space-time is a necessary detour if one is to become comfortable with the idea of an expanding universe. Then I tackle the heart of the scientific work, from the first maps that showed the distribution of galaxies to be unexpectedly clumpy, to the realization that galaxies over a large volume of space are moving, drawn by the force of gravity to a "Great Attractor." As in other studies, this conclusion carries with it the implication that most of the matter in the universe is invisible— what astronomers call *dark matter*. I explore further the big bang model and the cosmic background radiation, the most persuasive evidence that this astonishing event actually took place, and what this "roar from the past" can tell us about that remote time. I proceed to assemble what we know, what we think we know, and even what we think we don't know into a telling of the story of creation—how the universe has evolved from the big bang to today. Finally, there is my indulgence, an opportunity to spout off about why I think this kind of scientific odyssey is a vital part of human experience, and part of our destiny.

Our particular journey was one of many steps to a clearer view of how "what is" came to be. By taking you along, I hope to reveal the kinds of things scientists do and how they think, and so that you can share the *sense of discovery*, the supreme delight of the human mind. At times the scientific discussions may be challenging; I've tried to keep jargon to a minimum, but you may need to refer to the glossary of terms occasionally. Scientists believe that you don't really learn something just by being told about it, but only after you mull it over and, sometimes, not until you wrestle with it. So this book is not just about what, but how and why as well, a chance to sense the true flavor of science and, at the same time, read a good story.

# VOYAGE TO THE GREAT ATTRACTOR

# INTRODUCTION

I was five years old when my father hoisted me to the eyepiece of the small refracting telescope in Hyde Park, on Observatory Road in Cincinnati, Ohio. It was visitors' night, and we had shuffled through a serpentine line for what seemed like a child's lifetime. Then I saw Saturn, suspended, shimmering, pale-gold, its powerful ring slicing the blackness of space like the prow of a sleek ocean liner. From that moment my other haunts seemed small and tame. The world opened up for me that night. The new house we had just moved into, at first huge and exciting, had now lost its magic. The unknown streets that had seemed to stretch forever had become well-trodden paths, and the woods were a fright only when I pretended them to be. The other houses and other children were all familiar now. I had conquered my own world, but here was something more. The night I saw Saturn, I knew, instinctively, immediately, that my parents had led me to a new, wider realm with many mysteries. I knew I would find many uncharted lanes and hidden paths, of which I was so fond.

A cathedral sky of blue-black silence vaulted over our heads as we awaited dawn on this March night at Las Campanas Observatory. Here on the northern coast of Chile, the Andes mountains thrust charcoal profiles against a sky as dark as the one that watched over generations of our ancestors. Paradoxically, it is the darkness of the night sky that allows the collective light of thousands of stars to illuminate the way of a nighttime wanderer. Only on a cloudy night is the blackness overwhelming—our steps become wary as if they were taken in a dark closet.

On this late autumn night, I had entreated my former teacher, now friend and colleague, Sandy Faber, an astronomer from the University of California's Lick Observatory, to take an unhurried look at the Milky Way. At this time of year, the center of our galaxy, which astronomers call the Galactic Bulge, stretches directly overhead as radiantly as a matrimonial arch strung with stephanotis, far surpassing the relatively murky view of the Milky Way afforded to residents of Earth's northern hemisphere. Sandy was seized by a powerful reorientation of place, as I had first been years before. The Milky Way commands the sky so completely that we accept, without a moment's doubt, that we stand, pitched at some odd, arbitrary angle, on a small world at the far reaches of a colossal wonder. This reality is immediate and self-evident.

We could, at that moment, truly sense that Earth is round and that the sky does not parade over us. Of course, five centuries have passed since courageous voyagers proved once and for all that which even the ancients had known. And of course it is the Earth that turns on its axis and faithfully circles the Sun—four centuries had passed since Galileo and Copernicus passed on that bit of wisdom that the ancients had also known but the world forgot. But the difference is great between intellectual knowledge and actually sensing that these things are true. We still speak of the Sun *rising* and the stars *passing* overhead; maps of our countries are flat and people are still puzzled by why the New York to London flight takes them over the arctic.

But now a fundamental change in the consciousness of the human mind is taking place, leading us to become creatures of a vast realm of space and time—a hyperspatial domain. Pictures of the Earth taken from space are finally forcing us to absorb, at long last, these things we have long *known* to be true. As the image of *round* Earth finally becomes ingrained in the human psyche, new explorers—again only an infinitesimal fraction of the planet's human inhabitants—travel their minds and senses through the Milky Way galaxy and beyond, preparing the way for future generations to assimilate the role of creatures of the galaxy.

Perhaps the impact of the Milky Way on Sandy and myself was spurred by our astronomical knowledge: our studies have left us with a vivid mental image of a dense coalescence of a billion suns that forms the center of our galaxy, a remote 20,000 light-years away. After all, this sight of the Milky Way was available to millions of our ancestors, who drew a less cosmic perspective from the experience. The staggering insight of the true immensity of our galaxy has only crystallized in the last century, although, to be sure, many conceived of it centuries earlier. Yet, hardly had there been a decent interval to become comfortable with living in a universe 100 trillion times bigger than our Earth, when the universe "grew" again, this time by another factor of 100,000. In 1924 Edwin Hubble, working with the 100-inch Hooker telescope at the Carnegie Institution's Mount Wilson Observatory, demonstrated that many *nebulae*, fuzzy patches of light in the night sky, are indeed other galaxies—far, far away, and as splendid as the Milky Way herself. These galaxies stretched on and on, as far as our newly grown eyes could see.

Nor was it long before this cosmic revelation was surpassed by perhaps the most amazing revelation of all time: the universe is expanding. Again it was Hubble, combining his new estimates of distances to galaxies with measurements of their speeds from Lowell Observatory astronomer Vesto Slipher, who was at the focus of discovery. In 1929 Hubble presented evidence that galaxies are receding from one another at phenomenal speeds that are in direct proportion to their separations. In the years to follow, as new observations proved the point beyond doubt, theoretical physicists provided a simple but marvelous model whose import is still virtually incomprehensible: the universe began with a violent, explosive event—a "big bang." To appreciate how radical the notion of a universe in motion must have seemed in the first decades of the twentieth century, one has only to remember that the exceptional mind of Albert Einstein rejected even the possibility, when his own general theory of relativity virtually demanded it.

In the tumultuous decade of the 1920s, our mental image of the universe had evolved from a vast, but single and comprehensible, star system to a seemingly limitless profusion of galaxies carried along in the rapid expansion of a virtually endless abyss of space. It would have been remarkable for such a radical change in perspective to have been fully comprehended, even by the few people who followed such work. A key step was still necessary to move from this cosmic abstraction to the comfortable integration of the concept. Just as it had required a picture of Earth from space to establish firmly in our minds the true image of our planet wheeling through the blackness, so too we would need pictures of our universe, on a grand scale, to begin to assimilate our place in this newly found reality. It was toward this goal that Sandy and I were working that night at Las Campanas. We were helping to chart the distribution of galaxies and matter in our corner of the universe, to move from general to specific in the perception of the larger "world" in which we live.

When, as a child, I became aware of other houses beyond my own, my consciousness of my surroundings grew. Yet, my own ease with this enlarged world developed only after I recognized individual houses and could follow a path through the neighborhood that led reliably to my own. In the same way, images of a round Earth from space become "real" only because we recognize bits of geography from the maps of our childhood, like the Florida peninsula or the British Isles. Now, at the end of the millennium, we prepare for our next journey. Fully roused from our awakening early in this century, we are dressed and fed, ready to walk the neighborhoods of galaxies.

Already, our first glimpses on these cosmic strolls rival the startling revelations of Hubble's time. We have found that galaxies trace complex patterns—a lacy tapestry here, a web of galaxies there—that weave in and out of large, empty voids. Such a structure points to deliberate work, early in the history of the universe, of powerful natural forces that carved an intricate pattern into the distribution of matter. With advances in the under- standing of the physics of elementary particles, we now see the possibility of reading the cosmos as the result of a unique experiment in high-energy physics, an experiment beyond anything we could hope to repeat on Earth. Combining observations and theory, we aspire to the fundamental questions of cosmology: What is the universe made of? How was it born? How has it led to our own existence?

In pursuit of answers to these fundamental questions, it is the task of observational cosmologists to supply theoretical astrophysicists with an accurate picture of the distribution of matter in space. In practice, this means to map the distribution of galaxies and, since galaxies appear to amount to but a small fraction of all matter in the universe, provide a map of *all* matter, even that we cannot directly "see." At the close of the twentieth century, a few dozen people on this planet have chosen this as their principal occupation.

# $I$ • DISCOVERING A UNIVERSE

In 1980 I joined a group of scientists who were mounting an expedition into the universe of galaxies. Our group prepared maps and made plans, as would any band facing an ambitious campaign into uncharted territory. Before us, in Santa Cruz, California, were photographs of the entire celestial sky taken with telescopes in California and Australia whose specialty was to perform as giant cameras. Our job was to select from thousands of galaxies the few hundred our expedition would visit. Sorting through these charts, we ruled in and out whole galaxies with their billions of stars, as easily and yet as carefully as one composes a guest list for *the* social event of the year.

Our routine preparations seemed to pay little homage to the enormous philosophical implications of the journey—many might think it presumptuous or even arrogant to parse galaxies as if looking for ripe peaches. However, we were doing no more, and no less, than applying the human mind's remarkable ability to abstract. Galaxies are vast systems of stars bound together by gravity: they are the fundamental organizational unit of matter in the universe. Though each galaxy could be described by any one of a hundred parameters, the combination of which make it as unique as a snowflake, we were selecting just a few characteristics that would allow us to include or exclude it from our list. We would replace a whole galaxy with a few symbols or numbers, in that facile yet powerful stroke choosing to ignore a wonder that could in itself absorb many lifetimes—a detailed accounting of its billions of stars, an appreciation of its vast size and energy. A hundred years of work by our predecessors afforded us this luxury: we could start from their outpost and push the journey a bit farther into the outback of time and space.

Each of us knew a lot about galaxies, and each had long since found some personal accommodation with dimensions so staggering and power so phenomenal that human reason is pushed to its limit. Somehow this had left us free to relegate each galaxy to a file card and a number. On the card we would record each galaxy's address and, paring down our curiosity to just a few essentials, like a galaxy's shape, size, and color, decide whether to include it on our "must-see" list.

So on our itinerary we would designate a stopover of an hour or two with the telescope, hopscotching to the next galactic marvel without more than a moment's existential doubt. Our reason for this apparent disrespect was simply that our goal on this journey was to learn more about how *all* galaxies came to be, and what they might teach us about how *everything* came to be. Here, at base camp, we could not afford to linger and contemplate the spectacle of the ascent so far. We would set aside our sense of wonder in search of a greater one.

The dark of a moonless night would be fearsome if not for the diamond and ruby radiance of the stars. For hundreds of thousands of years our ancestors looked into this sky, to this other world, and saw these same stars. To be sure, they saw other wonders of a more temporal nature, like the careful meander of the planets, the flash of shooting stars, the awe of an eclipse, but these seemingly eternal fires—the stars—were undisputed lords of the night sky. In an Earth-world of complex rhythms and sporadic catastrophe, the fidelity of the stellar night sky played a central role in the development of the human psyche, providing a perpetual diary into which the account of generations of human experiences could be written.

Engaging as they are, these few thousand brightest stars tell a highly censored version of the story of the universe. Perhaps it is fortunate that a full view was withheld for so long from human eyes, since an early confrontation with the gargantuan size and ferocious dynamism of the universe might have rocked the secure cradle of anthrocentrism in which humankind was coddled through its infancy. Like the child in Wordsworth's ode "Intimations of Immortality from Recollections of Early Childhood," our species matured under the illusion of special importance, in a universe where nature endures and provides—a parent and guardian for life on Earth.

> Behold the Child among his new born blisses,
> A six years' Darling of a pigmy size!
> See where 'mid work of his own hand he lies,
> Fretted by sallies of his mother's kisses,

With light upon him from his father's eyes!
See, at his feet, some little plan or chart
Some fragment from his dream of human life,
Shaped by himself with newly-learned art:
   A wedding or a festival,
  A mourning or a funeral;
     And this hath now his heart,
    And unto this he frames his song:
      Then will he fit his tongue
To dialogues of business, love, or strife.

. . . The Clouds that gather round the setting sun
Do take a sober colouring from an eye
That hath kept watch o'er man's mortality.

After ten thousand generations of myth and superstition, the discovery of the real universe has been traumatic; coping with this awakening demands full use of our imaginations. Without question, it is well within the power of our minds to embrace this new cosmology—were we incapable, we would not have come this far—and it is surely to our benefit to widen our vistas in order to understand better human origins, human life, and the promise of human destiny.

When Galileo turned his small telescope to the night sky, he discovered that the universe did not live up to the crystalline perfection advertised by Aristotle. There are jagged mountains on the moon and violent storms on Jupiter. Galileo's discovery was indeed earthshaking, for with it came the revolutionary idea that the world beyond is made of the same stuff as the Earth. As our vision of the universe has continued to enlarge, this observation has been confirmed time and again. Perhaps the most stunning demonstration followed the invention of the spectrograph, an optical instrument that allows us to separate light into thousands of finely pared colors. With it, nineteenth-century physicists identified previously hidden spectral (color) patterns that belong uniquely to each atomic (chemical) element. When spectrographs were placed at the foci of telescopes and trained on celestial objects, they revealed the same spectral patterns, in this way proving that the distant stars are composed of the very same elements found here on Earth.

Galileo also observed that the Earth-Moon dance is copied by Jupiter and Saturn with their many moons. A century later, Isaac Newton described these minuets with a simple mathematical description of gravitational force that explains, with a single stroke, an apple falling to Earth, the orbits of moons around their planets, and the circling of planets around the Sun. Not only is the *substance* of nature the same throughout the universe, but so are the *rules* of nature: the four fundamental forces that govern all interactions of matter and energy—gravity, the electromagnetic, strong, and weak forces—are *universal*. This message too has been received again and again, until finally we have realized that these conclusions are little more than tautologies: the universe is exactly that realm over which substance and rules remain the same. Something wholly different would be unrecognizable and would be, by definition, excluded from our perception and understanding. There may be other universes, but we cannot hope to do more than dream or speculate about them within our present understanding of reality.

For the present, this universe seems to hold enough wonders to fend off any risk of boredom, as can be verified simply by taking a pair of binoculars and scanning around the night sky. Immediately one learns firsthand about another of Galileo's great discoveries—there are many more stars than were counted by our ancestors. A good place to look is toward the constellation Ursa Major—the Great Bear—with its familiar tail and hindquarters we call the Big Dipper. The brightest dozen stars are those which can be seen with the unaided eye, but even common binoculars will reveal hundreds more. This is because it takes but a small lens to gather and focus far more light than the pea-size aperture of the human eye, a fact that makes a far greater contribution to astronomy than a telescope's ability to make things look *bigger*. When astronomers build larger telescopes it is rarely to achieve greater magnification, but rather to gather more light so that fainter objects can be identified and studied.

With his deeper view of the sky, Galileo was probably the first human to realize that the stars number not in the thousands but in the millions, at least. In particular, he observed that the Milky Way, a diffuse, cloudlike glow that rings the sky, resolves into myriads of faint stars. We take the name, and also the word "galaxy," from the Greek description *galaxias*

Progressively deeper views of the universe. The top picture, taken by the author with a 35mm camera, shows the familiar part of the stellar constellation Ursa Major known as the Big Dipper. The middle picture, from the original Palomar Sky Survey (courtesy Caltech and The National Geographic Society), shows a small area in the top photo enlarged, with the deeper view afforded by the 48-inch mirror of the telescope. A prominent galaxy is now visible. (Other galaxies, invisible in the top photo but within its boundaries, are shown alongside, with the same enlargement as the middle photo. The object at lower right is not a galaxy but a true nebula, called "The Owl," of glowing gas in our Milky Way galaxy.) A yet smaller part of this middle picture, imaged with the COSMIC camera (designed and built by the author and Bill Kells) on the 200-inch Hale telescope, shows thousands of objects, most of which are galaxies. The bowl of the Dipper encloses about one million galaxies.

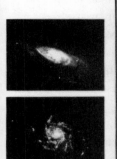

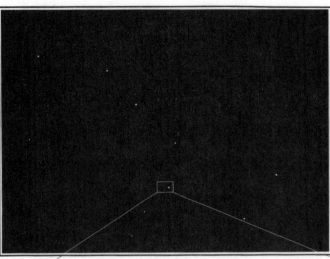

THE BIG DIPPER AND
ENLARGEMENTS OF
GALAXIES WITHIN

A SKY SURVEY
PHOTO SHOWING
MORE STARS AND
GALAXIES

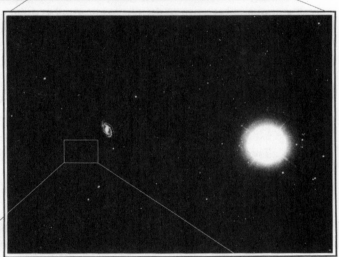

A DEEP "CCD" IMAGE
SHOWING THOUSANDS
OF FAINT GALAXIES

*kyclos* (milky circle). This luminous band girdling the sky is the combined light of distant stars collected into a flat, disklike shape, each star by itself too faint to be seen. The brighter stars, which in contrast appear in *all* directions in the night sky, are relatively nearby compared to the distant stars of the Milky Way disk; they seem to fully surround us because the disk of the Milky Way galaxy has some thickness and we with our Sun are at present near the middle of the slab.

The notion that our Sun is a member of such a vast star system had been popularized by eighteenth-century English naturalist and theologian Thomas Wright, but German philosopher Immanuel Kant showed astounding pre-science when he correctly deduced the true saucer-shape of our galaxy, and speculated that there might be other such "island universes." Only one of these other galaxies, the one peering through the constellation Andromeda, is visible to the naked eye, so the discovery of other galaxies had to wait for the widening of our eyes with bigger optics. Looking deeper into the Big Dipper's bowl uncovers more stars, of course, but previously hidden, ex-tended objects also begin to emerge. Whereas images of stars are pointlike, even the newly revealed fainter ones, these extended objects spread a dimmer glow over a broader area—an astronomer would say they have lower "surface brightness" than the sharp stellar images.*

These extended objects are galaxies, the huge systems of stars that Kant imagined. The abundance of extended objects in the sky became apparent soon after the invention of the telescope; by 1790, British astronomer William Herschel had carefully cataloged the positions of thousands with his home-made reflecting telescopes, the world's largest. Other observers had consid-ered these fuzzy blobs as annoying distractions in their searches for new comets and had recorded positions of nebulae only to avoid mistaking them for what they considered more worthy targets. Not so Herschel. His passion for the night sky extended with special devotion to these mysterious, ghostly glows. Herschel noted that his catalog of "false comets" included radiant globs like the Orion nebula, a gas cloud studded with luminous, young stars, but it also included thousands of objects with more symmetric forms, from perfectly round disks to very flat ellipses. Like Kant, he speculated that these might be wholly other star systems. To a person gazing into the eyepiece of a telescope, even a very large one, the images of such objects are faint and

---

*Stars appear to have size in a picture only because the image "grows" on a photographic emulsion when an area is flooded with light. Likewise, the famous cross structure on a star often depicted in artists' renditions is also an observational artifact generated by the diffraction of light around the structural crossbraces in the telescope.

indistinct—even today visitors to observatories grumble with disappointment at how little they can make out. However, photographic plates, first used in the late nineteenth century, can *accumulate* light and thereby show incredibly more detail: delicate spiral patterns were soon detected in many objects.

The story of how astronomers finally discerned the existence of other galaxies is often retold, probably because it includes an actual, public debate, for which two scientists were asked to serve as advocates for competing theories—a popular, but nonetheless false, conception of how science proceeds. The disagreement over the true nature of the extended objects was, by the beginning of the twentieth century, so well developed that it was almost a tradition. One group of astronomers contended that all such nebulae were glowing clouds of gas *within* the confines of the Milky Way. Some holding this opinion believed further that nebulae with spiral patterns were nativity scenes of individual stars and their surrounding planets. A contrary view was held by other astronomers who thought the evidence stacked up in favor of the view put forth by Kant and Herschel—that the more regularly shaped nebulae were entire star systems well beyond the Milky Way. These two very different cosmologies, one that the universe *is* the Milky Way, one giant star system, the other that the universe is a profusion of Milky Ways, each a vast star system, were the passionate positions of Harlow Shapley, from the Carnegie Institution's Mount Wilson Observatory, and Heber Curtis, from the University of California's Lick Observatory. The two debated the issue at the meeting of the National Academy of Sciences in 1920. Of course, neither side succeeded in convincing the other—if there had been decisive evidence the scientific community would have known about it and a debate would have been unnecessary. But in 1924 a young astronomer at the Mount Wilson Observatory, Edwin Hubble, provided new data that outflanked the battle lines of conflicting evidence that had held the subject in years of stalemate.

Over a period of four years, Hubble had used the new Mount Wilson 100-inch telescope to take dozens of photographs on glass "plates" of the Andromeda nebula, the largest of the spiral type. For the most part the image of the nebula was a soft, diffuse glow, but, upon close examination, Hubble identified faint starlike images peeking through. If the hypothesis that Andromeda was a giant, distant star system like our own Milky Way was correct, these faint points could be the brightest stars. The key discovery came when he noticed that some of these starlike images varied in brightness from exposure to exposure. Stars like these, whose light output repeats the same pattern over a period of anywhere from days to months, had been found in large numbers in the Milky Way. Henrietta Swan Leavitt, a research assistant at Harvard College Observatory, had further discovered that the intrinsic luminosities of these Cepheid variable stars (named for a moderately bright star of its type in the constellation Cepheus) are directly related to the *time* it takes them to complete one cycle from bright to faint and back again—brighter Cepheids have longer cycle times than fainter ones, by a predictable amount. This was a breakthrough, since once the *intrinsic* brightness of some light source is known, its distance can be calculated from a comparison with its *apparent* brightness, from the simple rule that light intensity falls off as the square of the distance.

The characteristic variation in light of a Cepheid star is a stellar dog-tag that allowed Hubble to identify this breed as the same pedigree found in the Milky Way. As a result Hubble knew immediately that these Cepheid variables lay well beyond the Milky Way (even before he applied the detailed method of deriving distances), because the Cepheid variables of the Andromeda nebula are *thousands* of times fainter than those in our galaxy. This extreme faintness could be explained only as a result of great distance—Andromeda is not just a "nebula" but a *galaxy*, a hundred times further than the most distant parts of our own Milky Way.

The argument was settled and the universe, for humans, would never again be the same. Four hundred years after the great explorers circumnavigated the globe, a new world had appeared on the horizon. Its galactic continents receded into distant seas seemingly beyond human reach, but voyages of wonder awaited those who would try.

The de facto leader of our group of six astronomers was thirty-six-year-old Sandy Faber. Her youthful enthusiasm and cheerful optimism, along with a complete absence of pretense, belied the fact that she was already considered a leading "extragalactic" astronomer (one who studies objects outside the Milky Way) and a world expert on the populations of stars that make up galaxies. Sandy's passion was to get to the bottom of things, and she had both the brains and drive to do so. Her gift for synthesis—bringing people together, merging ideas, combining observational data sets, and disciplining herself to master complex theoretical treatments—led to new insights and sometimes even resolution of major puzzles.

The careers of four of us had benefited from Faber's tutelage. David Burstein and I had been among her first doctoral students. We were well acquainted with her vigor and ambition, and long past the point of submitting second-rate work for her comments. Dave had a special talent for details—he knew the names and numbers of hundreds of galaxies and who had

The Great Spiral galaxy in Andromeda, the closest major galaxy to our own Milky Way. *(Photo courtesy of the Carnegie Observatories.)*

measured what about every one of them; the data he himself gathered was of the highest quality. He had a proclivity for developing involved explanations based on the hundred different facts and figures that only he could hold in his head, and was not shy about buttonholing a colleague for an hour or two to make sure he or she got the point. Few could follow, and fewer wished to endure, these complicated expositions. Nevertheless, as in oysters, these irritations sometimes brought forth pearls, of insight. Burstein and I had clashed regularly in our student days—I had found him pushy and presumptuous, he thought me arrogant and, later on, overrated. There was doubtless some truth in both claims, and worse head-butting lay ahead. Nevertheless, a certain mutual respect for abilities prevented this clash of personalities from seriously injuring the collaboration, and Sandy, who rarely let her own emotions show or interfere, had a knack for arranging post-skirmish peace talks along the scientific battle line.

Roberto Terlevich was a self-imposed political exile from dictatorship. He had transplanted his tempestuous Argentine soul to more stable English soil. A schoolboy smile and mischievous eyes would bring oft-needed humor to the long years of our project—for me, Roberto was like a Franklin stove in the corner, ready to provide warmth when it was most needed. His scientific endeavors were also broad-brush and colorful, often unconventional, and he delivered his ideas and conclusions with gusto. Our study had really been born when, in the summer of 1979 at Santa Cruz, Roberto had paraded in front of Sandy his Ph.D. thesis work on regions of copious star formation in galaxies. Following a chain of connections and analogies, the discussion had spawned a great swell of optimism that properties of one entire class of galaxies—those called "ellipticals"—might mimic the properties of these star-forming regions and be much more systematic and predictable than had been thought.

Roberto had come to visit Roger Davies, a recent recipient of a Ph.D. from Cambridge University in England, who had received a prestigious Lindemann fellowship for postdoctoral study overseas and had written Sandy and asked whether he might spend a year in Santa Cruz. Roger had been studying elliptical galaxies from the point of view of the motions of their stars, and he knew that Sandy and Dave had been making extensive observations to learn the history of star formation in these selfsame galaxies. If Roberto had at least some of the stereotypical Latin characteristics, Roger fit the well-educated, reserved, young British gentleman even better. His impeccable manners and gentleness offered no clue that he came from a solid working-class background in the industrial north of England. Meticulous and accurate in his work, Roger was a steadying influence—even his rare emotional outbursts were remarkably civil. (Since mine were not, I would come to find this annoying.)

This scientific salad of Terlevich, Davies, Faber, and Burstein blended ideas and talents in the quest of understanding one quirky corner of nature's quilt. They found a regular pattern in what had appeared to be incoherent data, and hoped that, with this new insight, a clue to the way galaxies had formed was within their grasp. Galaxies had long been recognized as building blocks of the universe, and many astronomers had concluded that understanding the conditions and processes under which they formed would bring us closer to understanding creation itself. Though it seemed hopelessly optimistic to believe that an entire galaxy, a complex confederation of hundreds of billions of independently evolving stars, could be characterized by a few simple parameters, many astronomers believed that elliptical galaxies offered just that hope. They had the simplest anatomy—just a round ball of stars, as well as the promise of an easier dissection—only old stars, no young ones. To try to understand the complexities of galaxies, this seemed like the place to start. The work of Terlevich, Davies, Faber, and Burstein would raise this hope further, but, as it was based on a pitifully poor sample of just twenty-four galaxies, few astronomers would take their conclusions seriously. Thus, the four would set out, as astronomers often do, to forage more deeply into space, to find *hundreds* of elliptical galaxies for which measurements could be made and the hypothesis tested. They agreed to embark on an ambitious enterprise that would test not only their hypothesis, but themselves.

It would require tremendous effort. When he returned to England, Roberto Terlevich persuaded Donald Lynden-Bell to join the project. While the others specialized in observations and their interpretation, Lynden-Bell's strong suit was theoretical physics and mathematical analysis of astronomical phenomena. He was the senior member of the group, and his thirty-year career had produced seminal work on the formation of galaxies: Donald had been among the first to model the complex dynamics of multibody systems bound by gravity and to understand their evolution through rapid contraction. Too many theorists reminded observers of campaigning politicians—they would as soon espouse one position as its diametric opposite, and would beam with pride if they could do so without the bother of a second breath. Fortunately, Donald was not one of these; he was the kind of theorist you could track and pin down—he followed his ideas with fervor and commitment, still wrestling with those he had first had as a college student. He was outspoken and a bit of a maverick, with a flair for bold strikes, such as his assertion that massive black holes—places where matter is so densely packed that intense gravity pulls in space itself—provide the energy sources for quasars, the most luminous beacons of the universe. His talents in mathematical methods would be invaluable to the group, though he would struggle to overcome his inexperience with modern computer analysis and observational techniques. A member of the "old school," Donald embodied academic formality; in social situations, he often appeared uncomfortable and stiff, but this awkwardness was well countered by abundant good humor and sincerity.

One day in late summer, 1980, a few months after the five had begun to work in earnest on the project, Sandy Faber described to me the group's ambitious plan. They would try to increase the sample of elliptical galaxies from a pitiful 24 to about 400, making accurate, telescope-intensive measurements for each object. A project of this scale was unprecedented in studies of galaxies. Furthermore, it was vital that a truly representative sample of the class was selected in order to assure that the first result was not just a statistical fluke. As in choosing citizens at random for a nationwide poll, great care is necessary in order to assemble a fair sample, one that includes, for example, those who live in heartland or suburbs as well as in cities. Likewise, it was imperative to select galaxies from all parts of the sky and choose them according to uniform criteria, since it was unknown if these properties might vary with location, environment, or history.

It was the goal of complete sky coverage that was to draw me into the project. Sandy worried aloud about gaining enough access to telescopes in the Southern Hemisphere. Telescope time is a scarce resource: astronomers typically apply six months to a year ahead, and usually get only a fraction of the time requested, perhaps a handful of nights over an entire year for one of the larger telescopes. To make matters worse, there are fewer telescopes placed to observe the southern sky. But, a few years back, the Carnegie Institution had commissioned a new 100-inch telescope at the Las Campanas Observatory in Chile, and I had been receiving about fifteen to twenty nights a year for various projects. Sandy knew that, with some of those nights, it was much more likely that the project could be completed within a few years.

However, I was reluctant to join such a large group. Mostly I had worked solo, a pattern that began in 1974, in the "youth" of my career, when my first thesis adviser, Joe Wampler, left the Lick Observatory and me, his student, to become director of the Anglo-Australian telescope. For two years I worked essentially alone. Though Sandy guided me through the completion of my thesis in 1976, Wampler's departure had given me a degree of freedom, unusual for a graduate student, that I had come to relish. I had been happy to continue working mostly on my own in the four years since, in part because I was trying to build a reputation based on my abilities alone, but more and more because I found that I genuinely enjoyed the responsibility of carrying through each phase of the work.

On the other hand, this was a good project and one that interested me. I wanted to get involved, but hoped I could play the role of a silent partner. So I suggested to Sandy that I contribute some observations gratis. But Sandy was persistent in her belief that my full participation would be a benefit, and she argued that, in particular, my experience in my own specialty—clusters of galaxies—would prove useful. At the time I thought this was simply flattery, but she was right: the cluster work would turn out to be crucial. Soon Dave Burstein was calling me, every other day it seemed, extolling the importance of the project and urging me to join the team. Ultimately, deciding I might as well, I rationalized that I could leave my contributed data with the group and exit gracefully if I became frustrated or dissatisfied with the collaborative aspects. It was with this rather tentative commitment on my part that the group grew to six.

With Hubble's demonstration that Andromeda is a galaxy as magnificent as our own, our perception of the size of the universe had grown immensely. Our view into space has continued to deepen because of telescopes with larger mirrors, more sensitive photographic emulsions, and, more recently, the development of electronic detectors a hundred times more sensitive than the best photographic films. Modern telescopes and instruments easily reach the most distant stars in our galaxy, so our deeper probes fail to add stars in burgeoning numbers. Instead, the night sky fills with the faint, fuzzy glows of more distant galaxies, whose limits we do not easily reach. As bigger telescopes open wider our eyes, more galaxies appear. Hundreds, then thousands of galaxies, each as magnificent as the Milky Way, float up from the depths of space. Finally, when probing a relatively small area with the most sensitive detectors and the biggest telescopes, the images of distant galaxies become so crowded they appear almost to touch. Our view has reached nearly the full depth of the universe as we know it.

These far-flung galaxies are only a few arcseconds in size.* It is sensible

---

*Each of the 360 degrees around the sky circle is divided into 60 minutes, and each minute is divided into 60 seconds, thus an arcsecond covers 1/3600 of a degree. The Moon and Sun each cover roughly 2,000 arcseconds, approximately ½ degree. The length of your thumb, with arm outstretched, covers about 5 degrees; to appreciate that the Earth is, in one sense, quite small, consider that Southern California, if propped upright so that it could be seen from New York, would appear this big.

to assume that these distant objects are actually as grand as the nearby Andromeda galaxy, which covers several degrees on the sky. If so, the fact that they appear 1,000 times smaller means they are 1,000 times farther away than Andromeda, a vast distance, to be sure. Furthermore, an imaginary sphere stretching to one of these distant galaxies has 1,000 times the radius, and with it 1,000,000,000 times the volume (a multiplicative factor of 1,000 for each of three dimensions) of the sphere that reaches only to Andromeda. If the number of galaxies per volume in our region of space is typical, we can estimate that the larger sphere encloses 2 billion galaxies rather than 2, and this is still at least a factor of about ten shy of the volume of the visible universe. In fact, the bowl of the Big Dipper alone frames about one million galaxies—something to remember when you're out for a walk under its generous scoop.

Our journey beyond the Dipper has uncovered another remarkable fact about the universe, one so obvious that we might not take time to appreciate it: the universe is transparent. It could easily be otherwise: visible light is easily absorbed by minute quantities of sootlike particles that astronomers refer to as *dust*. For example, visible light from the stars in the central regions of the Milky Way is absorbed before it reaches us, otherwise the bulge of our galaxy would treat us to a magnificent glow. Alas, we see instead dark rifts along the zone of the Milky Way; here gas clouds with a small amount of dust have blocked out all the light behind them. We can surmise, then, from the fact that galaxies are seen to such great distances, that the space between galaxies is quite empty, even emptier than the space *inside* the Milky Way. If stars were forming between the galaxies, there would be telltale signs of the accompanying gas and dust; also, dedicated searches have failed to turn up such galactic interlopers. It seems that stars form only *within* galaxies: this is a clue to how the universe we see today developed. Because of the remarkable clarity of our view, all the visible universe is laid bare before us and, as a consequence, all of its history as well.

To appreciate the scope and breadth of the universe uncovered in the 1920s is a challenge, even in the 1990s. A first step is a clear mental image of what the universe looks like on the large scale, from a point of view as vast as a galaxy itself. One starts with a truly empty, black void—there are no stars floating between the galaxies, in *intergalactic* space. One imagines floating above the vast pinwheel of the Milky Way, free to swoop around with ease from its central bulge of starlight to its outermost spiral arms. Close at hand, only as far away as the Milky Way is across, are two ragged collections of gas and stars. These are the two Magellanic Clouds, each a galaxy in its own right, though less than one-fourth as large as the Milky Way galaxy that holds them captive by its gravity. Though Andromeda looms no larger in the sky, it is obviously a galaxy as grand as the Milky Way, another disk of huge dimensions. From our godlike perspective, the Milky Way and Andromeda can be seen wheeling silently, two giant saucers separated by about twenty times their size. At a distance yet again as much is a smaller pinwheel, more delicate and almost frail in the thinness of its disk, called Messier 33, the name remembering the French comet-hunter who first cataloged the glowing nebula of the night sky. The darkness between these three giants is occasionally broken with lesser galaxies, some small, faint puffs of gas and stars, and a few barely visible round orbs.

The disappointing name for this magnificent collection of galactic wonders is *The Local Group*; similar associations of a few major galaxies, each with its entourage, can be seen not much farther out. Together, dozens of these groups combine to make the Local Supercluster; its core is a "town center" of galaxies called the Virgo Cluster, about twenty times as far away as Andromeda. Beyond, huge darkness looms, as empty as the Local Supercluster is full. Beyond this sprawls another great supercluster with an urban center or two, and another, and another.

How far does it all stretch? If our galaxy were the size of a small farm, with Andromeda and other galaxies neighboring farms just a hop down the road, then our universe would be about the size of the Earth. If we can imagine strolling to a neighbor's farmhouse, across the fields that represent intergalactic space, then we can also conceive of walking around the world— our universe. Scattered over the countryside are villages and towns where hundreds of these "houses" can be seen nestled together. Our model universe is like the rural countryside of the Middle Ages—great cities with their millions belong to the future.

The preparation that the six of us were making for our journey into deep, extragalactic space was a normal activity for observational astronomers. Of course, what an astronomer actually does with a telescope, and what an astronomer is doing when away from a telescope, is not a subject that preoccupies the daily thoughts of most people. It is not surprising, then, that when they meet an astronomer, most people shuffle with some embarrassment around the question they would most like to ask, "What is it you actually *do*?" (accompanied, if it happens to be after dark, with a look that always translates to "Why aren't you at your telescope?"). In fact, astronomers who specialize in observing with ground-based telescopes typically spend at most a few dozen nights per year doing so. For one thing, as I said, telescopes are expensive to build and operate, and, as a result, observing time is in short supply and has to be shared or competed for. Also, observatories tend to be at remote sites where weather and atmospheric stability is far better than average, and where skies are unpolluted by city lights or smog. Travel to such places usually takes one halfway across a continent, if not to another continent altogether.

Compensating for the scarcity of observing time is the wealth of information that now pours out of the electronic instruments at the telescope focus. Whether direct "pictures" of a tiny section of the sky, or a spectrum of the light arriving from a specific planet, star, or galaxy, each measurement is normally in the form of an array of millions of numbers that must be processed, extracted, and "reduced" using a computer, before the sought-after properties can be distilled. After an *observing run* of a half-dozen nights, an astronomer might spend a few weeks processing the data, many months understanding the information they contain, and as long as several years to fit the new observations into what is already known and to write a scientific paper reporting the results and conclusions.

Asked with less embarrassment, but more deserving of it, is the oft-heard "Discovered any new stars [or planets or galaxies] lately?" (Caution: May be accompanied by a sharp slap on the back.) This reflects the common misconception that astronomers peer into telescopes night after night, hoping to stumble across something never before seen. Of course, there are many transient phenomena—the brilliant appearance of an erupting or exploding star (a "nova" or "supernova"), or the discovery of a previously unknown comet or asteroid, that are important astronomical events. However, because they are so rare, their discovery usually requires a long-term, systematic search, or the good fortune of serendipity—regardless, these account for only a small fraction of astronomical research. As for the discovery of "new" stars or galaxies, by which the questioner usually means "unnamed," these are as plentiful as sand on the beach: naming sand grains or claiming to have discovered new pebbles in a quarry would make as much (or as little) sense.

What, then, are astronomers doing at their telescopes? They are recording the positions and analyzing the light of "families" of celestial objects—those with common features—and asking questions like: Why do some stars vary in brightness or even explode while most stay remarkably constant? (For that matter, why are there stars at all?) Why do some galaxies display a tightly coiled spiral pattern while others are smooth and symmetric? Why are some red nebulae studded with bright, blue stars, and others not? Questions like these are the wellsprings of scientific curiosity. They sound more approachable in an earthly context: Why are some mountains volcanoes while most are not? Why does rain pour from puffy low clouds and not from wispy high clouds? Why are plants green and animals not? But they are the same kinds of questions, a common feature of all scientific work. Such puzzlements surface from the gathering together of observed phenomena into groups; seeking their solutions is how we search for the underlying physical properties that made these things what they are. In astronomy, this process has been remarkably successful in uncovering fundamental laws of physics, because conditions in the celestial realm are far more extreme and varied than those found on Earth. With knowledge of these rules of nature comes the possibility of understanding how the universe, with its galaxies, stars, and planets, came to be, and provided the place and the parts for making creatures like us.

Like all great accomplishments in art and science, the discovery of galaxies was the result of many lives in pursuit of an idea; nevertheless, it is Edwin Hubble who will be remembered as the person most closely bound to this monumental step in our perception of the universe. Hubble's own description in ''Realm of the Nebula'' gives no details about his process of discovery (Hubble doesn't even use the first person when describing his own work) so, unfortunately, the reader does not gain any sense of what it must have been like to be the first human actually to know that the universe is vast beyond expectation. Unfortunately, this detachment was characteristic of Hubble, a legendary scientist but not a great humanist. To his colleagues and the world he presented a formal, controlled persona purged of emotion and uncertainty. Nor was he known for his humility, perhaps, as Dorothy Parker might have quipped, ''because he had little to be humble about.''

Because he kept his emotions private, we will never know whether Hubble was moved to the fidgety excitement so many scientists have felt when they realized they had taken a step forward in understanding nature. Perhaps the most profound reward for a scientist is to weather those nervous moments of near giddiness, a little like the first blush of romantic infatuation, that accompany the disclosure of some previously hidden facet of nature. For hours or days, sometimes even months or years, the scientist guards the secret, torn between a desire to share and a determination to hold off, until reasonably certain that there are no errors of observation, method, or interpretation. When released to colleagues, the new finding is likely to be stressed and pounded by worthy and unworthy challenges to its validity, in direct proportion to its perceived importance. If confirmed, it is set in place like a tile in the mosaic of human knowledge, further revealing the pattern and leading to the placement of other pieces. This artwork is nothing more and nothing less than a human representation of nature's own pattern. The goal of representing nature in human terms is the very heart of both science and art, a little-appreciated commonality that demonstrates the unity of human experience and enterprise.

The universe was redefined in 1929; the human-centered cosmos suffered its final collapse. The recognition of other galaxies, along with Harlow Shapley's discovery that our Sun resides far from the center of the Milky Way, made it clear that our location in this vast universe provides us little in the way of special credentials. Our Earth is "just another planet" orbiting the Sun—"just another star"—circling the perimeter of "just another galaxy." This knowledge has had a disturbing effect on collective human consciousness, but it is only a first, disheartening step in a discovery process that will, I believe, eventually recover a deeper sense of belonging. A child's discovery that his is not the only house, that there are other streets with other houses, even with other children, is the first blow to his sense of identity, but also the first step toward a *true* identity. Less than a century after humans discovered the universe, they are still unsettled by the implications of that discovery, but perhaps closer to something more valuable. At long last, we had learned the true "where" of our existence, and we would soon find the answer to "when" as well, clearing the way for us now to ask who and what we are, and, maybe someday, why.

Hubble himself had a next question, one whose surprise answer was so mind-expanding that even he could not believe it.

# 2 • BABY STEPS

An expedition usually begins with a map and, many a time, returns with a better one. The six of us, soon to be seven, were out to study the round balls of stars called elliptical galaxies, and hoping to visit all of them within a certain distance of the Milky Way. For this we needed a map, because pictures of the sky showed us only the directions of galaxies, not how far away they were. The ideal sample would then have been to scoop out all the elliptical galaxies in a volume of space, what astronomers call a "volume-limited sample," because it is often the best way to pull in a fair representation of the population. Consider how a political poll that surveys only apartment dwellers, or only those on small farms, might give a poor reading of an upcoming national election. Of course, even a volume-limited sample isn't very reliable unless it includes a representative range: a poll covering only the Loop in Chicago or a seventy-mile circle around Springfield would be much less useful than a poll sampling all of Illinois.

For our "poll" we needed to know how deeply we were probing into space. With no fenceposts or survey tracts to guide us, we would require the distance to each elliptical—the galaxies themselves would have to define the dimensions of the volume. To find these distances we would rely on another of Hubble's legacies, the *redshift-distance* relation.

In his measurements of the distances to the very nearest galaxies, using Cepheid variables, Hubble had pegged the first few markers on a three-dimensional map that showed the direction of each galaxy on the sky *and* its distance from the Milky Way galaxy. By measuring the apparent brightness of a member star of known luminosity, Hubble was using what astronomers quaintly call a "standard candle" to measure the distance. Unfortunately, this term was a little too apt: Cepheid variables are mere candles among stars—some stars are a hundred times brighter. Hubble wanted to push farther into space than the few nearest galaxies, but Cepheids were not bright enough to be individually detected in more distant galaxies, at least within the technological limits of his time. He had no choice but to try brighter standard candles, like the very brightest stars in each galaxy, in the hope that these all shine with about the same intrinsic luminosity. It was, as scientists say, a working hypothesis, one that would be justified if the strategy led to sensible results.

What Hubble was doing could be compared to trying to map your neighborhood without ever leaving home. Imagine that each house or apartment building has at least one large window and that the occupants can be seen walking in front of that window from time to time. In effect, Hubble was assuming that the tallest person to walk by is always about the same height. Actually, this is a pretty good working assumption, accurate to about 10 percent for male adults. Children have been effectively eliminated by considering only the tallest person in the house (unless the parents aren't home), so the "standard yardstick" is a fairly constant one.

The estimation of distance is now simple. One holds a ruler at arm's length and measures their apparent height, in millimeters, for example. The most distant people, those in the farthest houses, will appear smaller in direct proportion to how far away they are: those twice as far away will appear twice as small, three times as far away, three times as small. Now, not only the direction of each house but also its distance can be plotted on a piece of paper—this is a map. True, there might be no absolute scale to the map, for example, 1 inch = 400 feet, just "this house is two and one-half times farther than that one." However, these arbitrary units could be turned into *real* distances if the distance to at least one of the nearest houses is known or can be determined, since the map is *to scale*. (The house across the street is one hundred feet away, so the one twice as far away is two hundred feet away.) Since pacing the distance to Andromeda is out of the question, Hubble was betting he knew the actual height of his neighbor across the street, and the uncertainty that still exists today in the distances to galaxies can be compared to an argument about whether the people in the houses are residents of Chinatown or Little Norway. But, despite this uncertainty, the accomplishment of just making a map of the neighborhood with all the houses in proper order and spacing is a considerable one.

Mapping the neighborhood of galaxies has been done in much the same way, though brightness rather than size is more commonly used to estimate a galaxy's distance. Imagine making a map of your neighborhood at night using lights in the houses or apartments as standard candles. (It will help to imagine plenty of windows so that all the lights can be seen, or, better yet, totally glass houses.) In his historic measurement of the distance to Androm- eda using Cepheid variables, Hubble had, in effect, located many small "nite-lights" known to come in only a small range of wattage—other astron- omers had provided a "calibration" that was as good as actually reading the wattage right off the light bulb. As a result, for every "house" in which he could see one of the nite-lights, Hubble could calculate the distance by comparing them to the nite-lights in his own house, the Milky Way galaxy.

Unfortunately, because Cepheid stars are relatively faint, Hubble had to rely on brighter lights to reach more distant galaxies. These were not well calibrated, more like the other light bulbs in our houses that come in a variety of wattages from 50 to 300. With such a wide range, a large distance error would be made if only one light were chosen (was it a 75 or 250 watt bulb?), so Hubble relied on statistics: he measured a representative population of the brightest stars and assumed that the very brightest would be "the highest available wattage," much like the previous example of the tallest person in the house.

Two technical differences between the "standard rod" and "standard candle" approach are worth mentioning. Apparent size decreases in direct proportion to distance, but brightness falls as the square—a galaxy three times as far is nine times as faint. Also, measuring brightness requires an electronic device; our eyes sense subtle differences in brightness over a wide range, but are not good at establishing an accurate scale.* Hubble did the best he could using photographic films, which are only a little better in this respect than the human eye; today astronomers use electronic sensors called photometers that can with great accuracy measure the amount of light coming from a star or galaxy.

By 1929 Hubble had estimated distances to some twenty galaxies, using the standard candle technique. He had made the first map, albeit a crude one, of the positions of neighboring galaxies in space. The distance estimates were poor, no better than a factor of two. Such inaccuracy in mapping the locations of houses in the neighborhood would be a municipal disaster—with houses scattered so far from their true locations, street boundaries would become so wiggly as to be unrecognizable and all sense of organization would be lost. One might think that little could be done with such a poor map. Remarkably,

---

*As with most human senses, the eye's response to light is logarithmic: what seem to be even steps of 1 . . . 2 . . . 3 . . . 4 . . . are actually changes in sensitivity of 1 . . . 4 . . . 8 . . . 16. . . .

what Hubble did with his map led to one of the most significant advances in human thought.

Hubble's goal had been to measure the speed with which the Sun orbits the center of the Milky Way galaxy. Just as in our solar system where the Earth and other planets feel the force of gravity due to the Sun's great mass, bending their paths continuously into a closed track we call an orbit, the Sun is itself tugged by the combined mass of the billions of stars that lie within its orbit of the Milky Way's center. To measure the speed with which the Sun travels its near-circular orbit, Hubble imagined the galaxy to be like a merry-go-round, with the Sun as one of the horses. He intended to use the newly identified galaxies around the Milky Way as reference points, in the same way that people standing around the merry-go-round serve as fixed points that allow the rider to judge how fast she's moving. Hubble's procedure was to make a measurement of the speed of each galaxy relative to the Sun, from which he expected to find galaxies on one side of the sky approaching, and those on the other side receding, this a reflection of the Sun's romp around the Milky Way. He would look for galactic rotation, but he would find something else, something far more interesting.

Measuring the speed of a star or galaxy might sound like a formidable task, but it is one of the simplest and most direct measurements an astronomer can make. The key is to take a spectrum—spread the light into its component colors—and measure the colors of light coming from different atoms in the star (or many stars, in the case of a galaxy). Then one measures something called the Doppler shift and, *voilà*, the speed is known. Spectroscopy is the most powerful tool available to an astronomer. An encyclopedia of information is encoded in the intensity of light as a function of its color (which is synonymous with *energy*). This bounty is what allows an astronomer to pursue an observational science unlike any other: the astronomer has no direct contact with the subject of the research. Nor is information limited to visible light: the ''electromagnetic spectrum'' spans an enormous range in energy, from the minuscule packets of energy called radio waves to the dangerously energetic X rays and gamma rays. Familiar infrared, visible, and ultraviolet light comprise only a few octaves of an electromagnetic keyboard with many more ''notes'' than a piano.

The power of spectroscopy to change astronomy from a descriptive to an analytical science had become abundantly clear by the beginning of the twentieth century. Even before Hubble had shown that many of the nebulae were independent galaxies, much had been learned about them from spectroscopic analysis. The spectra of many resembled that of the Orion nebula, a womb of stellar birth that glows with the energy injected from its newborn stars. But other nebulae showed no trace of glowing gas, only the combined light of billions of stars. Even with these early spectroscopic measurements, astronomers were able to perform a kind of celestial taxonomy, and a rough assay of the chemical components of astronomical objects from the details of each spectrum. In time, they would learn to glean more specific information from spectra: the temperatures and densities of gas clouds, and how they are heated and cooled; the relative abundances of different elements; and the chemical reactions that prosper in the extreme environments of space. These would, in turn, be used to study the energy source of the stars, to infer distances to stars and galaxies, and to study the history of the production of the chemical elements by generations of stellar births and deaths.

However, well before this wealth of knowledge was mined, the earliest spectra had already surrendered a treasure—the *speeds* of stars—as easily as gold nuggets are plucked from a streambed, and just as valuable. In 1842 German physicist Christian Johann Doppler had shown that the pitch of sound waves, or the color (i.e., wavelength) of light, is altered by the relative motion of the observer and the source. In the tradition of a true impresario, Doppler had made his point by ferrying a chorus of trumpeters in the open car of a speeding train past a crowd of spectators. Those in attendance were perhaps the first to hear the melancholy drop in pitch that accompanies a passing train, a sound that became familiar and cherished when whistles and horns became the songs of the rails.

To be specific but not too mathematical, the "Doppler effect" is caused by the decrease in the time interval between the arrival of successive wave crests when an observer and source are approaching, or an increase in time between wave crests if the two are growing farther apart. For light, this change in frequency is observed as a shift in *color*: an approaching light source undergoes a shift to higher frequencies—a *blueshift*—while a receding source is *redshifted*. A familiar application using light is the radar "gun," which bounces radar waves (light with a much longer wavelength than our eyes can detect) off a baseball to see if a pitcher is "popping his fastball" or a Ferrari is traveling faster than the law allows. The change in frequency of the returning waves provides an instantaneous measurement of speed—it isn't necessary to clock the baseball's flight from the mound to home plate and then divide 60.5 feet by this time to calculate the speed. This feature is absolutely crucial in astronomy, since galaxies appear to move only an infinitesimal distance across the sky in a human lifetime. (It is not that they are moving slowly, but that the distances to galaxies are so vast that the apparent motion is extremely small.)

Furthermore, as will become apparent, the Doppler effect is invaluable as an astronomical technique because, if enough light can be collected, the measurement of speed can be made regardless of the distance to the object.

When astronomers began to record the spectra of stars on photographic plates, they found the familiar color patterns of starlight shifted to the red or the blue by small amounts. The Doppler shifts measured for these neighbors of the Sun indicated typical speeds of a few kilometers per second, with roughly equal numbers approaching and receding. (The Doppler shift measures only the relative motion *along* the line of sight from the observer to the object; the spectrum of a star moving rapidly *across* the line of sight shows virtually no shift in color.) However, when Doppler shifts were measured from the starlight spectra of roundish nebulae the results were shocking: speeds of *hundreds* of kilometers per second were the norm. In hindsight, one might wonder why this wasn't taken as strong evidence against the notion that such nebulae reside within the Milky Way galaxy—still a contending idea from 1910 to 1920, when these first measurements were made. However, such great speeds were just as puzzling to the proponents of the "island universe" model; great speeds were not what they expected for giant galaxies suspended in space. Even more perplexing was that almost all of these nebulae showed "redshifts," which indicated a mass exodus *away* from our galaxy rather than a mix of motions toward and away. Vesto Slipher of the Lowell Observatory, who had compiled the biggest sample, found that seventeen out of nineteen nearby nebulae showed large redshifts; only two had blueshifts. The systematic nature of these mysteriously large Doppler shifts threw even more doubt onto their interpretation as actual motions. No astronomer was bold enough to venture the correct interpretation, that these objects were other galaxies in full flight from our Milky Way.

So it was that these extraordinary data languished in what scientists are fond of calling "the journal of unexplained results" until Hubble came along in the late 1920s with his primitive galaxy map, again to take the giant step that consolidated scientific opinion. Remember that Hubble was looking for a systematic pattern caused by the merry-go-round spinning of the Milky Way—redshifts for galaxies on one side of the sky, blueshifts on the other. But when he plotted Slipher's velocities against his estimates of galaxy distances he found a much different effect, one far bigger than the one he had been looking for. Hubble's distance estimates revealed something Slipher hadn't seen—that the more distant galaxies had the biggest velocities away from the Milky Way. These motions reached a thousand kilometers per second, much larger than the rotation speed of the Milky Way galaxy. It was as if onlookers around the merry-go-round were caught running from the scene, with those on the perimeter fleeing the fastest while those closer in lagged. With these intended points of reference scattering even more rapidly than the merry-go-round was turning, the measurement of our galaxy's spinning was lost in the pandemonium, or, as scientists say, lost in the *noise*.

What Hubble had found was far more remarkable than what he was looking for. This is a common occurrence in science, reminding us that our imaginations are rarely up to the task of anticipating the universe. The relation Hubble found between galaxy distance and velocity was far from perfect, largely because his methods of estimating distance were crude, but the trend in the data was clear: velocities of galaxies increase with greater distance. Hubble's interpretation of what he had found, in keeping with his conservative style, was cautious: he described the trend by a straight line—the simplest mathematical form consistent with the data. By doing so, Hubble was assigning to each galaxy a speed in direct proportion to its distance, and assuming that the considerable "scatter" in the data—the failure of the points to fall neatly on a straight line—was due mainly to measurement error.

Hubble himself may have been hesitant, but theoretical cosmologists were quick to embrace the obvious interpretation: *the universe is rapidly expanding*. In fact, Hubble's observation appeared to solve a problem lingering since 1918. Einstein, in formulating the general theory of relativity—a mathematical model for the action of gravity on cosmic scales—had concluded that the universe could not stand still, but must either expand or contract. In order to achieve a static universe, Einstein had reluctantly added to his equations an arbitrary repulsion force to counteract gravity, but now the observed expansion of the universe seemed, happily, to make this ad hoc remedy superfluous.

The truly profound implication of Hubble's results, however, was grasped first by Georges Lemaître. In 1927, even before Hubble's announcement of a correlation between redshift and distance, Lemaître had found a mathematical solution to Einstein's equations for the case of an expanding universe. Lemaître was a priest as well as a scientist; perhaps this is why he was disposed to make the conceptual leap between the observation of an expanding universe and the idea of creation. He reasoned that, if the universe is expanding, its contents must have been more densely packed in the distant past. In particular, a linear relation between speed and distance implies a *uniform* expansion, that is, if the universe were thrown into reverse, spaces between galaxies would shrink in proportion until all galaxies virtually touched. Lemaître dared to extrapolate further, back to a time when all matter and space were crunched together at the density of an atomic nucleus—a "primeval atom." Some twenty years later, George Gamow would develop

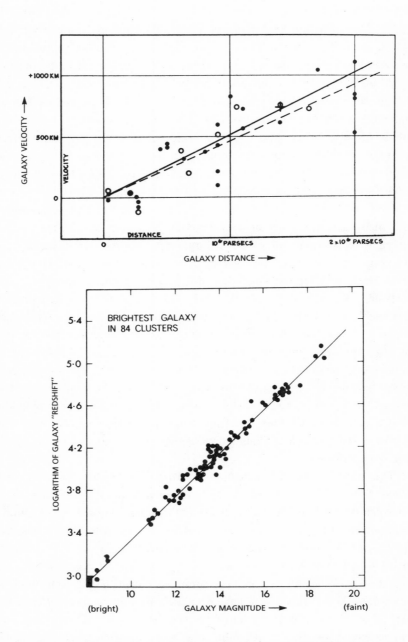

The relation of distance to velocity, which was originally found by Hubble in 1929, in its 1936 version (top). The bottom "Hubble diagram," made by Allan Sandage, shows a modern version that confirms the linear relation of distance and redshift on a scale vastly larger—the regime of Hubble's original data is shown as a box at the lower-left corner. A linear relation implies a universe that is expanding smoothly and evenly in all directions. *(Reference 1: 1929,* Proceedings of the National Academy of Sciences, *vol. 15, p. 168.) (Reference 2: 1972,* Astrophysical Journal, *vol. 178, p. 1.)*

Lemaître's idea into a physical model wherein the universe had been set in motion by a cataclysmic event, what Fred Hoyle would later name, derisively, a *big bang*.

The recognition of other galaxies beyond the Milky Way was seminal, but nothing compared to this amazing conclusion. The discovery of other galaxies had been forecast centuries before, but the dynamic expansion of the universe, and its implication that the universe arose in a cataclysmic moment of creation, was as unanticipated as it was revolutionary. Together, these conclusions of early-twentieth-century astronomy rank with the greatest achievements of human endeavor. Our universe had, in a real sense, finally been discovered.

The common denominator of our group was Sandy Faber. It was her expertise and knowledge in the blossoming study of the structure and stellar populations of galaxies, attached to an infectious energy and generosity with her time and ideas, that had drawn Dave Burstein and me to her as graduate students. In the late 1970s these qualities became appreciated far from her home base, drawing U.K. astronomers Roger Davies and Roberto Terlevich to Santa Cruz. Sandy's international reputation spread rapidly after the publication of a landmark review article she coauthored with University of Illinois astronomer Jay Gallagher that made the strongest case to date for vast amounts of "unseen" matter in the universe. Invitations to speak at conferences in England and Europe, and across the U.S. and Canada, came fast and thick for this new rising star in the astronomical world—it wasn't lost on anyone that Faber was one of a new generation of women astronomers who would begin to chip away at the serious gender imbalance that bedevils this field, like so many of the physical sciences.

Sandra Moore was an early enlistee of the postwar baby boom, born to an army colonel as he bade farewell to Okinawa, and a mother putting aside any thoughts of a career for the traditional role of wife and mother. Both were of Scotch-English descent. Their heritage endowed a crystal clear message reinforced by the Great Depression that, in Sandy's words, "Nothing came free in life, but hard work would be rewarded; education was the key to move upward—I was doted on intellectually but not spoiled monetarily."

The couple, already in their forties, had thought their chances of having children slim; to deal with the challenge of how to raise their only child, Sandy's father reached for a familiar model, the world of military discipline that had recently run *his* own life. Even the request to go to bed was delivered with a drill sergeant's cadence. Young Sandy thoroughly disliked this style, though it didn't negate the positive message and encouragement her father also offered.

"Despite his sternness, he was also inspiring," she says. "Financial difficulties had shaped his life and forced him into a practical vocation [civil engineering] where he could earn a living. Yet he never placed these limitations on me. Instead he told me I was wonderful and could do something exceptional someday. Exactly how I was to do this, he didn't say, though he clearly liked my interest in science. Practical questions like whether I needed to be self-supporting or could rely on a husband, how work would be melded with family, etc., never came up. It was a gilded future but at the same time awfully vague on specifics. . . . In retrospect, I think my father took the right tack. His approach reflected his own optimistic view of life, which I inherited."

Her mother had been attentive during her younger years; the crucial sense of unconditional love and acceptance was firmly rooted. Still, as she grew, Sandy saw her mother more as an example of what she did not want to become. "I admired men because they were active and in control of their destinies. Women seemed to me passive and weak, and their activities like housework and baby-tending were so much more boring than men's work out in the 'real world.' I had a lot of pride and did not relish the thought of spending my life waiting on other people's needs, which is what every woman in my world did."

Other children, Sandy remembers, were not tolerant of her attempt to fit somewhere between the traditional roles of little girl and little boy. She hung around more with the boys, adopting the tomboy habit and plunging into sports. And, in the evenings, she preferred eavesdropping on the World War II stories of her father and his buddies to the women's chats of families and personalities. But biology had plans of its own, and society had its traditional responses ready.

"As the years passed, there were more and more places I could not follow: Cub Scouts, for example, where the boys went hiking while the girls in my Brownie troop stayed home and embroidered place mats. I remember most poignantly standing on the sidelines watching bitterly while the same boys I had played baseball with that afternoon donned dashing uniforms and fought for glory on the local Little League team—no girls allowed."

No surprise that such a spirit found its own way.

"I began to discover a way to fight back. I saw that learning and studying could take me to faraway places that few of my peers could even dream of. I read incessantly, mostly about science, any kind of science. I had a rock collection; I learned how to identify every tree in the neighborhood; I spent hours reading about and watching spiders; I tracked the weather. I had a wonderful time escaping from people into the natural world. And I took delicious pleasure, I have to be honest, in acing academically the same boys who wouldn't let me play on the Little League team.''

But the world was changing. Doors could be opened, and teachers—the most important people on the planet—were the ones who could do it. The best saw in Sandy not just a girl, but a fertile and able mind. A sixth grade teacher invented a challenging spelling program just for her. A junior high teacher found a curious method of positive reinforcement when he called a conference with her parents to inform them that Sandy was destroying the atmosphere of his seventh grade general-science class by dominating the other students too strongly. As she was so exceptional, he asked, could not some other outlet be found for her brilliance? A tenth grade geometry teacher noted in the margins of a term paper on computers that she was gifted mathematically and that not to pursue this talent would be a great loss.

By the time she entered Swarthmore College, Sandy's strategy of finding a niche for herself, in a world where she hadn't at first been able to fit in, had succeeded brilliantly. And there she would find many other students who had chosen the same road; in their company she began at last to feel comfortable with who she was and what she could become. Soon competition gave way to collaboration, a newfound joy, and she developed a close working relationship with her physics lab partner. Two eager minds found common ground rediscovering the foundations of modern physics—measuring the speed of light and the charge on the electron, witnessing radioactive decay and rediscovering Newton's laws of motion. This Alice-in-Wonderland adventure, carefully crafted by the Swarthmore physics professors, was a turning point in Sandy's life. When Peter van de Kamp, an astronomer at Swarthmore, took Sandy under his wing, ''practically adopting me as his daughter, inviting me to his home, and promoting my career,'' her course was set. The nurture and encouragement of parents, the succession of kindly and supportive teachers, had lifted her from early struggle to high flight.

''I never felt I was struggling against the odds,'' she adds. ''On the contrary, the educational establishment seemed to me to be working actively on my behalf. I am an example of *how the system is supposed to work*, and for that I feel extremely lucky.''

By 1968 Sandy and her husband of one year, Andrew Faber, were both graduate students at Harvard University, he in applied physics and she in astronomy. Sandy, in her third year, had started a Ph.D. research project. She would be one of the first to study the systematic properties of galaxies as whole systems by measuring the colors and brightnesses of a large sample of ellipticals, using the telescopes at the Kitt Peak National Observatory in Arizona. But, despite a solid start on what would become a career-launching project, all was not well. Their ivory tower was under siege.

Andy Faber had a one-year military-service deferment for graduate school, and a low lottery number. Resolute in their opposition to the Vietnam War, the couple wrestled not so much with their own consciences but with two sets of parents who were dismayed at the possibility that Andy and Sandy might choose to move to Canada. At their urging, Andy found a draft-deferred job with the Navy that took advantage of his undergraduate experience in underwater acoustics research. It was a compromise, of principle and career, but it was a solution. They moved to Washington, D.C. While Andy waited out the storm at the Naval Research Laboratory, Sandy was marooned there. She wrangled a desk, and used the rather good astronomical library the astronomers at the Naval Lab had collected, but, far from her thesis adviser at Harvard and the company of astronomers who were interested in galaxies, her scientific harvest began to wither as a year and a half went by.

"However, my fortunes changed dramatically one day in early 1970, when I received a cheery telephone call from astronomer Vera Rubin, of the Department of Terrestrial Magnetism [DTM], Carnegie Institution of Washington [CIW], near Rock Creek Park. I had worked there as a summer student in 1966 with Vera and her colleague Kent Ford and enjoyed it immensely. Like all five labs of CIW, DTM is a rare place where scientists are given total freedom to pursue whatever interests them. The atmosphere is relaxed, honest, yet intense all at the same time."

With a new congressional law threatening Carnegie's tax-exempt status, its scientists were eager to direct more student research to reinforce Carnegie's posture as an educational institution. Months of isolation had brought a sense of scientific loneliness to Sandy and she eagerly accepted Rubin's invitation. She soon found what she had sorely needed and missed. At DTM, every one of the small group of astronomers and postdoctoral fellows was interested in galaxies: lively discussions began in the morning and lasted the day.

Looking back, Sandy realizes that Rubin and Ford were laying the groundwork for much of her own later scientific research. They were deeply involved in a study that would begin to convince the astronomical community of the reality of *dark matter*, the invisible mass that dominates the universe. She remembers the day when Mort Roberts, a master in the new art of radio astronomy, came to compare notes with Vera and Kent on the rotation of spiral galaxies and the monstrous amounts of unseen matter that must rule the very stars in their ceaseless circling. It was a discovery of such immense proportion that, twenty years later, its impact is not fully realized. As a young student Sandy sat and listened to history-in-the-making; she could not have dreamed of the role dark matter would play in her own research, and how it would come to capture the imagination of seven scientists whom she would one day bring together for a voyage into intergalactic space. Perhaps it was good fortune, or maybe it was the reward promised by her father to follow good, honest work. Whatever, Sandy found herself in full sail in winds that were carrying a few lucky and excited scientists toward a new universe.

Vera Rubin would become an influential role model for women astronomers, and Sandy as well, although ironically:

"I think that in my academic, and later professional career," she says, "I have actually had *fewer* obstacles to overcome than most, and this is paradoxically *because* I was a woman and not a man. The only girl in a class of twenty Advanced Chemistry students can't help but get noticed. Fortunately, the notoriety was almost always positive. I have no complaints: I'm playing Little League baseball after all."

In addition to the mind-shattering implications of the creation event, Hubble's discovery of a linear relation between a galaxy's distance and its velocity (redshift) has had important implications for astronomers. One is, of course, the provision of a simple way to estimate the distance to a galaxy. Distance and velocity are related by a constant of proportionality—this number of light-years* corresponds to this much speed—that came to be known as the *Hubble constant*. The relatively easy measurement of a redshift, then, gives a velocity that can be equated, through the Hubble constant, to an approximate distance. Compared to the difficult matter of finding Cepheid variable stars or other standard candles, which are useful, in any case, for only the closest galaxies, this technique of measuring a distance is elementary and extremely far-reaching.

Furthermore, the Hubble constant is related to the *age* of the universe in a very simple way: for a galaxy whose distance from the Milky Way is known in kilometers, and whose speed is determined to be some number of kilometers per hour, the two numbers can be divided to find how many hours it took to put that much distance between it and the Milky Way. In a universe with a linear relationship between distance and speed, this time is the same for all galaxies; for example, a galaxy twice as far away has twice the speed,

---

*Expressing the great distances to stars and galaxies in miles or kilometers would be awkward, like giving the distance from London to Athens in millimeters. A more suitable "unit" is the *light-year*, the distance a beam of light covers in one year (about six trillion miles).

so it covered the greater distance in the *same time*.* In this way the Hubble constant establishes the approximate age of the universe; present estimates for the "Hubble time" are between 10 and 20 billion years. This is a long time, of course, but it is not inconceivably long. In comparison, the Earth is about 5 billion years old, based on the ages of rocks as determined from the presence of unstable chemical elements (like radium or uranium) that decay into other elements. Our Sun and its planets are relative newcomers to the universe, but they are not *that* young.

The exact value of Hubble's constant has been difficult to pin down, decreasing by almost a factor of ten since Hubble's original paper in 1929. Hubble's first study grossly underestimated the distances to galaxies—the universe is much larger than he first surmised. However, even though *absolute distances* have been difficult to determine, the *linear* relation between *relative distances* that Hubble put forward has stood the test of time. The direct proportionality of redshift to distance—the straight line in what has become known as the *Hubble diagram*—has been verified out to distances hundreds of times greater than that of Hubble's first sample. Hubble himself, in collaboration with colleague Milton Humason, and later Humason, Nicholas Mayall, and Hubble's protégé, Allan Sandage, were able to push across a significant fraction of the universe by using whole galaxies as standard candles, rather than the individual stars in them. They showed conclusively that the expansion of the universe is a global phenomenon.

In the 1970s, Sandage used an even brighter standard candle to push to truly cosmic distances of billions of light-years. His choice was the luminosity of the brightest galaxy in the rare, rich clusters of hundreds of galaxies that dot the cosmic landscape. To borrow again from the analogy to real estate, he was looking at distant neighborhoods and selecting the biggest or the brightest house in each. These measurements validated Hubble's linear relation to even greater distances, but they also promised another prize that Sandage coveted. Looking far into space is looking far back in time as well, since light arriving at the telescope left each galaxy billions of years before. Sandage's aim was to measure the expansion rate as it was when the universe was much younger and compare it to the present expansion rate, to see if and how much the expansion might have slowed. No less than the fate of the universe was at stake: would its outward flight be so strong that it would overcome the pull of gravity and expand forever, or would the universe

*This and other features of "uniform expansion," including a discussion of why the universe has no "center" and no "edge," will be explored in a later chapter.

succumb to gravity and slow to a stop, or even fall back together into a big crunch?

Alas, this worthy quest was frustrated by the very "look back in time" that made it possible. In the end, it became clear that the standard candles were themselves evolving: some astronomers pointed out that galaxies should grow *dimmer* as they age, as their younger, more luminous stars die out, while others suggested that the largest galaxies might grow *brighter* in time as they gobbled up small neighbor galaxies that strayed too close. No one could be sure which of these or other related processes would dominate, and by how much, but it was becoming clear that the standard candle was not likely to be a steady beacon for all time. Judging even the sense of the correction—brighter or dimmer—let alone how much of a correction to make, seemed beyond current understanding. Unfortunately, this uncertainty made all the difference in reading the fate of the universe from the Hubble diagram. The technique was, for the time being at least, put aside. The universe had yielded something about its past; it would be some time before it would disclose secrets about its future.

The map given to us by our predecessors was a stunning achievement, like the first maps of the Americas made by the great European navigators of the sixteenth century. Just like their efforts, however, the charts of the new universe were incomplete and, as we would later learn, not very accurate. Our attempt to include all elliptical galaxies within a given volume was frustrated from the start. If redshifts had been measured for all nearby elliptical galaxies, of which there are many thousands, selecting a *volume-limited* sample would have been easy: limit the distance sampled by accepting only those galaxies with redshifts smaller than a certain amount, in other words, choose from the three-dimensional map those galaxies within an imaginary sphere centered on our galaxy. Our target volume was a sphere with a radius of 240 million light-years (about 100 times farther than the Andromeda galaxy)—at this distance the redshift amounts to a 2 percent change in the wavelength (color) of the light emitted by the galaxy, a shift not discernible with a human eye, but easily measured with a spectrograph. Unfortunately, only a few hundred redshifts had been measured for elliptical galaxies by 1980; like the maps of Renaissance mariners, our maps had huge areas marked "unexplored." In these areas there were no redshifts, and therefore no distances.

Denied the ideal, unbiased sample, we sought the next best thing: a sample that was biased in a way we thought we understood. As astronomers often do, we set out to construct a sample limited not by distance but by brightness, which is the most immediate property of any astronomical object and the one most readily determined. Even though redshifts were not available for elliptical galaxies all over the sky, their apparent brightnesses had all been measured, or at least estimated. For each galaxy we considered there was a catalog entry listing a *magnitude*, the awkward brightness scale inherited from the Greeks and refined by modern astronomers. Unlike a volume-limited sample, a *magnitude-limited* sample would not contain all ellipticals within a certain volume, because intrinsically brighter galaxies would be seen to much greater distances than fainter ones. But at least this bias could be applied equally in all directions. To make sure that it was, we tackled the job of reconciling heterogeneous data from different catalogs so as to apply the brightness limit fairly over the whole sky—as uniformly as possible. Our map would be incomplete, we knew, but predictably so, the way a sailor's map might show *all* islands larger than ten miles across, but only one out of ten islands smaller than two miles across—the same all over the world. It would be incomplete—biased—but in a predictable way.

When we had completed our examinations of catalogs and charts, we had chosen about 500 elliptical galaxies brighter than thirteenth magnitude. The Greeks had divided the brightness scale into categories: the brightest stars visible to the human eye were designated as ''stars of the first magnitude,'' the faintest, stars of the sixth magnitude. Thousands of years later photometers— electronic ''light meters''—were invented, and measurements showed that a first-magnitude star is actually about 100 times brighter than a sixth-magnitude star. This relationship, that five ''steps'' of the magnitude scale (there are five steps counting from one to six) correspond to a factor of 100 in brightness, was adopted as exact, and extended to objects fainter than naked-eye stars. (A single magnitude corresponds to a change in brightness of a bit more than 2.5.) The faintest galaxies in our sample were seven magnitudes fainter than sixth-magnitude stars, i.e., a factor of about 600 times less light incoming. At the time we were assembling our sample, other astronomers were detecting galaxies at twenty-sixth magnitude, 100 million times fainter than the faintest star one could see by eye. These were at the far reaches of the universe, however; our little romp was organized to stay closer to home.

Sandy gave Jesus Gonzales, one of her graduate students, the tedious job of making ''finding charts'' for the ellipticals in the northern sky; Roberto Terlevich and Roger Davies were asked to prepare the charts for the south. For each selection they made a photocopy of the patch of sky containing the elliptical galaxy. An area about 10 percent of that covered by the Moon was sufficient to show the galaxy in question, the pattern of stars *in our own Milky Way*, which lay in the foreground, and any companion galaxies that might indicate whether the elliptical was a member of a group or cluster of galaxies. Each photo was mounted on a card on which was written the galaxy position in celestial coordinates, to a precision of about ten arcseconds, various catalog names, the magnitude of the galaxy, its redshift (if one had been measured), and other sundry details. These charts would be taken to the telescope and used to verify that the correct galaxy was visited, since many galaxies look alike, but the star field onto which each is projected is unique. The cards would also serve for record keeping: the dates and times of observations could be written on them, and special features, noted during the locating of the galaxy and during data taking, would be entered.

Miserable Xerox copies were made of these handsome charts so that each of us could have a copy. They were adequate only because these were relatively bright galaxies and therefore easy to identify. Applications for observing time for the coming seasons were made to six different observatories, and many of the requests were approved. Our planning complete, we were ready to set off on our journey into intergalactic space.

Dave Burstein's parents were moving from the house he grew up in, and his mother had asked him to return home and clean out the stacks of paper he had accumulated through school. As Dave sat on the floor among the mess, he came across a stack of mechanical drawings he had made ten years earlier in Mr. Lush's ninth grade class. He smiled as he remembered the fun the students had with that name. Lush had been a good guy, though; he had let Dave work on these drawings in homeroom—he had Lush for homeroom too that year—and while Dave concentrated over the large drawing table, Lush would delight in his morning ritual of reading homilies from the *Reader's Digest*—little two-page stories with a moral or message.

Now, as Dave revisited the first of his drawings, Lush's voice flooded into Dave's head—the very story Lush had read the day he made the drawing replayed with incredible fidelity. It was frightening, a nightmare. Dave felt his mind's tape recorder racing ahead, daring him to keep up. Panicked, Dave turned over the page. The voice stopped. He turned over the next drawing and Lush's voice picked up in midsentence, as if a switch had been thrown. But this time he was telling a different story, the one he read on the day *that* drawing had been made! Dave flipped through a few more and the same haunting voice cut in and out; finally Dave tossed the drawings down and sat stunned on the carpet. He had known about this strange memory "ability" all his life, of course, but this time it seemed to have gone too far, and it left him shaken.

Dave's remarkable memory almost defeated him in his quest to become a scientist. It might seem that the ability to recall vast amounts of information with great accuracy is a considerable help in succeeding in school. That is absolutely true, and exactly the problem. It was that same year at Theodore Roosevelt Junior High School in West Orange, New Jersey, that Dave participated in a College Bowl–type contest, on stage in front of the whole school. Dave had read most of the *World Book Encyclopedia* by that time, and he answered almost every question before any of the other kids had a chance, getting all but thirty-odd points out of 500. He remembers specifically the question "Who returned from Germany in 1938 claiming 'Peace is at hand'?" Instantly Dave shot back, "Neville Chamberlain." Trouble was, thirteen-year-old David had absolutely no idea who Chamberlain was. He had a head full of facts, but little *knowledge*.

The memory ability came from his mother's side of the family, and manifested itself in all degrees up to complete photographic memory. In her later years Dave's mother could complete the *New York Times* crossword puzzle in fifteen or twenty minutes—she had seen all the clues over the years and the answers now popped up unconsciously. Perhaps this is why she was sympathetic to Dave's "problem," at least it was a problem for *him*, because from an early age Dave had managed to turn this extraordinary ability for remembered facts into a snot-nosed arrogance that infuriated most everyone with whom he came in contact. Dave's father, an educated man with a law degree, would be explaining something to Dave and his two older sisters at the dinner table and Dave would break in, dispassionately, to inform them that the dates or events in question were wrong. And Dave was right, always. In his own words, Dave recalls himself a "difficult child," and remembers how his mother chortled with glee when a teacher commented, on one of Dave's report cards, that he was doing very well in school. What a relief, she thought—he has found *something* he likes.

Remembering a lot of facts may pass for being smart, but there is not a lot of satisfaction to go along with it. Dave would try to watch their little five-inch television, but he got bored. Radio was worse. He began to read books, everything he could get his hands on, including the encyclopedia, nearly cover to cover. And, of course, with all those facts, and the ability to parrot back much of what he heard, elementary school was easy. But when he entered junior high his academic performance took a nosedive. For the first time something other than total recall was required.

"My first few months in seventh grade were not happy for me," explains

Dave, "as I was not doing well in school. What I realized many years later was that this pattern of initially doing poorly was a product of my memory. Having an excellent memory such as mine is a blessing and a curse. It's a blessing in that the elementary, junior high, and even high schools placed a premium on memorization (they still do), so people like me do very well. But it's also a curse, because many of us never learn to think analytically; all that they do is essentially conditioned response. I could just remember how to do some task, rather than figure it out from general principles.

"Another curse of this memory ability is that those who have it more often than not don't realize it. Our society is so structured that people [with exceptional memories] can do very well in life in a large number of occupations. However, the physical sciences are not among them."

Unfortunately, this is just the direction Dave was headed.

"My interest in science, in general, and astronomy in particular, came at a very early age. I was told (and I also remember) that, at age six, if an adult asked what I wanted to be, I replied 'an astronomer.' My parents were quite puzzled and somewhat disapproving, attitudes they retained until I was well into my career (and in a permanent job, some thirty years later). The only other scientist in the family—my mother's cousin, who was at the time a successful physics professor at Columbia University—was considered an outcast and somewhat of a failure because of his unwillingness to attend or mix in at family functions.

"I went to public schools through twelfth grade. In the ninth grade, we took a kind of 'occupational preference test,' from which we had to choose a profession on which to write. I wrote an application for a fictional job at the Jet Propulsion Laboratory. As I was very good in mathematics in high school, the idea of becoming a theoretical mathematician began to supplant that of becoming an astronomer. In my junior year in high school, 1964, I applied for summer science programs sponsored by the National Science Foundation [NSF]. One program was in astronomy at Ojai, California, and two were in mathematics, at Ohio State and Flagstaff, Arizona. I had greatly enjoyed my trip to Tucson with my parents the previous summer—the first time I had ventured west of Philadelphia—and this heavily influenced my choice of the school in Flagstaff.

"The NSF program reinforced my desire to be a theoretical mathematician. If I had gone to Ojai, that probably would have reinforced my decision to be an astronomer. Either way, I wanted to go to a small, liberal arts school that had a good science program. Wesleyan University, in Middletown, Connecticut, was pointed out to me as such a school, and I was accepted there early, in January 1965.

"After three weeks at Wesleyan, the idea of being a theoretical mathematician vanished from my thoughts. My ability was with numbers; I had never learned to think analytically. I did dismally the first three months at college in both math and physics. I realized many years later that during those three months I had to learn how to think all over again, basically from scratch. I can still remember the pain of having done it that way."

The faculty at Wesleyan was good, and after Dave pulled himself up by his mental bootstraps, he found plenty of opportunities to work with scientists. One of the astronomers at Wesleyan's Van Vleck Observatory gave him the job of measuring the rotation of spiral galaxies from photographic spectra taken by Margaret and Geoffrey Burbidge, British astronomers who had moved their careers to the United States and made fundamental contributions to theoretical and observational astronomy. With physicist James Faller, Dave learned about sophisticated astronomical instrumentation; Faller was one of those scientists who was going to try bouncing a laser beam off a small, prismlike cube that would be left on the Moon by Apollo astronauts. Dave had never considered himself mechanically inclined—as a youth he enjoyed countless hours with an Erector Set, and tried to build his own telescope, but he was consistently disappointed with his efforts. With Faller, Dave would learn that he could do precise, careful work with scientific apparatus, even if he himself was not a master builder. The great care he learned in making accurate, reliable measurements would become a trademark of his astronomical career.

Dave became an enthusiastic user of telescopes when he did his senior thesis with the twenty-four-inch reflecting telescope at Van Vleck. When he looked around for graduate schools in astronomy with good observational facilities, he realized that several schools of the University of California had access to the excellent Lick Observatory on Mt. Hamilton, near San Jose, at the south end of the San Francisco Bay. The Lick Observatory headquarters had recently been moved to the new U.C. campus at Santa Cruz, and though Dave had never heard of the place, he sent in an application. Within a few months, he was offered an NSF fellowship and, pleasantly surprised, immediately accepted.

It seemed that all was going well for Burstein at this point of his life— he was on a straight course to a career he relished, and he had met the woman he would eventually marry, a "townie," as the Wesleyan students called the residents of Middletown, by the name of Gail Maureen Kelly. He had overcome "his memory," the single biggest obstacle to his becoming a scientist. But life threw another curve at Dave that almost spoiled his chance to become an astronomer. It came, as it did for Sandy Faber, courtesy of the Vietnam War.

Dave was in full stride in his first quarter as a graduate student at U.C. Santa Cruz in 1969, doing well in his classes and developing skills on computers, a very new addition to astronomical data processing. But, like many others, he was on borrowed time. Classified by his draft board 1A, and having passed his physical, he knew he would be called to active duty before the year was out. By the New Year he had decided to join the National Guard, but this required that he move back east; he dropped out of school in the middle of his second quarter. Dave finished the active duty period of four months, but he couldn't easily transfer to the National Guard at Santa Cruz and so decided to remain near his fiancée, Gail, and work again at Wesleyan. It was a year and a half later when Dave returned to Santa Cruz to resume his graduate studies in astronomy.

Some things had changed. He was married now, soon to start a family, and he brought back an injury from the Guard that was about to affect him in ways he wouldn't understand. Dave is a big man, 6'2" and around 240 pounds, and when he fell off a National Guard truck in winter 1970, it did irreparable damage to the muscles of his lower back.

"I was treated for this injury by a doctor-friend of my father's, who prescribed the new drug Valium as a muscle relaxant. From fall 1970 to late summer 1972 I took Valium daily, in dosages from 10 to 40 mg per day.

"Unknown to me, the Valium was adversely affecting my mind. I found I could not do graduate course work that I had done before. I could not think as clearly as before. I never associated this with the Valium—it was a gradual effect. It was only after a two-week National Guard summer camp at Fort Irwin, near Death Valley, in the summer of 1972, that I decided I had to stop taking the Valium. It was at this camp that I was taking 40 mg of Valium a day; even I realized this was too much."

Dave went through a mild but noticeable withdrawal period with chills, shaking, and the like; he found out a couple of years later that Valium had been shown to be addictive and to have an adverse long-term effect on mental abilities.

"It took me about four years to recover from taking the Valium; after 1977 I felt I could do as well as I could before I started taking the stuff. Unfortunately, this was a very critical time for me. Many of my graduate student colleagues and most of the UCSC faculty formed their opinions of my scientific abilities during this time."

This was also the period when I got to know Dave. I had entered U.C.

Santa Cruz the year when he was gone to the National Guard. We had little interaction at first because I was finished with my coursework by the time Dave took his up again. But within a few years we were both working on thesis projects that involved taking deep astronomical photographs and scanning these images with a machine to recover the intensities of light coming from the galaxies. Sandy Faber had just come to Santa Cruz when Dave was looking around for a thesis adviser and project, but she already had two students under her wing and turned Dave down. This left Dave working fairly independently of Merle Walker, the astronomer who agreed to be his thesis adviser, and I had been cut loose of mine, Joe Wampler, who had just left for Australia. When it came to finding our way through the technical difficulties of our theses, we had little choice but to talk to each other.

"Alan and I began interacting scientifically much more when we started our Ph.D. theses: we had both chosen photographic surface photometry as the main observational parts of our theses, and there really wasn't anyone at Lick who could help us with the nitty-gritty details. Sandy knew a bit about the subject, but not at the depths we needed to probe. Hence, Alan and I learned much more from each other than from any of the faculty about how to analyze our data. I remember that I would typically first come to Alan, having discussed such-and-such an effect, and he would come to me later, having analyzed that effect quantitatively in his own data, giving me back his comment and then discussing another effect. This back-and-forth educating of each other went on for about a year and a half."

This give-and-take worked fairly well in spite of the fact that Dave and I didn't get along very well. Dave seemed to want more of a personal friendship than I was interested in—he thought we had a lot in common, but if we did, they must have been things I wanted to leave behind. Dave was not very confident of his performance in graduate school in this period, and he compensated, as people often do, by highly touting his abilities and talents. My parents had raised me to believe that this was the epitome of bad manners. (A smug satisfaction indicating that one thinks oneself a bit better than someone else—a behavior I myself have been accused of—was bad enough; to actually tell a person how good one is was intolerable.) This behavior surfaced especially in sports: we were both involved in intramural softball and played volleyball and tennis. Dave was surprisingly good at sports for such a big, bearlike man. At first glance he appeared rather clumsy, but he could hit a ball a mile and spike the hell out of a volleyball. As we were both very competitive, there were more than a few occasions when verbal feathers flew, particularly in the heat of "combat." What I found particularly enraging was that Dave, even when he performed well, had to remind me and others of how much better he was before his back injury—if only we could have seen him then.

I remember particularly the way he bragged about his tennis prowess, for years, until I couldn't stand it anymore and I challenged him to a match. I was a decent tennis player—at least I was fairly quick and coordinated enough, I thought, to outhustle him. I was wrong. Dave was much better than I could have imagined; he had played for his high school team and almost made varsity at Wesleyan, and he wiped the court with me in the first set, 6–2. By then his back was giving him considerable pain and we had to stop. At that moment I might have gained new respect for Dave, had he not taken the opportunity to lecture me on how much more severely he would have trounced me if it weren't for his injury.

This unfortunate incident set the tone for several years to come. I was finishing up my thesis and on my way to a Carnegie Fellowship at the Hale Observatories in Pasadena. Our contacts became few and strained. The next year, when I returned for a visit, Dave was trying to find a postdoctoral fellowship himself, and having little success. He forced a conversation, asking why life was so unjust: the two of us were alike in so many ways, according to him. We had both finished our theses with Faber; she had been a very good influence on each of us. He demanded that I explain why things were working out so well for me and so poorly for him. He was looking for sympathy, or support, or something, but I was stone cold to his supplication.

That was the low point of our relationship; things would improve as we both gained confidence in our careers over the next three years. Dave's road to a permanent job in astronomy was much rougher than mine, but he eventually landed a good one. We began to talk from time to time—Sandy was a common point in both our research lives, and she played a parental role in this sibling rivalry, to good success.

"Over a period of time," recalls Dave, "Alan and I developed a kind of not-so-easy relationship that recognized the learning that we did from each other, but still chafed at our personality differences. It is in my nature to be at once critical of things around me, but also to want to try to help to change the things that are wrong to those that are more in keeping with how I perceive the right. A ninth-grade job-aptitude test I had taken spit out two options for me—scientist or social worker. This trait of mine is obviously designed to get me into trouble, especially since I am not very diplomatic with words, and it does so regularly. As a result, we have had our blowups over the years, with some being more notable than others. Yet, through this all, we have still interacted scientifically, have still gained knowledge from each other, and have both profited from the relationship. Given the very strong personalities that exist among those of us who worked on the elliptical galaxy project, it was remarkable that all of us worked as well together as we did.''

∽

By 1960 the new view of the universe had matured. The universe had been recognized as dynamic—the product of a cathartic genesis, immense in scale—though apparently finite, and old, but not eternal. Much of what had been learned lacked detail: the big bang was, for the most part, an unspecified event. The cause of the expansion was unknown, the process of galaxy formation continued to be largely unexplained, and the destiny of the universe remained a mystery. But a basic picture had been sketched out, and the next generation of astronomers would face the task of seeing if these were simple outlines of a grander picture, or just a mirage.

Our imaginary journey of floating among the galaxies had finally come to include a ruler and a clock. With these there is, of course, a natural desire to relate the size of the universe to *human terms*, those appropriate to everyday life. This will not work for the realm of cosmology. The range of size we humans experience on Earth is far too limited to favor analogies with the scale of the universe. Let me explain why. If the Earth were the size of a pinhead, about the smallest thing a person can comfortably see, the visible universe would still be 6 trillion miles across, which is still solidly outside the grasp of human experience. Conversely, if the entire universe were reduced to the tiny size of our planet, about as large a distance as humans have a feeling for, the rescaled Earth would shrink to a sphere smaller than an atom—not much help here either. There simply isn't a big enough *range in size* in our world to build a useful scale model of the universe.

There is no choice but to break out of this mode of thinking. We must adopt a unit of measure limited only by our minds, not by our bodies. What could be better than following Einstein in his most famous ''thought experiment,'' by accelerating to the speed of light and expressing distances in terms of time? It takes only eight minutes for light from the Sun to reach the Earth; the entire solar system is but a few light-hours across. We now traverse this planetary realm with ease. Journeys on a light beam to the nearest stars require a career change—it is eight years to Sirius—but at 100,000 years, a pilgrimage to the outer bounds of our galaxy is another matter altogether. Still, one can at least imagine such a journey. A sizable piece of our galaxy could be crossed in the time that humans have recorded their history, a tiny fraction of the nearly 5-billion-year lifetime of planet Earth. We could even imagine sailing to the next galaxy, Andromeda, some two million light-years away, about the length of time the kin of *Homo sapiens* have lived on this planet. It should be no surprise that a voyage across the entire visible universe would last the age of the universe itself: its size is limited by just how far light itself has traveled since the big bang, about 15 billion years ago.

These numbers are overwhelming in terms of a single human lifetime, but they are within the context of *humanity's* lifetime. Human lives are brief and human scale is small, but our minds are able to hurdle these bounds and relate the universe to the lifetime of our species and even of our planet. Once this conceptual leap is made, the universe appears as vibrant rather than a lifeless snapshot. Galaxies, which before appeared frozen, wheel majestically through the sky. Our own Milky Way spins like a giant saucer, each of the billions of stars in its disk orbiting our galaxy's central bulge. One such traveler is our own Sun, which has circled the galactic center some thirty times since its birth, entrained with its family of planets like a mother goose. Galaxies scintillate as if sewn with sequins. Their stars, no longer paradigms of constancy, are born in bright-blue flashes, become red and puffy in middle age, and perish in blinding explosions. Each galaxy sparkles with the rhythms of star birth and death, a vast ecosystem of generation and regeneration.

The very fabric of the universe—intergalactic space—stretches as galaxies continue their outward flights from the big bang. The galaxies themselves are not expanding (nor is anything inside them, like the solar system), because gravity has locked stars in their orbits; similarly, the mutual attraction among galaxies in dense clusters gathers them into ever-rising castles surrounded by ever-widening moats. The great voids are growing and they will continue to do so as the universe slides toward an icy, still death in the inaccessible future.

This is a world beyond the dreams of our ancestors of thousands of centuries, uncovered only in this century by humans of relentless curiosity and perseverance, with the opportunity and privilege bestowed by the wealth of our civilization. What great fortune to be alive when the universe was discovered.

# 3 • ISLANDS IN THE SPACE SEA

Our mission was to learn more about galaxies, the colossal star systems that are the defining element in our universe. How had they formed, and when? What had caused the visible matter in the universe to organize itself in such a regular, consistent way? How could galaxy properties be used as yardsticks and clocks to measure the universe and chart its evolution? These questions formed the basis for our expedition, in the same way they had guided the quests of the two generations of extragalactic astronomers before us.

Astronomy is a field of investigation led by observations. Rarely have the predictions and speculations of theoretical astrophysicists been far ahead of the evidence gathered at the telescope. Nature's imagination has proven far richer than our own. It is unlikely that the answers to these questions will fall like gentle rain from the sky of pure thought; finding them will require gathering vast amounts of information about galaxies.

What are the "observables" of galaxies? The two most readily measured are size and brightness. Brightness is recorded as the galaxy's magnitude, and size as an angle measured on the sky. These can be measured accurately, but luminosity and physical size, the crucial parameters that derive from them, depend on the *distance* and are known only as accurately as it is. Spectroscopy delivers many additional properties, though much work is required to extract these properties, and the derivations depend to a greater extent on certain assumptions and a physical model for the system. With a spectrum, the collective motions of stars that make up the galaxy can be measured (stellar *kinematics*), the ages and types of stars (stellar *population*), and the rate of star formation can be inferred as well. A spectrum will also yield the redshift, which, by application of the Hubble relation, provides an estimated distance. And a spectrum allows a kind of chemical assay of the stars and gas, in particular the measurement of how much material is present in the atomic elements heavier than hydrogen and helium, which is a measure of the history of chemical evolution in the galaxy.*

From the size of the galaxy, determined from the apparent size on the sky and a distance estimate, and motions of stars, measured by Doppler shifts in the spectrum, the mass of the galaxy can be calculated. This is done using Newton's law of gravitation, which relates how much force is produced by an amount of mass at a given distance, and Newton's famous formula $F = m \times a$ (force equals mass times acceleration), which describes the motion of a body in response to a force. Stars orbit a galaxy much as the planets in the solar system orbit the Sun, and Newton's laws can be used equally well to "weigh" the Sun or the Milky Way galaxy. Comparing the luminosity of a galaxy with its mass produces a parameter called a *mass-to-light* ratio, an interesting measure because it describes "galaxy efficiency" in the sense of how much light is produced for a given amount of mass (analogous to horsepower per pound of weight in an automobile engine).

Size, luminosity, mass, mass-to-light ratio, abundance of heavy elements—these are some of the important parameters used to describe a galaxy. The search for understanding galaxies is propelled to a large degree by the discovery of correlations among these parameters. For example, we know that galaxies with greater mass are bigger, shine more brightly, and their stars and gas move more rapidly than in smaller galaxies. We have learned that they shine more brightly because they have more stars (rather than the same number of brighter stars), that their greater size demonstrates nature's preference for keeping approximately the same number of stars per volume (rather than cramming a greater number of stars into the same volume), and that their stars and gas move faster in response to the increased gravity that goes along with greater mass. These are already helpful clues to how the galaxies and their stars formed, but there are more subtle and probably more telling relations that are not so well explored and whose implications are poorly understood. We wonder, for example, whether bigger galaxies have, over their lifetimes, formed stars at the same rate, or faster or slower, than smaller galaxies. Has this anything to do with the observation that bigger

---

*Astronomers refer to all heavy elements as *metals*, collecting into this one broad category not only the traditional metals of the chemist, but also the building blocks of life—carbon, nitrogen, and oxygen—as well as chlorine, silicon, neon, etc. The gathering together of everything heavier than hydrogen and helium under one label is an astronomical shorthand based on the fact that nearly all of this material has been synthesized by successive generations of stars.

galaxies have a higher proportion of heavy chemical elements (for example, iron, magnesium, and carbon)? The answers to such questions will help us understand the birth processes and life cycles of galaxies, and illuminate conditions in the early universe when galaxy formation began.

In 1980 members of our group believed that elliptical galaxies were likely the simplest galaxies, and, because of this, the best place to start looking for such correlations. Ellipticals have but one structural component, a roundish ball of stars astronomers call a *spheroid,* whereas spirals have both this and a flat disk. Stars in the disk of a spiral galaxy move on more-or-less circular orbits, in a flat plane, but stars in a spheroid move on orbits that are more *isotropic*—in all directions. In this way the elliptical galaxy is simpler: it is made up only of stars on elongated, randomly oriented orbits. Another way in which elliptical galaxies appear to be simpler systems is that they are no longer forming stars; on the contrary, they appear to have birthed most of their stars early in the universe's history. This is again a simpler behavior than for spiral galaxies, where there is a wide range of star-forming rates and a complex history of star formation. It was reasonable to hope that the best chance of cracking the case of galaxy evolution lay in these simplest of galaxies. This is where we would start.

Fifteen-year-old Roger Llewelyn Davies, awake in the middle of a cool summer's night, peered with wonder approaching disbelief at the little black-and-white television screen. He was on holiday with his family, in the coastal resort of Southsea, England, at a modest guest house 200 miles from the family home in Scunthorpe, Lincolnshire, 100 miles north of London and not far from the North Sea. It was as far as Roger had ever been from home. But that night, as the rest of his family slept, Roger slipped away and traveled as far as any human had ever gone—he went to the Moon.

On this night, July 20, 1969, Roger joined Neil Armstrong as he pushed from the bottom step of the spaceship *Apollo 11* and fell softly, silently to the powdery surface of Mare Tranquilitatis. Who could resist the thought that humankind had crossed a threshold, that something genuinely new had happened? Certainly not Roger, who remembers that night as the moment when his boyhood fascination with astronomy became a direction for his life.

For a young person to be swept up in the mysteries of outer space is not unusual; indeed, in our country a surprising number of children are firmly committed to a career of astronomy—at least, for a week or two. But for Roger to plan such a future would have been most extraordinary. Born and raised in a small steel town in rural England, an only child whose father, along with every other male member of his family, labored in the steelworks, the most he could hope for was employment there as a chemist. Even that was a stretch—no one in their family had ever attended university.

Scunthorpe was a stronghold of the British Labour party that led the nation after World War II; Labour's nationalization of the steelworks in the 1960s and its subsequent prosperity represented a success story for the socialist government. Ironically, this Labour island floated in a Conservative sea— the surrounding farmers with their plentiful harvest of wheat, oats, barley, potatoes, and sugar beets had little use for progressive politics. But in Roger's world in Scunthorpe the socialist perspective ruled: the Labour Council saw to it that the schools and parks of these working-class families were well cared for, and the town enjoyed a strong sense of community and security. Roger played contentedly with his "mates," passed enjoyable hours with his family, and sang with the school choir, traveling to music festivals around the north of England to compete. He was a sensible, confident, modest, well-mannered, and good-humored young man.

But a scientist? After all, this was no utopia, it was still a working-class community where grains of iron ore spoiled the wash on the line whenever the wind blew the wrong way, where hard work and simple values—not dreams about the origins of the universe—were the daily pudding. Yet Scunthorpe produced not just one but *two* people whose lives engage such ethereal pursuits. As he was growing up, Roger would hear again and again of Wallace Sargent, the first student from the John Leggot Grammar School to attend university. When Roger entered the school, Sargent was already a distinguished professor at Caltech, an explorer of distant galaxies and quasars who worked with the world's largest telescope, the 200-inch at Palomar. Sargent was the son of a gardener and a cleaning woman. The furthest his father got was to be a "fitter's mate" at the steelworks; an academic life was about as far from the family tree as a shot at the Prime Ministry. How had he made his way?

Through the schools. What this working community did have to offer was an education, an opportunity to any and all to broaden their horizons beyond Scunthorpe. It was part of a social revolution that occurred near the end of World War II, when government leaders realized that returning troops would be unwilling in the future to accept the class system that had ruled British society for centuries. Where good health care and education had previously been reserved for the privileged and the wealthy, laws were passed and monies allocated that began to provide, for the first time, these crucial benefits and opportunities for all English citizens. This product of the socialist revolution and the war was a fresh breeze that carried Sargent, and fifteen years later, Roger Davies, to previously unreachable destinations.

Roger remembers these early schooldays fondly: "My elementary school education concentrated more on instilling the fun of learning and understanding rather than packing me full of facts and figures. In the long term this approach has proved to be a great asset to me, and it was in primary school that I first gained an interest in astronomy through school projects—I particularly recall one on the planets. My yeargroup was amongst the last in Scunthorpe to be subjected to the 11+ examination, where at age 11 pupils were examined to determine whether they would proceed to the grammar school (the more academically able) or to the secondary modern school (for more vocational emphasis). Fortunately, I passed this examination."

The rigors of Roger's early education may sound extreme compared to the practices of U.S. public education today. "At John Leggot the teaching was intensive, for the five years from age 11 to 16 we took ten academic subjects, each of which were examined twice a year with typically two exams of 2–3 hours per subject . . . about 200 exams over 5 years.

"I developed a natural inclination toward mathematics and my interest in astronomy grew. My parents bought me a small telescope and I ran the school astronomical society. All this was in addition to the usual practical school subjects such as woodwork, art, games. In the last year of my grammar school education (1968–69) I joined in an extracurricular scheme known as the Duke of Edinburgh Awards for school students that involved meeting a basic level of competence in a wide range of mostly outdoor activities, such as sports, hiking, and camping. This scheme involved doing a project, for which I chose the Apollo Missions 8–11 which were underway from Christmas 1968 to summer 1969. Those missions left an enormous impression on me; I recorded every detail for my project. The Moon landing occurred while we were on our annual family holiday. I remember getting up in the early hours of the morning and being the only person in the TV lounge of the small guest house where we were staying to watch it live. The Moon landing cemented my interest in astronomy."

Roger took his first summer job at the steelworks in 1969, as a lab assistant in the quality control laboratory of the coke-oven plant where his father worked as a maintenance engineer. But he was beginning to understand that his proficiency in math and science might carry him away from all this, to a university education, perhaps even to pursue the astronomy he enjoyed so much. In the early 1970s the steel industry began to slump, and the likelihood of finding a secure and satisfying job there ebbed away. Roger realized, as did most of his friends, that his future did indeed lie far away from the comfortable home of his childhood. He selected two universities with strong coursework in physics and research groups in astronomy, and ultimately chose University College London (UCL) over Manchester University.

"My first term in London was a disaster, so many distractions, so many fascinating diversions. After the Christmas break the department organized exams to evaluate the performance of first year undergraduates; I recall my interview with the senior tutor to discuss my performance in these exams. It was abysmal—he gave me a rocket and it worked. I had a lot to catch up on, one term behind and a full lecture load in the second and third terms to absorb. I worked very hard through the first half of 1973 and enjoyed it; physics at university is quite different from school physics and I found it challenging and stimulating."

Roger finished with first class honors, and, with summer experience at the Royal Greenwich Observatory at Herstmonceux, the pipe-dream of pursuing a career in astronomy became a reality as he was offered graduate research positions at both UCL and Cambridge University. He chose Cambridge, partly for its excellence, and partly because he had grown weary of the bustle of London life and welcomed the peace of a more pastoral setting and a calmer pace. He fell in with a talented research supervisor who built his own instrumentation for telescopes, instrumentation used to study cosmology through observations of distant galaxies that are strong sources of radio waves. Roger chose to investigate elliptical galaxies for his thesis; like many other young astronomers of the day, he was hoping to learn about how these simplest of galaxies (or so it was thought) formed and evolved, as a clue to all of galaxy evolution. His task would be to measure the motions of stars in each galaxy, from the center to their periphery, in order to learn how much mass the galaxy contained and how this distribution of mass over the body of the galaxy might have been initially laid down.

It was an ambitious undertaking, and Roger was fortunate that Don Morton, the new director of the Anglo-Australian four-meter telescope, shared his enthusiasm for the project and invited him to come use a new spectrograph built by Joe Wampler. In six months Roger was able to gather excellent data for his project, and to broaden his vista again, to the jazz clubs of Sydney, the great beaches of the north coast, and the wild countryside and mountains of the interior.

Roger returned to England, finishing his thesis in December 1978. His timing continued to be excellent—there was now great interest in the structure and stellar populations of elliptical galaxies, so Roger had little trouble in obtaining a number of offers for postdoctoral fellowships. He chose the

Lindeman, which allowed him to take a year at U.C. Santa Cruz, where he could work with Sandy Faber before returning to finish his fellowship at Cambridge.

"It seemed like all the important astronomers were there at the time. I arrived in Santa Cruz and got the bus up to the glorious campus in the redwoods. I found the Natural Sciences II building and staggered with my suitcases to Sandy Faber's office; I had, of course, never met her. I have rarely been so relieved—a friendly face wearing octagonal spectacles and a lift in an enormous estate car [station wagon] to a college room where I could stay for a while."

Soon afterward, Roberto Terlevich dropped by to visit, and, with Sandy, Dave, Roger, and Roberto, our project was born. Roger's journey had been a series of openings to wider and wider worlds. He had reached his dream of a life in astronomy. Though far from Scunthorpe, it was very much part of his story and responsible for where he now found himself; the schools of the industrial town of his birth had launched him. Through his own pluck and hard work he had become an explorer of cosmic space. How many such minds go undeveloped? How might they change the world?

Galaxies are the principal wonders of the universe-at-large. They blaze against the darkness of space; the concentration of visible matter into these behemoths is stunning. Mass that was once in the form of a thin gas has been transformed into stars, an average of 100 billion in each galaxy—a dozen stars or so for each person on Earth. Though they occupy only one-hundredth of one percent of the volume of the universe, galaxies are the sites of all the action, the crucibles of stellar evolution and change. The space between them is, by comparison, empty and still.

Almost as striking as the degree of concentration of visible matter into galaxies is their remarkable uniformity. Galaxy shapes vary, but these variations are familial, like the difference between a giraffe and a horse rather than the difference between a giraffe and a jellyfish. They come in a variety of sizes and brightnesses as well, but these represent only scaled versions of what is basically the same animal. This simplicity is repeated over huge distances (and thus back in time as well): a few basic shapes, a wide range in size, a lone galaxy here and there, many doubles and triples, a few groups of several large and many smaller galaxies, perhaps one populous cluster. This is the way the universe looks, in neighborhoods a hundred million light-years across.

Finding the underlying causes of this remarkable regularity is a preoccupation of current astronomical research. As is frequent in science, taxonomy has been a first step: grouping subjects into classes that share some characteristic can uncover clues to the physical process that is responsible for the trait. For example, the classification of nebulae into regular and irregular types was the first hint of two different phenomena. The rarer disk-shaped nebulae were later found to be external galaxies—very massive, old systems well organized by gravity. In contrast, the irregular appearance of the more common amorphous nebulae betrayed their relative youth and the chaotic interplay of forces not in equilibrium; eventually the suspicion that these were smaller star and gas clouds in our own galaxy was confirmed.

Some of the early work in galaxy classification was again done by Hubble. He identified three primary types—spiral, elliptical, lenticular (lens-shaped)—distinguished by shape and morphology rather than size or brightness. Spiral galaxies, like our Milky Way, are the most common type; they typically appear in the sky as stretched ovals with spiral bands in their outer regions. Astronomers soon recognized that a spiral galaxy is basically a flat disk—those that are seen edge-on are 20 to 30 times as thin as they are across, while those viewed face-on are nearly perfect circles.

Not all spiral galaxies are completely flat; many also swell in the middle. This *bulge* is known to be more-or-less spherical because the central regions appear roundish regardless of the angle at which we view the galaxy. (Only a sphere looks round from every angle.) Even though the view of our own galaxy is obscured by our location in its dust-laden disk, we can see that it has both these characteristics: the thinness of the disk is evident as the narrow band we call the Milky Way that rings the sky, and the round bulge of our galaxy shows as a striking widening of this band toward the constellation Sagittarius.

Elliptical galaxies, in contrast, are simply roundish balls of billions of stars. After many years of research, it is still unclear whether they are all egg-shaped (prolate) or onion-shaped (oblate), or whether there are some of both. This has been a difficult issue to settle because we have only one view of the cosmos. We see each galaxy from only one direction: in our lifetimes, instantaneous by cosmic standards, we have no chance to travel to a place from which we might view a galaxy's "better side," and no time to let it turn beneath our gaze. Instead we must use our senses and wits to sculpt the true shape of these glowing silhouettes. This is how we have been able to conclude that spiral galaxies have flat, circular disks, by observing that those seen face-on look round, those viewed edge-on appear flat, and those tipped at arbitrary angles describe near-perfect ellipses. For the roundish ellipticals, however, whether they are egg-shaped or onion-shaped remains ambiguous.

These are different types of galaxies, from the flat disk spirals seen at the top, to the round elliptical galaxy seen at the bottom. Spirals are full of gas and dust, the raw materials of ongoing star formation, while elliptical (E) (bottom left) and S0 (bottom right) galaxies are, in comparison, nearly gas-free. *(Photos courtesy of the Carnegie Observatories.)*

Members of the S0 (S-zero) class of galaxies (Hubble called them lenticu-lars) look like a cross between an elliptical and a thin-disk spiral. The round bulge in the middle dominates the light of the fainter disk that surrounds it, and there is no spiral pattern seen in the disk.* It was the spiral pattern prevalent in so many galaxies that allowed astronomers to move from taxon-omy to a rudimentary understanding of the physical processes that give rise to the different types. As more was learned about the life cycles of stars, it became clear that the spiral pattern is delineated by young stars forming in thin disks, and that this spiral pattern, the thinness of the disk, and the formation of new stars are all intimately related.

We are now certain of the basic picture: the disks of spiral galaxies are mainly made up of stars moving on near-circular orbits around the galaxy's center. Floating among the stars is a lesser mass of gas from which the stars themselves have formed. Mostly this gas is made up of hydrogen atoms. Helium atoms make up most of the rest, but about one percent is in all-important heavy atoms, including carbon, nitrogen, oxygen, silicon, magne-sium, and iron. Most of these heavy atoms are bound together in tiny grains of hydrocarbons, silicates, and ices—a thin mist of the same material of which the Earth and we ourselves are made. The average density of this gas and dust between stars is only one atom per cubic centimeter, a better vacuum than has ever been produced on Earth, but 100,000 times more dense than the gas *between* galaxies, and dense enough to be a starting point for star formation.

Along the spiral arms in a galaxy are thicker clouds of gas where new stars are condensing and lighting their nuclear fires. In at least some cases the spiral pattern is itself the agent for this activity. A spiral arm is a pressure wave that sweeps continually around the galaxy, like a water wave sloshing back and forth in a bathtub. As it passes, the wave squeezes the cold, dormant gas, compressing it to where the pull of gravity takes over and drives an avalanchelike collapse. As each gas cloud contracts, its temperature rises; this causes it to glow more brightly, and the loss of this radiant energy causes the cloud to contract further. Only fragmentation into stellar-size globules can stall this inexorable collapse, for in this new form central temperatures rise to millions of degrees, igniting the enormous power source of nuclear fusion. The energy released by a star's central nuclear fire releases tremen-

---

*There is also a class referred to as "irregular," which, as might be guessed from the name, is a catchall for the few percent that are not spiral, elliptical, or S0 galaxies. Some appear to be galaxies too small to sustain a stable, well-organized shape; others seem to be galactic equivalents of train wrecks, the results of disruptive collisions.

dous heat—in a gas this translates into pressure, enough pressure to counter-balance the staggering weight imposed by gravity.

Stars like these, born in small clusters by the hundreds or thousands, will eventually settle into stable adulthood for tens of millions or even billions of years. During this time they will steadily convert the hydrogen atoms in their cores into helium atoms. In their twilight years they will fuse this helium into carbon, nitrogen, and oxygen atoms, and the most massive stars will heat their central furnaces to tens of millions of degrees, temperatures sufficient to fuse these lighter elements into heavy metals like cobalt, nickel, and iron. All this will be a futile attempt by the star to find more fuel to stave off the relentless crush of gravity, but this desperate activity is doomed. Eventually, nuclear fuel will be exhausted and gravity will rule once again.

Stars with a modest mass, like the Sun, however, will escape destruction by gently puffing their outer mantles into space (forming one of the kinds of nebulae) and then, relieved of the great weight, cool quietly to a diamondlike, crystalline form known as a white dwarf. But the weight of the outer layers of the more massive star is too great; eventually the core collapses catastrophi-cally, and moments later this triggers a violent explosion called a supernova. In the chaos, all the heavy elements formed by nuclear fusion are spewed back into space. The debris from these explosions is the sole source of all atoms other than hydrogen and helium. It scatters through space, eventually mixing with other drifting gas clouds. Someday this material will be driven into collapse and form new stars with encircling planets, and with it the stellar life cycle will have completed another generation.

The moral of this story is as follows: 90 percent of your body weight, everything but the hydrogen atoms, was once inside a massive star that blew up as a supernova. The 10 percent that is hydrogen came directly from the big bang. Perhaps your ancestors were not among those who sailed on the *Mayflower*, but you could hardly have wished for a more distinguished family tree.

Like a snapshot of a high jumper clearing the bar, the process of birth and regeneration in spiral galaxies is frozen before our gaze. Swirling clouds of glowing gas pumped by the energy of star formation must once have been the scene in elliptical and S0 galaxies as well, but now these kinds of galaxies are nearly quiet. Gas expelled by their dying stars is heated and blown into space; without a thin disk to collect and concentrate this gas into dense clouds, no new stars are born. It seems that elliptical and S0 galaxies, and in general all galaxies with large central bulges, formed most of their billions of stars long ago, in quite a fervor. Now they are slowly fading, as succeeding generations of dying stars grow fainter and redder. The big-bulge galaxies are furthest along the path to oblivion that all galaxies, even the still-young, thin-disk spirals, will eventually take.

Astronomers have long sought the origins of different galaxy types, reasoning that this variation is a direct result of the way in which galaxies formed and evolved. To recount some of the ideas one needs a sketch of the universe when it was very young, probably about one billion years old. As the universe expanded, its smoothly spread gas cooled and fractionated into large clouds. At first slowly, and eventually very rapidly, these were contracted into comparatively small, dense clouds by gravity—a larger-scale version of what is happening in star-forming regions today. We don't yet know whether these clouds collapsed as single large units or grew from the aggregation of smaller chunks. Furthermore, the causes of differentiation into spiral, S0, and elliptical galaxies are not detailed in this simple outline. Clues to the "what and when" of the processes involved might emerge from comparing galaxies—their sizes, shapes, and star-formation activity—in relation to the conditions where they are found today.

A clear picture has not yet emerged, but headway has been made in identifying processes that played a major role, either at the time galaxies were born or later in their lives. The discussion is usually couched in terms once owned by social workers: do galaxies acquire their traits by *nature* or *nurture*? In the astronomical context this translates into: did variations in *initial conditions* cause some protogalaxies quickly to develop into either thin-disk spirals or large-bulge elliptical and S0 galaxies, or were galaxies created in a single form that then slowly evolved into the types we see today, perhaps due to the influence of the different environments in which they live?

The answer does not appear to be a simple one, just as it is not in the case of the "nature vs. nurture" debate in behavioral psychology. There is evidence that both conditions of birth and environment play a role. In the 1970s, when I was a graduate student, the fashionable notion was that environment, not initial conditions, was the principal factor. The basic thinking started from a point of view emphasized by astrophysicist Jim Peebles at Princeton: conditions all over the universe must have been very uniform at the time galaxies were forming—calculations showed that for many hundreds of millions of years after the big bang the density and temperature in the still gaseous universe must have varied little from one place to another. Peebles's model was based on his realization that large density contrasts, from the great clusters of galaxies to the huge voids, developed slowly and relatively recently. The force of gravity would have been the agent that amplified the slight undulations in density left by the big bang, contracting denser regions so they became denser still. Likewise, regions that initially had been only slightly underfilled would be emptied out in the process. These fluctuations in density can be compared to the high- and low-pressure systems that control the weather, but in our atmosphere these regions first grow but later dissipate, whereas in the universe, gravity is unopposed, so the fluctuations grow and grow in amplitude. This notion that structure arose by gravitational amplification of slight density undulations in an originally smooth distribution of matter still forms the basis for most models of the formation of structure.

Peebles's favorite version of this model, which he called *hierarchical clustering*, worked from the bottom up: first to form were star clusters; these then clumped together to form galaxies; finally, these galaxies aggregated (while retaining their individual identities) to form clusters, and clusters of clusters (superclusters), with voids between them. According to this picture, initial conditions shouldn't have played much of a role in determining galaxy type, because at the time galaxies were forming all of them shared a similar environment. On the other hand, we know that differentiation into types is not random—elliptical and S0 galaxies are today found preferentially in more crowded regions, and spirals favor more suburban environments. This means that galaxies could not have been formed willy-nilly without regard to location, so the proponents of the hierarchical model concluded that environment must have worked on galaxies to change them into different types—later.

There was a fairly large body of observational evidence that supported this view; my own Ph.D. thesis provided a small bit of it. The question under consideration was the distribution of galaxy "size," generalized to mean not only the physical dimensions of galaxies but also their luminosities and masses, all of which are strongly coupled. According to George Abell, then an astronomer at UCLA, the distribution of galaxy brightnesses varied little from place to place in the universe—it had a *universal form*. Abell had come to this conclusion by comparing the populations of clusters of galaxies, dense groupings of several hundred galaxies bound together by gravity. By choosing clusters for his study, Abell had secured the favored volume-limited sample, simply by roping off relatively small regions of space. For galaxies in these clusters he measured both the range in brightness and the number of galaxies in each interval of brightness. This distribution, referred to as the *luminosity function*, is much like the distribution of the heights or weights of the members of a church congregation.

Working with clusters gave Abell an additional advantage: all the galaxies in a cluster were at approximately the same distance from our galaxy. Therefore, he could take relative brightness as a measure of luminosity—the true brightness. In this way, the additional uncertainty due to different distances was avoided.

Not that Abell's task was an easy one. Measuring even the apparent brightnesses of galaxies is not a cinch. It is relatively easy to determine the magnitudes of star images—because they are pointlike, nearly all the light lies within a circle several times bigger than the blurring caused by turbulence in the Earth's atmosphere. But galaxies are not so easily wrapped up, because they are extended—generally their boundaries are indistinct. True, the *surface brightness* of a galaxy drops steadily as one looks farther from the center, but, since the area grows larger and larger as one looks farther out, the fainter glow of the periphery continues to contribute a significant amount of the total light. Deciding at what point one has "gotten it all" has proven a thorny problem for astronomers.

By the time I had taken up measuring galaxy magnitudes, there were computer-controlled machines to scan a photographic plate and measure the brightness at each point of the image—for my thesis research I helped develop such a machine for the Lick Observatory by cobbling together an old measuring machine with a new computer, one tiny by today's standards. With the intensity map produced by such a machine, all the light in a galaxy image could be added together, or at least a consistent procedure could be adopted to get some large fraction of it. The scanning technique was enough of an improvement over the method Abell had used ten years earlier to justify a new test of his conclusion that the distribution of galaxy brightness in populous clusters is universal. I chose twelve of the "richest" clusters from a catalog Abell himself had compiled for *his* Ph.D. thesis, when he had searched by eye each newborn photograph as it emerged from the darkroom of the Palomar Sky Survey, a massive project for which he was a principal observer. This catalog, compiled from the same atlas of photographs that our new group was using to examine and choose elliptical galaxies, was still the basic source material for studies of clusters.

The experiment Abell had done, which I was now to repeat with a better technique, was analogous to rounding up a random sample of people in different countries around the world to test if there were significant differences in the distributions in height or weight, as might result, for example, from differences in genetics or diet (nature or nurture). Eventually my three years of work would lead me to the conclusion that Abell had been right—the variance was amazingly small. I remember that at the time I was rather disappointed with merely confirming something that was already known, but it would be an important part of my education to learn that testing earlier ideas with better data is vital to the process.

The Hercules cluster of galaxies. All extended objects are galaxies. The sharp round dots are stars in the foreground, in our own Milky Way galaxy. *(Photo courtesy of the Carnegie Observatories.)*

In fact, I did find some differences in the luminosity of the average galaxy from cluster to cluster, but the variation in the luminosity distribution was never larger than a factor of two, a small amount compared to the total range of several hundred. This would be like finding a range in people's heights of 1 to 8 feet, but an average height that varied by less than an eighth of an inch from country to country. So little variation compared to the range would argue against any hypothesis correlating height with diet or climate, for example. The result for galaxies seemed to be saying something similar: from place to place galaxies have the same distribution in size and brightness, as if they were all born in pretty much the same way—this echoed the prevalent view of the time promoted by Peebles and others.

This constancy is a bit misleading, however. Though the variation in size or luminosity may be small, the relative proportions of different morphological *types* varies radically from place to place. In crowded regions like rich clusters, elliptical and S0 galaxies dominate and spirals are rare, while in the thin spray of galaxies between groups and clusters the spirals are the most common type. Taken together, the universality of the luminosity function and the variation in morphological types is like taking a closer look at the height distribution and observing that the ratio (number of women)/ (number of men) is varying widely even though the height distributions remain the same.

This connection of galaxy type to environment had been recognized as early as the 1930s, and by the 1950s some astronomers were citing it as evidence that nurture rather than nature was the key to the different galaxy types. In these extremely crowded regions, they argued, galaxies would suffer glancing or even head-on collisions that would change their appearance forever. Later, it would be discovered that the galaxies in rich clusters move in a sea of extremely hot gas that reaches relatively high density in the cluster's core. It was suggested that the effect of this strong, hot wind blowing through a spiral galaxy might be dramatic, sweeping from its disk the gas and dust that were necessary to sustain star formation.

It was in this area of relating galaxy type to environment where I first felt the thrill of discovery that can reward a scientist for years of struggle with ideas and measurements. I found a new and unexpectedly strong connection between the mix of galaxy types and the crowdedness of their immediate surroundings. The experience filled me with a joy, enthusiasm, and satisfaction I have achieved few times in my life. It charged me with an energy for scientific work that I still draw on.

The work began in 1976 when I came to the Carnegie Institution, then linked with Caltech astronomy under the name Hale Observatories, as a postdoctoral Carnegie Fellow—a two-year appointment. A strong letter of recommendation from Sandy Faber, whose reputation was soaring at the time, seemed to have been instrumental in my receiving what for me was a prized opportunity. Even as a child I had heard about the Mount Wilson and Palomar Observatories; to join scientists at what was probably the most important center of observational astronomy, and to use instruments like the giant 200-inch telescope at Palomar, had not figured into my dreams, let alone my plans.

Applicants for Carnegie Fellowships were asked to submit proposals of their intended observational programs. This was not required by any of the other fellowship applications I had made, and it was a task I dreaded. After several years focused on a Ph.D. thesis, a student can suffer tunnel vision; becoming an expert on a single topic seems to do a remarkable job of erasing the broad picture of the field painstakingly assembled by the faculty in a dozen or more academic courses. This scientific amnesia is temporary, but I was still in its grip when called upon to propose a new project for the Carnegie Fellowship. It was Sandy who supplied the needed treatment. One afternoon I paced nervously in her tiny office while she asked a series of simple yet pointed questions that led me to review what was well known about clusters of galaxies and what was still poorly studied. I was stunned by how easily the discussion converged on the paucity of information on morphological types of galaxies in clusters, and the realization that Carnegie's new Irénée du Pont telescope, with its ability to take high-definition, wide-field photographs, was exactly the instrument to collect the needed data. I never forgot the process. Since then, I have never been at a loss for project ideas, and have been faced with the more pleasant challenge of how to choose the most interesting and promising problems and cast off the rest.

My project was to photograph large regions of the sky containing clusters and classify the galaxies by type. It was an ideal application of the new facility; no doubt this played a role in the offering of the fellowship. The du Pont telescope had just been commissioned when I began in August 1977, and not all of the instruments and their detectors had been completed. So for those of us who wanted to use the telescope as a camera, by placing giant 20″ x 20″ glass photographic "plates" at the focus, generous allocations of time were forthcoming. After two and a half years and about thirty nights of observing, I had good plates of fifty-five rich clusters. These were deep photographs of clusters of galaxies, like the plates I took for my thesis project; however, each photograph covered four times as much area of the sky and, most importantly, the images were three times enlarged. This allowed me to distinguish the spiral arms, disks, and bulge components that separate galaxies into their different morphological types.

Back in my office in Pasadena, I peered zombielike through an eyepiece, six to eight hours a day for nearly half a year, until I had recorded the position, assigned a morphological type, and estimated the magnitude for some 6,000 galaxies in these clusters. The data were entered into computer files (by the now arcane route of punch cards) so that the large number of entries could be rapidly cross-compared and manipulated in order to look for patterns or correlations.

It had taken only a few days of looking at the plates before I was convinced that, as my predecessors had discovered, elliptical and S0 galaxies are more common in clusters, particularly in their dense central cores. But my goal was to look for more specific relationships that might give a clue why this was so. With smaller samples of lower-quality data, some of my colleagues had found intriguing clues. In his 1976 Ph.D. thesis at Caltech, Gus Oemler had suggested that different types appear in near fixed ratios in different clusters, for example 2:1:2 for spirals to ellipticals to S0's in one type of cluster, with two other types of clusters with other fixed ratios. I could find no such regularity for my sample, but worried that my failure to find such an effect occurred because these ratios were very sensitive to where I set the outer boundary of the cluster. Clusters, like galaxies themselves, have a high central density but thin out farther from the core, until they blend inconspicuously with the background of galaxies in the general field. In fact, Caltech astronomer Wallace Sargent and then student Jorge Melnick had found a "radial gradient" in population: starting in cluster centers, elliptical and S0 galaxies are most common, but with increasing distance from the core spirals become slowly but steadily more dominant until they are nearly as common as they are in the general field. Because of this, Oemler's ratios describing the "global" population mix were sensitive to where the outer boundary was drawn, but the irregularity of many clusters meant that any adopted prescription would be both important and rather arbitrary. Nor could I find any consistency in the radial gradient model of Melnick and Sargent. I found that gradients were strong in some clusters, particularly those which have a roundish, symmetrical structure like a huge ball of galaxies, but the gradients were weak in less well organized clusters. These and other trials left me frustrated, wondering whether there was any strong, repeatable correlation of galaxy type with any position, location, or environment.

There was. The key move was to break away from parameters that had to do with the *global* properties of the cluster, like the total population ratios or the distance from the cluster center, and look at a *local* parameter. During those months of visually inspecting the plates, I had often been struck by the fact that, no matter how chaotic the distribution of cluster galaxies, dense clumps were dominated by elliptical and S0 galaxies. This was true even if only a few galaxies were involved, regardless of such global parameters as the distance from the cluster center or the total number of galaxies in the cluster. I began to wonder if purely local environment might be the deciding factor, but was well aware that this was a rather heretical point of view. It was only after I exhausted every other approach I could think of, and hadn't found anything convincing, that I returned to this idea.

I remember telling fellow Carnegie astronomer Stephen Shectman about my idea of writing a computer program that would find the density of the immediate environment around each galaxy, to see if that would correlate better with morphological type. It was his opinion that looking for such a correlation was probably a waste of time. I was taken aback by Shectman's dismissal of the idea—I valued his opinion because I knew he was very smart. After a day's hesitation, I decided to give it a try anyway, mostly because I was fresh out of other ideas.

Just before midnight, several days later, the computer began to spout the results for one-half of the data sample. I had programmed the computer to calculate how crowded each galaxy was by its ten neighbors, then to record how often different galaxy types were found at each level of "crowdedness." Computer graphics were primitive in 1979, so from a printout of columns of numbers I began to plot results on a piece of graph paper. The first point represented the lowest-density regime—on average, just one galaxy in an imaginary cube 3 million light-years on each side. In such sparse locales about 80 percent of the galaxies were spirals, about 10 percent were ellipticals, and 10 percent S0's. As I plotted subsequent points for more and more crowded places, I saw a steep decline in the percentage of spirals, dropping to 50 percent by a density of 10 galaxies in the box, to 20 percent by 100 galaxies in the box, and approaching zero beyond. As the percentage of spirals fell, the percentage of S0's rose first, followed at higher densities by the ellipticals, which finally caught up with the S0's when each contributed about 45 percent of population. These percentages were reached in very crowded conditions, when galaxies were separated by only one or two times their size.

I was elated to have found such good correlation—the dependence of morphology on *local* density was much stronger than any of the global parameters that had been tried before, and it was by far the cleanest result I had obtained in any of my projects. The next day I felt that I was floating, in a state of private euphoria as I nervously awaited the computer's judgment

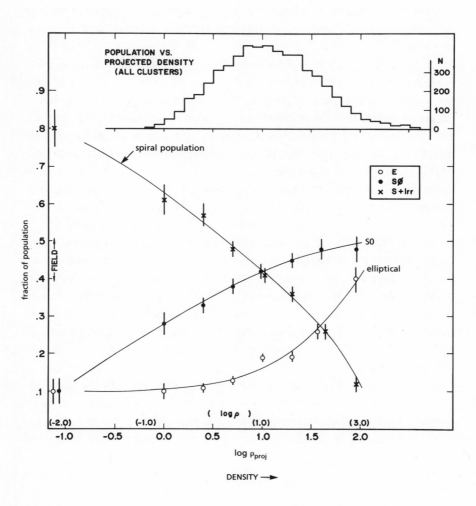

The correlation between galaxy type and environment, the "morphology-density" relation. The percentage of spirals drops, and percentages of elliptical and S0 galaxies rise, with increasing crowdedness of the region in which they are found. *(Reference: 1980* Astrophysical Journal, *vol. 236, p. 351.)*

for the remaining half of the data. That night I had these as well, and I could barely keep from trembling as I plotted the new points. They fell almost exactly on top of the earlier ones. From then on I couldn't contain my excitement about finding this vivid, previously unrecognized pattern in nature, and I buzzed around the Observatory's offices and down at the Caltech astronomy department for a couple of days. I found both interest and encouragement from my colleagues, who seemed almost as surprised and delighted as I had been.

This result, which is occasionally called the Dressler effect, to the gratification of my ego, added an important clue to the nature vs. nurture debate, but it didn't come close to settling it. I was able to argue, persuasively in the minds of many, that these differences in population reflected the extreme sensitivity of galaxy formation to the density of the *early* environment. One might think this is nurture, but for galaxies environmental differences at the time of conception is really what we mean by "nature." This sensitivity to early environment probably has something to do with how much spin a galaxy acquires and how this depends on the proximity of its neighbors (spiral galaxies acquired a large *angular momentum* while ellipticals got very little), or the rate at which the collapse of the initial gas cloud takes place in a higher-density region (ellipticals collapsed more rapidly and thereby turned their gas into stars sooner), or both. More recent, promising work, in which galaxy formation has been simulated in a computer, indicates that coherent collapse of the gas cloud will lead to a spinning spiral galaxy. However, if the contraction is chaotic and develops many subunits, the spins of each of these cancel as they subsequently merge, resulting in the birth of a slowly rotating elliptical.

My discovery of a tight correlation between galaxy type and the crowdedness of the surroundings couldn't by itself answer the question "nature or nurture?" but it did decode another word in this cosmic game of *Jeopardy*. In the not too distant future, the formula for galaxy-making will be spelled out in its entirety.

Roberto Terlevich thought he had found another key piece of the puzzle of how galaxies form. Ironically, the clue came not from the study of anything as grand as an elliptical galaxy, but from pygmy star systems that are tiny in comparison, and often far removed from any major galaxy. These mini-galaxies are regions of intense star formation—like the Orion nebula, but hundreds or thousands of times bigger. Terlevich liked the name *H II galaxies*: H II is the designation for ionized hydrogen gas, and emission lines from glowing hydrogen gas are the strongest features in the spectrum of a region heated by young stars. He believed that these, rather than elliptical galaxies, might be the simplest examples of galaxy formation, miniature galaxies caught in the act of forming their first and perhaps only generation of stars.

Terlevich was passionate about his work, as he was about all matters in his life. His path to the study of the birth of galaxies had been as difficult and unlikely as that of any of the astronomers I knew, and the experience seemed to have left him as an unusual mixture of gentle soul and rebel. On personal matters there were few among our astronomical colleagues who were more good-natured, thoughtful, and supportive. It was as if Roberto, having been through a lot himself, had become more tolerant of his friends' foibles, more adaptable and more resilient. One could so easily elicit a warm and engaging smile from his broad friendly face, crowned with its handsome mane of graying hair; his handshake was firm and true, and followed by a Latin embrace if you gave any encouragement at all. He was an easy man to like.

But Roberto was also an iconoclast. When he was engaged in scientific or political discussions, his eyes would light up and he would speak ardently about the matter at hand. "Party-line" explanations seemed to bore him; his ideas were never conventional, and confirming or supporting somebody else's view of the world was clearly of much less interest than breaking new ground. It would be no accident that Roberto would be the spark of our project. This was a spirit that needed to find its own way and rouse others to follow.

"Both my parents have very modest origins," says Roberto. "My father was born just before the great war, in San Lorenzo, a very old and small village in the Istrian peninsula—an Italian community, though at that time part of Austria. In 1929, after struggling for many years to make a living, he left for Argentina, where he met my mother, a destitute widow with two children who had emigrated from Portugal.

"I was born in 1942, and spent all my childhood in a working class neighborhood. My father worked as a taxi driver and my mother stayed at home. All my primary school was very boring, and I became sort of a rebel. My taste for physics and astronomy developed during my early teens. But I had to wait until university to have fun with my studies.

"One important influence during my teens was a teacher of mathematics who completely changed my approach to science. He showed me how simple and how much fun math and physics can be. I was fourteen and about that time I started to develop an interest in astronomy. I joined the amateur astronomy association in Buenos Aires, and soon I was building my first telescope and reading all sorts of literature on astronomy. One big problem was my poor knowledge of English. I remember having to rely on translations from other amateur astronomers to learn about the recent developments.

"When I finished my secondary education I was undecided on what to do. On the one hand I was very keen about astronomy, but there was no astronomy in Buenos Aires. The only places with astronomy degrees were La Plata [forty miles east of B.A.] and Córdoba [400 miles west], and I could not afford to live away from my parents' home. At the same time I was pressed by them to study engineering or economics, though I did not want to do so.

"I spent one year working in an insurance company. This was interesting and I earned a lot of money. But the following year I had my compulsory military service, which I did in the elite group called the *Granaderos*—the presidential guard. I participated in one coup, as a loyal defender, fought in combat and was bombarded by the Argentinean Air Force. We lost, and the president had to go, but we stayed. It was during this time that I realized how fragile life is and I decided to follow my vocation of astronomy.

"The main reason I went to university to follow a career in astronomy was because of the influence of my eldest brother-in-law, Norberto. I believe he was the only person who recognized my potential. This is probably connected to the fact that he was the only one among all my relatives, and even distant relatives, who went to university. He had a very successful career as a top manager in a very large finance corporation, and he was always pressing me to develop my intellectual capabilities. I remember that Norberto was always interested in my schooling and always willing to help, in many ways. At university, it was really the first time I felt I was doing something difficult and worthy.

"So, after leaving the army, I did the introductory course at La Plata University and passed the admission examinations. My first year was a revelation for me; it was so exciting—all the new knowledge and the feeling that your brain is somehow growing at a very fast pace. The big surprise came when I passed all the first-year examinations with distinction—I was not expecting that. After my first year I ran out of money and I got a job at La Plata Observatory as a telescope operator. This included a salary and, perhaps more importantly, a room for me to live in (I had been commuting every day from Buenos Aires). I built all kinds of equipment from photoelectric systems to domes for telescopes. These were good times.

"After finishing my degree, I was offered a temporary job at the newly formed Instituto de Astronomía y Física del Espacio, IAFE [Institute of Astronomy and Space Physics] in Buenos Aires. The salary was not too good but the group of researchers was excellent. So, although my wife Elena and I were living in La Plata, forty miles east of Buenos Aires, I decided to accept and therefore spent about four hours every day traveling to and from work. I devoted most of my time to developing new instrumentation. My first project was to design and build an automatic two-channel photometer.''

Roberto and Elena, who was also a student in astronomy at La Plata, were struggling, but they were fulfilled in what they were doing. It was a tense time in Argentinean history, when, after thirty years with twelve different governments, most of them imposed by military coup, populist Juan Perón had returned from exile to try to reconcile the bitter struggle between the urban and agrarian factions. Roberto and Elena became involved. They were appalled at the unfair working conditions for young scientists and became active in trade union activities in order to do something about it.

Like others before it, this attempt at reconciliation was short-lived. ''With the Junta coup of March 1976,'' Roberto recalls, ''people like myself became the target of paramilitary groups that rampaged through the country. Elena, who had a permanent position as lecturer at the La Plata Observatory, was dismissed from her job without any explanation. I lost my job a few months later; again no explanation was given. We had to stop asking for explanations, we were told—if we continued, our safety was not guaranteed. At the same time many of our friends started to 'disappear.'

''We decided to leave the country and continue with our work abroad. We wrote to many places asking for help. The only answer we got was from Martin Rees, inviting us to visit Cambridge. [Rees is an impressive intellect in theoretical astrophysics with catholic tastes in astronomical topics.] He was aware of the difficult times in Argentina because of one of his students, Juhan Frank, who had worked with me at IAFE before going to Cambridge to study for a Ph.D.''

Exiles from their homeland, not knowing the fate of their less fortunate *desaparecido* friends, Roberto and Elena knew that they would probably never return home. The first months in Cambridge were a time of extraordinarily difficult adjustment for them and their two children, but the academic community in Cambridge gave them a warm welcome and a secure new home. In short order they were both admitted to the Ph.D. program. For the first year they borrowed money from their relatives to pay the fees, but eventually both Roberto and Elena received grants. They would go on to finish their doctoral work and find postdoctoral and permanent positions in England.

Roberto chose Jorge Melnick, a young Chilean astronomer who had received his degree at Caltech working with Wallace Sargent, as his thesis adviser. Together they would look for the regularity of intrinsic properties of both very young and old star systems to find clues as to how they had formed. In particular, Terlevich had focused on three parameters he measured from the spectrum he took of each of these systems: the velocity of the galaxy in the Hubble expansion (its redshift), the typical speeds of stars and gas *within* the galaxy (what astronomers call *velocity dispersion*), and the fractional content of heavy elements (the *metal abundance*, or *metallicity*). He measured each object's brightness with a photometer placed at the focus of a telescope. Then he used the measured brightness to calculate a luminosity (the true energy output) by combining it with an estimate of the distance that he derived from applying the Hubble relation between redshift and distance.

The very young systems, the H II galaxies, gave the most interesting results. First Roberto found a rough correlation of luminosity with both the velocity dispersion and metal abundance. That there might be such a correlation between luminosity and velocity dispersion was not particularly surprising, because it made sense for a more luminous system to contain more gas and stars, and the greater gravitational pull exerted by this greater mass would require stars and gas to move faster. As might be expected for such a simplistic line of reasoning, the actual correlation Roberto found between velocity dispersion and luminosity was not very tight. The relationship he found between luminosity and metal abundance had even greater spread, but there was at least some discernible trend. What surprised Roberto was that he found a correlation between the deviations from the two relationships— any particular H II galaxy departed in the same way in both relationships.

Scatter is the deviation of data points from an ideal relationship like a straight line or a simple curve. It owes to measurement error or the failure of one variable alone to completely predict another (the way a person's height is not a complete predictor of weight). In this case, Roberto knew that measurement errors were considerably smaller than the scatter—they were not the culprit. Therefore, when he idealized the relationship between luminosity and velocity dispersion and drew a straight line through the widely spread data points (predicting a unique value of velocity dispersion for each value of luminosity), he was attributing the scatter from this simple line to real physical differences in the H II galaxies. He did the same for the relation of metal abundance and luminosity. What had caught Roberto's attention was that when an H II galaxy's velocity dispersion was higher than predicted by the simple straight line, so was its metal abundance higher than predicted by the straight line fit to the relation of metal abundance with luminosity. The converse was also true, of course; when the velocity dispersion was lower than expected, so was the metal abundance.

Roberto tried to explain why this might be so. He hypothesized that a higher-than-average velocity dispersion indicated an H II galaxy smaller in size than the typical one of that luminosity, figuring that in a system in which stars are jammed closer together, their pull of gravity on each other is stronger, forcing them to move more rapidly. Astronomers would say that such a star system is more "tightly bound." If these same systems also had a higher metal abundance, might this be because the enriched gas from supernova explosions would be better held in by the greater pull of gravity?

In fact, this was not a completely new idea. Metal abundances in very small galaxies—astronomers call them dwarf galaxies*—are extremely low compared to giant systems like the Milky Way or Andromeda. Since these are sparse stellar systems with a very low density of matter, their gravity may have been too weak to hold on to gas expelled by supernovae, gas that would enrich the content of heavy elements, the metals. Roberto's new observations were particularly intriguing since they appeared to offer direct evidence of this idea: other things being equal, more tightly bound systems were better able to hold on to expelled gas, and less tightly bound systems less so.

Though just an extension of conventional thinking in this respect, Roberto's interpretation contained one radical element that was classic Terlevich. He was suggesting that these H II galaxies were gravitationally bound systems—this was necessary if the speeds of the stars were to correlate well with the total mass in the system. This idea met with immediate and widespread skepticism, as it ran counter to the idea that regions of intense star formation usually blow themselves apart with the energy released by the new generation of stars.

When Roberto came to Santa Cruz to visit Roger Davies in the summer of 1979, he realized that Roger, Sandy Faber, and Dave Burstein had in hand all the necessary data for elliptical galaxies to see if such correlations applied to them as well. In particular, Roger in his Ph.D. research at Cambridge had paid special attention to deriving reliable velocity dispersions—the typical speeds of the stars—of elliptical galaxies by comparing the best measurements of several researchers, including his own. Sandy and Dave had been

---

*Dwarf galaxies are actually more numerous than the giant galaxies classified as ellipticals, S0's, spirals, or irregulars. However, their masses are so tiny in comparison—hundreds or thousands of times less than giant galaxies—that their total contribution to the mass in all galaxies is small. This group includes no spirals or S0's, only chaotic-looking irregular systems and spheroidal systems with extremely low density of stars. Because dwarf galaxies are smaller and simpler than a giant spiral galaxy, many astronomers regard their study as crucial to understanding the evolution of giant galaxies.

taking spectra of many of these same ellipticals; in these spectral rainbows they had singled out particular absorption features—dark bands at specific colors produced by magnesium atoms—with which to measure the metal abundances of stars in the galaxies. Finally, they had magnitudes for the galaxies that could be used, with estimates of distance from the Hubble relation, to calculate luminosity. The parallel to Roberto's study of H II galaxies was complete, and Roberto's drive and eagerness to extend his radical new result to an even more global application pushed the three to examine the data for elliptical galaxies. Sandy remembers being rather embarrassed that, with some of these data "in the can" for so long, it required a push from Roberto to get down to cases.

The four assembled a sample of twenty-four elliptical galaxies with trustworthy measurements. As with Roberto's H II galaxies, these ellipticals also showed a correlation of luminosity with velocity dispersion, a trend reasonably described by a simple straight line. This relation had been found four years earlier by Sandy and another of her graduate students, Bob Jackson, and had come to be known as the Faber-Jackson relation. Again, the correlation was far from perfect—at a given luminosity, velocity dispersion varied by about 20 percent. As had been the case for the systems Roberto had studied, metal abundance also correlated with luminosity, more poorly than did velocity dispersion, as he had also found. Another straight line was fitted to these data. Most importantly, as for the H II galaxies, the scatter from the straight lines describing these two relationships was not random: at a given luminosity, a higher-than-average velocity dispersion (above the straight-line fit) usually went along with a higher-than-average metal abundance, where "average" was again defined by the value predicted by the straight-line fit at that luminosity. The converse was also true, of course: elliptical galaxies with velocity dispersions that were "too low" for their luminosities were usually found to have too low a metal abundance as well.

In the paper they would eventually write and publish, the four referred to this as the "delta-delta" relationship, "delta" referencing the Greek letter Δ that mathematicians and scientists use to connote a difference between two quantities. In this case the "delta" was the *deviation* between the expected and observed value. The delta-delta relationship meant that when there was a positive deviation in velocity dispersion—the observed value was greater than the expected value—there was also a positive deviation in metal abundance, "observed" minus "expected." Such deviations are also commonly called "residuals," that is, something left over. The delta-delta relationship was a case of *correlated residuals*.

The correlation of residuals implied that the properties of elliptical galax-

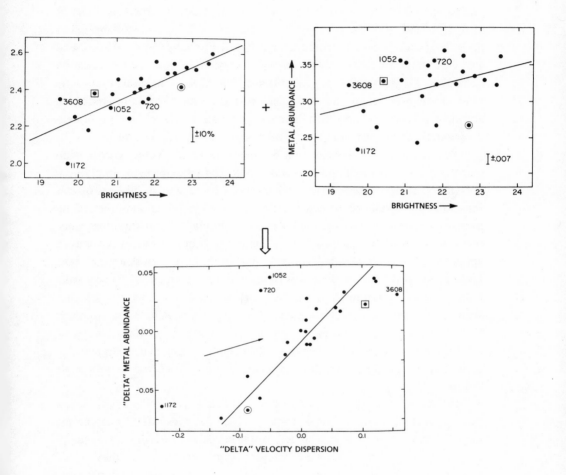

The relations found for elliptical galaxies by Terlevich, Davies, Faber, and Burstein between velocity dispersion and brightness, and between metal abundance and brightness. The top diagrams go into making the bottom one, the "delta-delta" diagram showing that the deviations from these relationships, approximated as straight lines, are themselves correlated. The point with a circle around it shows a galaxy that is low in both relationships; the point with a box around it shows one that is high. The four points with numbers were cases that did not follow the delta-delta relationship well. Their importance was minimized but they would turn out to be important.

ies depended on more than just the total luminosity, though it was clear **that** this was the "first parameter"—this means that a change in luminosity predicted the biggest change in the other measurable quantities: velocity dispersion and metal abundance. But the remaining scatter suggested a "second parameter," one on which both velocity dispersion and metal abundance also depended, though more weakly. This was the message of the delta-delta diagram. As they prepared their paper, Terlevich, Davies, Faber, and Burstein worked to identify this second parameter in their list of measured quantities. They decided that galaxy shape, described by its elongation—the ratio between the long and short axes in the football-like contours of an elliptical galaxy—was the best candidate, though the correlation of elongation with the residuals from the relationships was far from perfect.

The possibility of a correlation between "residual" velocity dispersion and "residual" metal abundance was a potential breakthrough, implying a connection between how tightly bound a galaxy is and its history of star formation that produced the heavy elements. The identification of the second parameter with galaxy elongation was an additional, tantalizing clue, since one could guess that the shape of a galaxy was connected to the way it collapsed from a gas cloud—a process that could have both determined the final speeds of the stars and influenced the history of formation. The data that fed this hopeful interpretation relied on simple correlations that, remarkably, seemed as good for these giant galaxies as they were for the tiny, simple H II galaxies. This was more than could have been hoped for. But a sample of twenty-four galaxies was too small to make a convincing demonstration, so, amid great excitement about the new result, the quest for a very large sample began.

Ironically, both of these initial steps may have been false ones. A few years later Terlevich's original result for H II galaxies would become the target of criticism—many researchers would continue to question his assumption that the H II galaxies were actually bound systems, and they would provide counterevidence. This would in turn raise considerable doubt as to whether these young star systems could be compared to the older elliptical galaxies which were, without doubt, tightly bound systems.

But even at the time there were misgivings about the interpretation of the delta-delta relation for the twenty-four elliptical galaxies, doubts that sprang from a different direction altogether. There was a simpler explanation for the correlation of residuals than one involving the tightness of bound systems, evolution of generations of stars, and the ability to hold on to gas expelled from supernovae. If distance estimates for many of the galaxies were significantly in error, luminosities for these galaxies would also be in error. For example, if the luminosity of a certain galaxy had been underestimated due to an underestimated distance, then the velocity dispersion would appear too high for that luminosity, and so would the metal abundance. That is, these quantities would have had values appropriate for the true, higher luminosity, the one that would have been found if the distance had been determined *correctly*. In this way, the delta-delta relation could be nothing more than a reflection of errors in distance estimations; it might have nothing at all to do with galaxy formation.

In their paper, Terlevich, Davies, Faber, and Burstein discussed this possibility as well. They had considered but rejected it because they judged that the two sets of residuals were not affected equally, as they should be if the error was in another quantity, the luminosity. This was a technical reason, one that required precise knowledge of the source and size of all errors in the data. In retrospect, it would become clear that these were not well known; rejection of the alternative explanation that errors in distance were the cause of the delta-delta relationship would turn out to be unjustified. More fundamentally, however, Terlevich et al. rejected this explanation because it carried with it a scientific bombshell, one they were unwilling to accept. Distances had been estimated with the standard method of using the Hubble relation between redshift and distance, the one expected in a uniformly expanding universe. In order to account for the delta-delta relationship, distances for these relatively nearby galaxies would have to be in error by as much as 30 percent.

So what? Why was this so hard to believe? The tightest Hubble relations between expansion velocity and distance had a good deal of scatter, so might not true variations of distance at a given recession velocity "hide" in the considerable scatter? Perhaps, but ever since Hubble, conventional wisdom had held that the scatter in his relation was due to the difficulty of measuring accurate distances for galaxies—typical methods provided estimates with a factor of two uncertainty—instead of indicating a real spread in the correspondence between distances and velocities. In other words, it was believed that if truly accurate measurements of distance to nearby galaxies were available, the scatter in the measured Hubble relation would be reduced to something much smaller, perhaps 10 percent or less. The possibility was not taken seriously that redshift might not be an accurate indicator of distance, as would result if, for example, galaxies had large velocities *in addition* to that arising in a smooth, universal expansion. If such motions were comparable in size, say one or two thousand kilometers per second for this sample, serious mistakes would be made in estimating distance to nearby galaxies when it was assumed that their velocities were entirely due to the expansion of the universe.

It is understandable that this possibility seemed too fantastic to be seriously entertained. What could have accelerated entire galaxies to such enormous speeds? What indeed.

# 4 • A NIGHT AT LAS CAMPANAS

Along the ancient highway of the Incas, in the Andes mountains of northern Chile, lies the Las Campanas Observatory, an earthly portal into intergalactic space. Here I had come, on February 26, 1981, to begin observations of the elliptical galaxies chosen by our group. It was an inauspicious beginning— summer skies are usually clear at Las Campanas, but on this night a billowy blanket smothered the heads of the Andes. When the clouds parted long enough for a peek through to the sky, the telescope revealed cottonball images instead of sharp points—the air above churned and boiled. It was not a good night.

As was my habit, I had arrived at the Observatory a day early to recuperate from the long trip to Chile. By almost anyone's standards, the trek from California to Chile was a wearing one. The flight from Los Angeles through Miami and Buenos Aires to Santiago, Chile, had appropriated nearly eighteen hours of my life, nearly all of it spent wedged in airplanes or confined to an airport waiting room.

These journeys are not without aesthetic redemption, however. Crossing the equator in the gut of the night, the plane sails toward the southern tropics above the steaming Amazon jungle, often through raging thunderstorms. Arrhythmic pulses of light strobe the cabin, the scene resembling nothing more than an episode of *The Twilight Zone*. These are violent storms, but in a jumbo jet the buffeting is slight—a price worth the opportunity to peer into nature's eye as it flashes in anger. It is a rite of passage appropriate to crossing into the Southern Hemisphere, crowned the next morning as the jet skims the high Andes before descending into Santiago. These stark, sharply chiseled mountains thrust broad glaciers miles into the sky, an apt reminder of the primal Earth and the reason for this journey—to search for clues to the origin of our world.

Such Valhalla-like visions fade during the seven-hour bus ride to La Serena. The bus pulls out of Santiago, a sprawling city ringed by a guard of high Andes. Santiago is about the size of Chicago, but the look is much more European: I watch men in proper blue suits bustle through ample pedestrian streets. But when a horse-drawn cart, driven by a withered tradesman probably not fifty years old, cuts through my view of the Metro station, I am reminded that this is a culture in which the present cohabits with the past.

The bus is a giant Mercedes with enormous windows and plush, generous seats. A slender stewardess, whose coal-black hair frames a portrait face only slightly darkened by native ancestry, serves drinks and sandwiches as the bus sways its way north along the Pan Americana, the singular highway that bounces between the cobalt-blue Pacific and the wall of white-tipped mountains confining this string-bean-shaped country. A few hours north of the fertile agricultural valleys surrounding Santiago, the countryside turns as dry as the desert southwest of the United States. We are heading toward the Atacama desert, one of the driest places on Earth. The Observatory lies just on its edge, 100 miles north of the rich agricultural valley of the river Elqui. This 300-mile trip covers less than one-tenth of the length of Chile, which, if laid out across North America, would stretch from Alaska to the middle of Mexico, with the same wide range of climate and terrain.

By sundown I reach the Observatory offices on the hill El Pino, named for the lone pine that commands the promontory overlooking the crescent bay of La Serena. After a congenial but customarily too-late supper with my friends who work at Cerro Tololo Observatory—the U.S. national facility for astronomy—I am all too happy to yield to my body's plea for some sleep.

The following morning brings the two-hour drive to the Observatory. I ride in front with driver Leonardo Peralta; my suitcase shares the bed of the truck with fish, bread, and fruit for the commissary, and a wooden crate of electronics the size and weight of a safe, which has just arrived from Pasadena. Leonardo is warm and friendly, as always, but his knowledge of English is as limited as mine is of Spanish, and soon the conversation wanes. We speed ahead until once again we are locked in an involuntary caravan of buses stuck behind an overloaded, oil-belching truck that is struggling up the next hill.

Passing miles of rocky foothills whose cactus and shrubs are kept green by perpetual fog, we return briefly to the frigid sea at a ragged coastal fishing village that is called, somewhat inexplicably, Caleta Hornos—"Oven Cove." From there we climb a long steep grade named Cuesta Buenos Aires that elevates us to a desert plateau and the mining town of Incahuasi (Inca House, in Incan), once prosperous from copper mining, now a ragtag of shanties, but with dirt floors swept clean by proud Chileans. Beyond, the grade Pajonales slices through the mountains, looking like a geological birthday cake oozing Neapolitan ice cream, crumbling rocks streaked with pastel layers of purple, yellow, orange, and gray-green, and red rifts that reveal a now-abandoned mineral treasure. Finally, the Observatory in view, we leave the Pan Americana and slowly climb the twenty-six-kilometer dirt road that leads to the 8,400-foot summit of Las Campanas, "the bells." We suspect that the mountain was named by the native Aquitas tribe for the huge angular boulders that ring when struck by stones or buffeted by high winds.

We arrive in time for lunch with several of my sleep-deprived, bleary-eyed colleagues. In the heyday of Mount Wilson, Edwin Hubble and his colleagues dined in coat-and-tie—in those days the customary uniform for a night's observing. Seating protocol was assured by engraved napkin rings that recorded the pecking order—by telescope size: service began only after the astronomer who would that night observe with the legendary 100-inch telescope rang the dinner bell. Things have changed. My friends kibbitz around the table in casual joviality. They wear blue jeans and shorts, T-shirts and sweats; this generation seems determined to distance itself from the stuffy stereotype of its distinguished scientific ancestry. Soon the talk turns to a favorite but often dreaded subject: *the weather*. The weather had been pristine for the first few nights of their observing *run*, but has turned sour the last two, with high humidity, blustery winds, and—the astronomer's nemesis— clouds. This night doesn't promise to be much better; their disappointment isn't hidden.

I walk from the lodge, a handsome chaletlike structure built of indigenous stone and knotty pine brought from the lush forests of southern Chile, out a mile and a half to the long ridge where telescopes rest quietly in their domes, awaiting nightfall. The promontory named Manquis juts northward like the prow of a ship: the valleys beyond frequently fill with fog by morning, leaving the ridge afloat in a sea of puffy, gray-white cloudtops. Away from the late-afternoon sun the sky is piercing blue, almost indigo—a vivid contrast to the cleaved, sharp faces of the boulders thrusting from the crest, the dusty rose shoulders of the nearby peaks, and the far eastern horizon with its sawtooth of 12,000-to-18,000-foot slate-gray peaks straddling the border between Chile and Argentina. This wall of mountains that encompasses the entire horizon is but one vertebra of the backbone of South America—the *cordillera de los Andes*. Its winter snowpack has all but disappeared from view, squeezed drop by drop through fissures in the massive stones. It will

**Manquis Ridge, Las Campanas Observatory.** *(Photo taken by the author.)*

be several years before an infinitesimal fraction of this last snowfall will gurgle from *Las Brisas* (the Breezes), the spring at the base of Las Campanas that irrigates our mountaintop colony.

Las Campanas has no trees, just an abundance of scrub and bushes that is a haven for thousands of birds. The greenery is a testament of life's determination; delicate flowers of yellow and purple peek carefully from chutes that appear to be rooted in solid rock. Even the few bushes of size do little to hide the foxes and burros that roam the ridge, or the free-ranging goats that often climb from the now-parched valleys up the steep, rock-strewn slopes. Manquis is utterly still this day, except for the flutter and chirp of birds. The mountains are burnished blades that slice away the world beyond—one might as well be on another planet.

The night comes and it is cloudy and pitch-dark—there are no stars to light my way. I turn in early; the next night is the first of my allocated run of seven, and sleep will be catch-as-catch-can until the run is over.

Fernando Peralta, Leonardo's big brother, pushes quarter-sized control buttons that cause the 100-ton du Pont telescope to swing around the sky to the point where, according to the coordinates I have given him, our first elliptical galaxy is to be found. We sit in a control room that is sealed off from the dome that keeps the telescope in an undisturbed bath of nighttime breezes. Fernando is at the console, where dials and banks of radiant numbers report information about the telescope, and I am in front of a computer terminal that controls an instrument for collecting light at the telescope's focus. We hear the whir of electric motors and gears as the telescope slews to its instructed destination; Fernando remains absorbed in his task as he peers over his impressive hook nose at the rapidly changing computer readouts. His dark skin is stretched taut over a large, muscular frame; a crop of thick black hair is graying in distinguished fashion around the ears. There's a lot of native South American in Fernando; his face reminds me of a certain Toltec statue I once saw in the Mexican National Museum.

Fernando has been with the Observatory from its prehistory. In 1967 he hauled crates of equipment up the mountain with Observatory founder Horace Babcock on a mule-train expedition that first investigated the site. Now he and fellow night assistant Angel Guerra were alone entrusted with the nighttime operation of the Observatory's most valuable piece of equipment, the 100-inch du Pont telescope, named for chemical magnate Irénée du Pont in recognition of the family's generous gift that supported its construction. Fernando's story is reminiscent of that of Milton Humason, a mule-driver during the 1912–18 construction of the Mount Wilson 100-inch telescope who went on to become Hubble's assistant and a highly productive astronomer in his own right. Fernando's aspirations stop with the equipment, however; his intimate acquaintance with each of the telescope's technical systems will help us survive some as-yet-unforeseen crisis, but he has only a passing, polite interest in the astronomical work.

The telescope coasts to a stop and Fernando's fingers are busy tapping the green-glowing buttons to make fine adjustments in position. The coordinates I have supplied, *right ascension*, the celestial equivalent of longitude, and *declination*, celestial latitude, ring up like cherries on a slot machine as the telescope reaches its destination. Fernando swivels his chair toward the controls of the television camera that we will use to look through the telescope, and turns up the sensitivity control. The images of a few stars appear, but they are soft and jumpy, not the crisp, still pinpoints we were hoping for. I sigh.

A large, faint patch appears slightly left of center screen, probably the elliptical galaxy we are looking for. I reach over and switch the camera to "integrate" mode and the light is accumulated in a digital memory for a couple of seconds before it is displayed. This is certainly a galaxy: it floods the center of this now-deeper picture with a diffuse glow. I hold up my "finding chart" and compare the patterns of stars and the location of the galaxy. There is no doubt: this is the elliptical galaxy E019-G013.

We have both seen galaxies before, countless times, yet we still react with satisfaction when the first one appears. We have plucked a galaxy out of the sky.

Light that has traveled without interruption for 100 million years has just hit the 100-inch-diameter mirror of the du Pont telescope. It seems strange that, last night, after an equally long, epic journey, *photons* like these were wiped out in a cloud floating in the Earth's atmosphere, just a scant 1/50,000th of a second before they would have been collected by the telescope. Tonight they complete their journey, to my way of thinking, but just as they do, other photons that left E019-G013 that same second (some 100 million years ago) are falling futilely on the dome, in the parking lot, on the Andes, into the ocean. They are extinguished; the million points of light collected each second by the 100-inch mirror are the only ones that will be "listened to." As far as we know, this eight-foot slab of glass is the only deliberate interception of the light from E019-G013 that at this moment is crossing an imaginary spherical shell 200 million light-years across. If only we had a

**The du Pont 100-inch reflecting telescope.** *(Photo courtesy of John Belke and the Carnegie Observatories.)*

bigger mirror, we could catch more. But it is no easy matter to build a telescope even as big as the du Pont.

Light from distant objects (in astronomy, all objects fall into this category) arrives in near-parallel bundles. Beams from different objects are all mixed together, but in this apparent jumble each beam arrives from a slightly different angle. I remember my amazement at age three or so about the way the Moon followed us all the way to Cincinnati on a drive home from Dayton. I asked my parents for an explanation of why the Moon wasn't left behind, but I don't remember if they were able to explain to me that the Moon was so distant that light continued to come from the same direction (one might tell a child that the Moon is too far away to drive *around*).*

A lens or mirror sorts out these parallel beams, producing a focused spot for each bundle of rays—the lens sorts different angles into different positions. By rearranging this melee of information into a spatial format, the optic has produced an image. Bad optics, like eyes whose lenses can no longer accommodate, cannot precisely sort out the rays from different directions: a fuzzy image results. Also, as the air rising from a hot desert road bends and distorts the exact parallelism of the light, so too do air currents in the Earth's atmosphere spoil a sharply focused image. This was responsible for the fuzzy stellar image on this night, what astronomers call "bad seeing."

Mirrors have replaced lenses as the main light-gathering elements in telescopes because light passing through a lens is broken up into a rainbow of colors, each focused to a slightly different point. In contrast to the mess of a *chromatic* image, reflection from a mirror is color-free and consequently sharp. To do its precision work, the surface of the mirror must be curved with extraordinary accuracy. In order to convert angle into position with absolute fidelity, the bumps and wiggles on the mirror's surface must be kept to a millionth of an inch. Yet, it is surprisingly easy to achieve this level of accuracy, at least in small optics, as I found out when I was thirteen years old.

As a "bar-mitzvah boy," I received the customary surplus of neckties and fountain pens, but my teacher for the ceremony sought a more lasting gift. Told of my childhood interest in astronomy, student-rabbi Stephen Forstein chose a book on telescope making, a lead my parents followed when they purchased a kit for making a four-inch-diameter mirror. Actually, like

---

*What I do remember is that, whether or not they had answers, my parents always gave me the impression that they were pleased that I asked—they never seemed to run out of patience. I am certain that this encouragement had a lot to do with my becoming a scientist. If nurtured by parents, the curiosity we feel as children can last a lifetime.

one of those boat or airplane "kits" that comes with a set of plans and some flat balsa-wood, this kit included just a few simple pieces: a four-inch disk of Pyrex about an inch thick, a Coke-bottle-glass disk of the same size, a dozen bottles of sandlike grit, and some rouge. But this and the book were enough.

Glass has extraordinary properties. Neither a simple solid nor liquid, its molecules are easily dislodged by the abrasive action of fine particles; microscopic chunks of the material readily reattach at a new location. In making a mirror, one glass disk is passed over another, a soapy slurry of grit separating the two. Normally the surfaces begin flat, but this soon changes as the top disk, destined to become the mirror, is passed back and forth. During each stroke the abrasive loosens tiny glass bits, which are then pulled down by gravity to the bottom disk, called the tool, where they stick. After many hours of work, the mirror begins to be hollowed out, and the tool becomes convex. A smooth curve develops because the center of the top disk spends more time in contact with the tool than its edge, so the amount of glass removed from the mirror depends on the distance from the center. Furthermore, because high zones will be ground down and low zones filled in, both surfaces tend naturally to spheres—two spherical sections are the only curves that can be in perfect contact as they are moved over one another.

After this *rough grinding*, the mirror surface is approximately the right shape, but it is marred by tiny pits that are huge mountains compared to the goal of lakelike smoothness of one-millionth of an inch. The grinding process continues with a series of finer grits which leave smaller and smaller gouges, until the glass can be polished smooth by a powdery substance—*rouge*, for example—embedded in a waxy substrate. This step returns the shine to the glass and it becomes possible to test the surface for smoothness and shape.

Probably the most remarkable part of the process is that a simple test can be made to see if the mirror follows the desired curve to a millionth of an inch. Light squeezed through a pinhole is beamed to the mirror and reflected back to a point just an inch or so to one side of the pinhole. The beam starts from a perfect point, but returns to a point only as accurately as the figure of the mirror allows. As one peers at the return beam and cuts across the point of focus with a knife edge, the mirror will go instantly dark if the rays all come together perfectly, but, if not, irregularities in the mirror's surface will stand out in stark relief, greatly exaggerated like the shadows of mountains cast by a late-afternoon sun. Cycles of polishing with specific strokes and retesting will eventually converge to a finished glass surface of phenomenal accuracy. The precision glass disk is then turned into a mirror by placing it in a vacuum chamber and depositing an extremely thin layer of aluminum on its surface.* Although professionals use more sophisticated techniques for larger optics, conceptually the process is much the same.

My father and I found an old fifty-five-gallon oil drum for a polishing stand; we set it up in a small room in our finished basement, filling it quarter-full with water for stability. (Many years later we found that the drum had pressed an inch-deep depression into the vinyl floor, and that the mix of residual oil and tap water had turned into microbial soup, the stench of which reaffirmed my decision to leave biological investigations to my older brother David.) Over the better part of a year's spare time I circled this drum, guiding the four-inch mirror across the tool, hour after hour, until the shiny surface was a near-perfect shape. Most of the other parts for this telescope were bought from a scientific-supply house—the tube that held the optics, the two-axis mounting that pointed the telescope anywhere in the sky, and the "eyepieces"—these were the small lenses that took the light focused by my mirror and re-created a parallel bundle needed by the tiny lens in my eye. In effect, the whole telescope was an extension of my eye: it gathered a four-inch parallel beam of light from an object and collapsed it down to "eye-size"—a gain of several hundred over what the unaided eye could see.

Early one evening in the winter of my thirteenth year I pointed the telescope at the Moon and gawked with amazement at the clarity of the image my handiwork had fashioned. There were indeed mountains on the Moon! Soon I had rushed my parents to the eyepiece and delighted as they beamed with pride. They were no less proud, but a tad less enthusiastic, when I awakened them at 3 a.m. to look at Saturn—it had been eight years since the four of us had been together at the observatory in Hyde Park. The next morning my father rolled his bleary eyes up from a three-minute egg and admitted that he had forgotten to put on his glasses for the wee-hour show-and-tell—he remembered only a bright orb flanked by two smaller blobs. The crispness of Saturn's rings had bounced perfectly off my telescope mirror, but had not survived his myopia.

The beautiful gaseous nebulae of stars caught in birth and death are faint; it wasn't long before I began polishing an eight-inch mirror that would make a brighter image by gathering a bundle of light four times as large. With my experience this mirror took only three months to complete. However, this time I decided to make the mounting myself, a tricky task since it must hold

---

*Household mirrors are coated on the back surface with silver, but for astronomical mirrors the light must bounce off the precisely figured front surface rather than pass *through* any glass. Silver tarnishes when exposed to air, so front-surface mirrors are commonly coated with aluminum—a less-good but more durable reflector.

the telescope with a viselike grip, yet move almost effortlessly in order to track smoothly these celestial moving targets.

I tried many combinations of pipe fittings, machined shafts, sleeves, and bearings, aluminum trusses, steel angle, and plywood frames, to achieve smoothness, rigidity, and precision, but I never achieved a really successful design. My undoing always seemed to be the achievement of exact balance in any position. But, in the process, I learned (in an enjoyable way—by *trying*) some basic principles of mechanical engineering that can lead to structures that are simple, light, and strong.

All this went on during the six years I attended Walnut Hills, a public "college prep" middle and high school that was as challenging socially as it was scholastically. I spent less and less time working on telescopes, but the one part that I kept after, because it was the most fun, was building and rebuilding the control system that moved the telescope, a combination of gear trains, motors, electric clutches, brakes, and selsyn encoders. This system enabled the telescope to slew quickly to a point on the sky and then turn precisely at the glacial rate of one revolution in 23 hours and 56 minutes.* Such a peculiar ratio can be approximated by combining gears with a large, odd number of teeth, random versions of which were common in old bomb-sights that were floating around in war-surplus stores in the 1960s.

With Bruce Block, my best friend then as now, I spent an embarrassing number of hours poring over bins of equipment sold at twenty cents a pound at a surplus store called Gadgeteers, then tearing into these virtually indestructible chassis in order to extract some particular gear buried deep in the works. This form of play was very important for my development, but it was also wonderful that Bruce's interests and talents were in theater and the arts. The time we spent building (and destroying) things was complemented by time spent writing skits and "doing comedy bits." I even assisted for a while in Bruce's marionette company, which included stage, lights, recorded music, and special effects. I realized, however, that this might not be my best career move the Saturday we were presenting Hansel and Gretel at some eight-year-old's birthday party. In my ineptness I improvised a novel but abrupt conclusion to the show by dropping the controls for the evil witch squarely on her head. But the range of experiences these hobbies provided was, for both me and my friend, essential.

*The length of a *sidereal* day, which is reckoned to the stars instead of the Sun, is about four minutes shorter than the *solar* day. This is because as the Earth turns, it also moves on its orbit around the Sun, so it takes an extra four minutes for the Sun to appear to return overhead.

In time there was a platform on the roof over the second-story porch of our house. Not only did my parents have this built as a birthday present, they even tolerated occasional treks through their bedroom in the middle of the night, since this was the only avenue to Dressler Observatory. From there I could look over most of the trees, at the dark and mysterious winter sky with its light show of star clusters, nebulae, and planets. I cannot claim to have used it often—there were basketball games, buddies, girls, the piano, school-work, and girls—but I think I used it well.

I doubt that as a boy I ever imagined using a telescope as big as the du Pont. This machine, weighing as much as a bus, glides across the sky with a smoothness I never achieved even for my five-foot-long, hundred-pound telescope. Though 1,000 times as heavy as my eight-inch mirror, the quartz mirror of the du Pont, floating on dozens of stress-free supports, offers a surface no bumpier than a millionth of an inch to the long-traveled light waves from E019-G013. Locked on target this night, the telescope was tracking so effortlessly that the galaxy scarcely moved on the television screen. It was time to start observing.

I reached to the spectrograph control box and hit a button to open the shutter, then swiveled to the computer terminal, typing the command to start recording light entering the spectrograph. Fernando called out "Doctor!" accenting the second syllable, and pointed to the instrument control box. *"Espejo!"* On first nights I usually forgot to flip out the *mirror*, which was still deflecting the beam to the acquisition TV camera. The unfortunate photons of the moment were still being squandered; nothing was reaching the spectrograph. "Will he ever learn?" he must have been thinking.

Finally, the mirror backed out of the beam, light from E019-G013 streamed past the open shutter into the spectrograph. Fernando picked up the guide paddle, a flat metal box with small buttons for tweaking the position of the telescope, and tapped them occasionally to make sure that the light from E019-G013 was funneled exactly down the slit. For the first time a human being was taking a spectrum of E019-G013.

We commonly think of a telescope as an instrument that makes something far away look closer. Yet magnification is only a small part of what a telescope can do. The combination of telescope and spectrograph is much more powerful, because it does more than take the astronomer closer to the distant star or galaxy, it takes him *there*.

By separating light by color, sometimes as finely as a million colors from violet to deep red, a spectrograph reveals extensive information about the conditions right at the light's point of origin, or at places along the way where it filtered through intervening matter. This is not just a closer view: the astronomer is on site, able to dig a shovelful of material and subject it to laboratory analysis to deduce its chemical and physical properties. Without spectroscopy, astronomy might be little more than a kind of space zoology. With this tool, it has evolved into a study of the biology and genetics of celestial inhabitants.

Even at a very coarse level of inspection, a spectrum contains fundamental information, for example, by its shape, which peaks in intensity at a given color and falls off both to the red and blue. As customarily plotted by astronomers, this shape is something like the profile of the Rock of Gibraltar—sharply rising on one side, slowly declining on the other. The color at which the intensity peaks is precisely determined by the temperature of the material giving off the light. Hotter objects, those in which the atoms are vibrating or moving more rapidly, have more energy to get rid of, so to speak; therefore they emit more energetic packets—higher energy photons. This kind of radiation is called *thermal radiation*, because the amount and color of the light is simply related to the temperature of the material that's emitting it.

In daily life we are familiar with light coming from sources with temperatures of thousands of degrees Celsius or degrees Kelvin,* like the Sun, the filament of an incandescent light bulb, a stove burner, or fire. Objects at thousands of degrees produce light in the visible part of the spectrum, i.e., that which our eyes can see. This is no coincidence: eyes evolved to make use of the light that bathes our world.

Hotter objects with temperatures of tens of thousands to hundreds of thousands of degrees are not a normal part of everyday life, a good thing because they give off photons so energetic that they can tear apart living tissue. One of the reasons that nuclear weapons are so destructive to life is that, with temperatures of millions of degrees, much of their enormous release of energy comes out in gamma rays, ultra-high-energy photons that are truly deadly. On the other hand, such potent light can also be a lifesaver. The X-ray machine spreads an extremely low dose of high-energy light that penetrates soft tissues of the body. Because they absorb X rays, bones, which are

---

*The Kelvin scale has the same units as the Celsius scale, but its zero temperature is at $-273$ degrees Celsius. This is the so-called "absolute zero," never reached, where atoms are frozen still.

made mostly of heavy calcium atoms, and other harder tissues cast shadows that are recorded on film. This provides a simple but powerful tool for checking what's going on inside the body.

On the cooler and safer side of the keyboard are objects with temperatures of hundreds of degrees Kelvin; these emit infrared light. For example, room temperature is about 300 degrees Kelvin; all objects in your house are releasing large numbers of photons, so you could see in pitch dark if human eyes were able to detect these feeble packets of energy. Some snakes do just that, with unusual eyes evolved over the eons. Because the body temperature of a living creature is typically a bit higher than room temperature, my cat Andromedus and I stand out against the walls and furniture with a more intense, warmer ("bluer") infrared glow. The possibility of "seeing" prey in the absence of visible light has given snakes a distinct survival advantage, and humans, in their own brand of reptilian behavior, have developed similar means that allow them to make war by night. In a more natural way humans can "see" infrared light, after a fashion—our skin is a sensitive absorber of what we commonly call *heat*. The sensation of warmth results from the absorption of infrared light coming from, say, the embers of a dying fire or an ardent companion.

Cooler objects, with temperatures of only tens of degrees Kelvin, radiate the lowest-energy light, radio waves. Because the Earth and everything on it is much hotter than this, there are no *thermal* sources of radio waves in our world.*

In outer space, thermal radiation is also the most common type, so an astronomer usually needs only a rough measure of the intensity of light at a few colors to determine the temperature of the source. The light from astronomical objects spans the entire range of the *electromagnetic* spectrum. (The name derives from the description of a photon as oscillating electric/magnetic fields, their strength rising and falling like bicycle pedals.) Stars glow mainly in visible light, though some are so cool that they are much brighter in the "invisible" infrared or so hot that they shine mainly in ultraviolet light. The frigid gas clouds floating between stars in the disks of spiral galaxies, whose

---

*However, light can also be generated by *nonthermal* processes, so there are both man-made and natural sources of radio energy in the Earth realm. The rapid acceleration of electrons, as happens in a thunderbolt or in the Van Allen "belts" circling the Earth, releases copious amounts of radio energy. Present-day communications systems still rely heavily on radio waves; photons for radio and television broadcasts are created by wiggling the electrons in a metal rod up and down millions of times per second. (Visible light, whose frequency of oscillation is a billion times faster, cannot be produced in this fashion.)

temperatures are only thirty or forty degrees above absolute zero, are observed with radio telescopes, giant antennas that electronically sense this low-energy light. On the other side of the tracks, temperatures can reach hundreds of thousands of degrees in the explosion of a star, or in the hot spot a star acquires when its companion star dumps on it huge amounts of matter—*it happens*. The X rays that are released can be detected with specialized "satellite telescopes" orbiting well above the Earth's atmosphere, a shield that absorbs X rays and harmful ultraviolet light before they reach the ground.

The Sun, a typical star, is a particularly instructive example of the relation between temperature and the color of emitted light. First, a necessary digression about the source of the enormous energy, the process of nuclear fusion taking place in the Sun's core. Tremendous pressure caused by gravity's attempt to compress the Sun results in a central temperature of millions of degrees; at such outrageous temperatures, protons—the nuclei of hydrogen atoms—bash together with enormous ferocity. Some of these collisions result in *fusion*, the sticking together of four protons to produce the nucleus of a new helium atom, according to the prescription 4 protons = 2 protons + 2 neutrons. In each such "nuclear reaction" a small amount of mass is converted into a great deal of energy according to Einstein's famed formula $E = mc^2$. The energy released is just enough to counteract the crush of gravity. The Sun remains stable because the process is self-regulating: if too much energy were produced, the Sun would expand and cool, thereby reducing the fusion rate; if too little, it would contract and heat up, increasing the rate. The process corrects itself and thereby achieves balance.

The relation of the color of light with temperature has everything to do with why we are able to live so close to this phenomenal power source. With so much energy produced in nuclear reactions, nature reasonably chooses to dispatch it in high-energy photons called gamma rays. Yet the Sun illuminates our world with a warm yellow light, not gamma rays. As the gamma ray photons diffuse out from the Sun's core, they are absorbed and re-emitted countless times in the half-million miles of gas surrounding the core; each succeeding layer is cooler than the last, so the photon energy drops slowly through X rays, to ultraviolet light, and finally to benign visible light. The "surface" of the Sun, to the extent that a gas layer can be considered a surface, is simply the stratum where the density has dropped so low that photons can finally escape without further absorption. Thanks to the relation of color and temperature, the outer layers of gas shield us from the solar core by serving as a light *transformer* that "steps down" lethal gamma rays to beams suitable for tanning.

Such simple analysis is very revealing about the temperature of the source of radiation, but much more is to be learned from a finer dissection of the light. This is because atoms, specifically the electrons that orbit the nucleus of the atom, react strongly with any light they encounter, adding or removing light of specific hues. These modifications to the spectrum are called spectral *lines* (a spectrograph spreads the rainbow of colors into a band, and these alterations appear as cuts across the band); they can be bright like the letters of a neon sign or dark like spaces between white piano keys.

Glowing gas produces bright lines of various hues, an immutable and unique combination for each of the most common elements in the gas cloud—hydrogen, helium, carbon, nitrogen, oxygen.* Each of these atoms can absorb energy from a collision with another atom, energy it quickly releases in discrete, premeasured quantities—these are the photons. The exact energy determines the color of light: bluer for higher energy, redder for lower energy. The number of lines and their colors are predetermined by the individual structure of each particular kind of atom—how many electrons orbit the nucleus of the atom and how tightly they are held to it by the attraction of opposite electrical charges. With accurate measurements and the application of well-understood physics, the density of the gas (how many atoms per cubic centimeter), its temperature (the typical speed of the billiard-ball-like motions of the atoms), and the relative proportions of the different atomic species can be deduced.

Bright emission lines are the rule if gas is the source of the light, but often a gas cloud merely intervenes between a source and observer. In this case a spectrum of continuous colors is interrupted by dark lines, which are the specific colors where the light has been *extracted* by atomic interlopers. Stars are excellent examples of such *absorption-line* spectra. For example, in the Sun the continuum of colors we call a rainbow (what you see when a spray of water droplets disperses sunlight into its colors) comes from a layer of gas at a temperature around 5,000 degrees and, therefore, by rules of thermal radiation, yellow-green light is the most intense. As this light leaves the Sun, it must traverse the cooler, lower-density solar atmosphere, where atoms of carbon, magnesium, calcium, sodium, and iron are waiting to devour their favorite colors. Some molecules, combinations like cyanogen (carbon coupled to nitrogen), magnesium hydride, and titanium oxide, are also floating in the solar atmosphere. With their more complex electronic configurations, they have ravenous appetites and can swallow up huge swaths

---

*An everyday example of this is "neon signs," in which tubes of glowing gas are excited by electricity. Neon is used to give the red-orange hue, and other gases like argon and krypton supply the other colors.

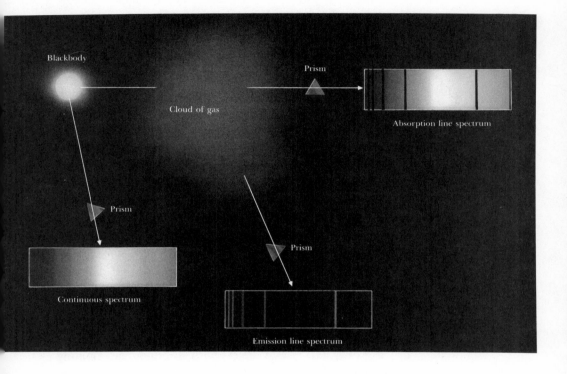

The production of absorption-line and emission-line spectra by gas. The "blackbody" is a source of light with a broad, continuous range of colors. Atoms in a cloud of gas absorb light at certain colors, leaving gaps—absorption *lines*—in the continuous spectrum. If viewed from another angle away from the continuous source, the light these atoms subtract can be seen re-emitted as a spectrum of bright lines at the same colors as the absorption lines. *(Reference: W. J. Kaufman, III: Universe, Third Edition, 1991, New York: W. H. Freeman & Co., p. 94.)*

of color from the continuum. As a result of these deletions, the spectrum acquires a series of dark lines or bands at the colors where most of the light has been extracted. The missing teeth in this solar grin are called absorption lines; like the emission lines of the glowing gas, they offer a direct way to study the chemical composition, density, and pressure of a stellar atmosphere.

From both absorption- and emission-line spectra, conditions *at the source* can be derived. This capability to examine circumstances of matter so remote from physical contact amazed nineteenth-century physicists, and is still a marvel to astronomers who rely on this gift to learn virtually everything that is known about the universe.

The spectrograph on the du Pont is mounted at the back of the telescope, behind the 100-inch *primary* mirror. As light hits the primary it bounces back up toward the sky, where, at the top end of the telescope, it reflects again off a convex *secondary* mirror thirty-eight inches in diameter. This sends it back toward the primary in a long, thin cone through a three-foot hole in the center of the primary mirror, to the spectrograph, where it focuses at a tiny metal plate. In this plate are cut a pair of small rectangular apertures a fraction of a millimeter in size. Their purpose is to pass only the light from two small regions—a few arcseconds across the sky—into the machinery below.

At the moment, the elliptical galaxy E019-G013 straddles one of these slits in the metal plate; on the television image, now showing light reflecting from the aperture plate, a rectangular chunk appears to have been carved from the galaxy's center. The second aperture is far enough away from the center of E019-G013 that it collects mostly light from the night sky. Earth's atmosphere, like that of a star, is also replete with absorbing and emitting atoms, so the night sky produces its own spectrum, one that adds to the light of E019-G013. The signal from the galaxy is perhaps ten times as strong, but the contamination by sky light is enough to hamper the intended spectral analysis; therefore, light from the sky will be collected separately and simultaneously, and later subtracted from the total signal that includes galaxy and sky added together.

Now that light from E019-G013 and adjoining sky has been selected, to the exclusion of all other sources in the area, the instrument will make and record the two spectra. Past convergence at the telescope's focus, the light beam, now inside the spectrograph, is diverging. A few feet down it finds a concave *collimating* mirror that tucks the beam inward so that it is again a parallel shaft of light, and bounces it back the way it came, but at a small angle from the incoming rays.

This parallel beam is a mixture of all colors, but a prism—a glass wedge—can spread this single beam into a fan of parallel beams of different colors, each one departing at a slightly different angle. This particular spectrograph uses a *grating* to disperse the light; it looks like a flat mirror, but, on closer inspection one sees its surface is meticulously scored by fine parallel grooves, 1,200 to the millimeter. These grooves act in concert to redirect the light according to color. Parallel light does not make an image, however, so to put this information into a recordable format, one again needs a lens to do its job of translating parallel light coming from different angles into different positions at a focus. Each color is focused to a unique position—*voilà*, the image of a spectrum appears.

In Hubble's day, a photographic plate would record the spectral image, but with miserable efficiency. Most of the photons would tunnel with a dead thud into the film emulsion, never to be heard from again; only about one percent left their mark as detectable silver grains. Modern detectors make use of one or another technique related to television. The image formed by the spectrograph—in this case two bands of light ordered by color—is converted into electrical signals, that is, the intensity of the light at each point registers a minute electrical current, voltage, or charge, which is recorded as a number. An array of these numbers, stored in a computer, is the "picture" that the astronomer will analyze; the picture "scales" in a simple linear fashion—an intensity of light twice as great is recorded as a number twice as large. This is another advantage over photographic emulsions, which have a limited range of sensitivity and respond in a very nonlinear fashion,* making it difficult or even impossible to recover the true intensities.

This particular detector, built by my colleague at the Observatories, Steve Shectman, is about twenty times more efficient than a photographic emulsion. It has two components, one a chain of four *image tubes*—sort of light amplifiers—and a double array of photodiodes that generate electrical signals when struck by light. The image tubes make up most of the length of the package, roughly a cylinder six inches in diameter and thirty inches long. At the front of each image tube is a photocathode, a one-and-one-half-inch-round glass disk coated with a special material that sprays out electrons

---

*Nonlinear means an output that is out of proportion to the input, like the avalanche that happens when you try to shove just one more thing into the closet. Scientists are fond of extending this terminology to human behavior, accusing someone of "going nonlinear," flying off the handle in response to some insignificant event.

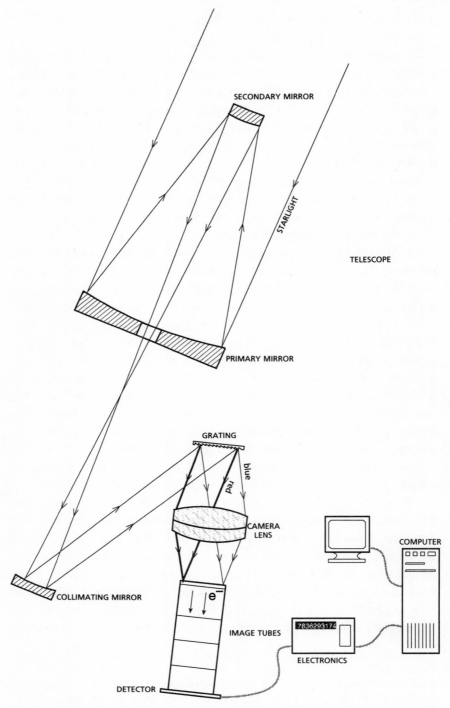

The layout of an optical spectrograph, such as is used on a telescope.

to the rear when it is hit by photons. Accelerated in a vacuum by electric and/or magnetic fields, the fast-moving electrons hit a phosphor screen—like that in a TV—at the back surface of the image tube, producing a glowing spot of light.

The careful shepherding of electrons in these tubes results in an image on the phosphor that matches the input picture, only brighter. By ganging several tubes together, each one multiplying one input photon into many at the output, the signal is amplified millions of times. Although only one out of every five photons hitting the front photocathode unleashes such an avalanche (unfortunately, the process is not 100 percent efficient), a single photon ultimately cascades into a flash of light bright enough to be seen by a human eye.

The photodiodes, arranged in a line along the rainbow of the spectrum, produce measurable electrical signals when they absorb the amplified light pulses. Two rows of diodes, one for each of the two apertures, record the spectra of galaxy plus sky and sky alone. Each flash of light generated from an incoming photon is "counted"—a number is added into a computer memory which keeps sums of the total light at each position. These diodes are not able to make fine distinctions of color (if they could, we might not need a spectrograph), so the color of light is known by which diode collected it, i.e., the position along the spectrum. Among the various calibrations that will be taken during the night is the spectrum of a lamp containing argon gas that projects along the detector an emission-line spectrum with discrete, cataloged colors. The color corresponding to each diode will be determined with great accuracy.

Since the photon "counts" are accumulated in a computer memory, it is a simple matter to subtract the sky signal and plot the spectrum of the galaxy. As the exposure continues, the counts build up; E019-G013 is delivering about one counted photon to each of the nearly 1,000 bins of the detector every second. After waiting half a minute I instruct the computer to plot the spectrum, and a traveling dot draws a squiggly line on the screen. The line is made up of dots placed according to the color and the intensity at that color: color runs horizontally along the computer screen; the total number of counts determines the vertical position of each point.

At this time each color bin has accumulated only about twenty-five counts; the general shape of the spectrum and hints of a few strong absorption lines are visible, but the line on the screen jitters up and down like Pinocchio's polygraph. These are statistical fluctuations—astronomers call this "photon noise," the result of there being only a small number of photons detected at each color. The variation expected due to pure chance is the square root of the number of counts, in this case $\sqrt{25} = 5$ counts, so the number in each bin is uncertain by about $5/25 = 20$ percent. With such large fluctuations, we describe the spectrum as *noisy*—no subtle absorption or emission lines will be visible. But the signal adds up as the exposure continues and the noise becomes proportionally less; after about fifteen minutes there are roughly 1,000 counts in each bin, so the noise fluctuations are down to about 3 percent. Now the subtle pattern of absorption lines due to calcium, magnesium, and iron atoms in the atmospheres of the stars in E019-G013 have become distinctly visible, and the observation is complete.

Astronomers fight the uncertainty of "small number statistics" more than most scientists, since it is at the frontier of weak signals where the unanswered questions lie. Although such fluctuations are not a part of most people's daily routine, our intuition about them has been well honed by experience. For example, if asked to prove that a roulette wheel wasn't "crooked," few people would expect thirty-eight spins to result in the ball falling once and only once into each of the thirty-eight numbered bins. By the time the wheel had been spun 3,800 times, only an average of 100 occurrences are expected for each number—the variation from bin to bin is still large, $\sqrt{100}/100 = 10$ percent. To establish the honesty of the wheel to one percent would require 380,000 spins. Considering the time that would take, there may be a better way to check a roulette wheel, but for galaxies all we can do is count.

I instruct the computer to store the array of numbers on a magnetic tape, and then move the mirror so that light from the lamp of glowing argon gas illuminates the spectrograph for the color calibration. Fernando slews the telescope to the next object, but when we turn on the television camera, we see only one faint star and a smudge where the next elliptical galaxy is supposed to be. "*Nubes*, Doctor!" he exclaims. Clouds. When I poke my head out the door that leads to the catwalk around the perimeter of the dome, I see he is right. The clouds are back with a vengeance.

I hand Fernando the position of a Milky Way star that is much brighter than our target galaxies. I need spectra of stars like this to calibrate my observation. Perhaps this one is bright enough to be seen through the clouds. It is a giant, a phase all stars pass through in their later stages. Its hydrogen fuel exhausted after 10 billion years of nuclear fusion in its core, the force of gravity again took over, shrank the star's core and heated it to higher and higher temperatures, until fusion could once again generate enough energy to forestall collapse. Now helium nuclei, the heavier kernels containing two protons and two neutrons, are fusing into carbon nuclei, the next step in the chain of heavy-element manufacture. The star is shining hundreds of times more brightly than before, but it will do so for a much shorter time, only a hundred million years or so. The greater energy released from helium fusion pushes out the surrounding layers of the star to a thinner broth. Somewhat paradoxically, this puffed-up envelope is cooler (and thus *redder*) than before, even though the star's center has become much denser and hotter. Stars like this are truly huge; when the Sun goes through this phase, its "surface" will push out beyond the orbit of the Earth, engulfing the three inner planets in the process. But not to worry, there's a good 5 billion years before this happens, plenty of time to move to another neighborhood.

Fernando has found the ninth-magnitude star, and although it appears to fade and flare continually as clouds come and go, it remains bright enough for a spectrum to be taken. These unwelcome water bags are capturing more light than the telescope, but at least the distribution in *color* is unaffected—light of all colors passes equally poorly. Spectroscopy, then, can be done on a partly cloudy night, when many other kinds of astronomical observations cannot. The trouble is, the objects of research programs are often near the limit of what the telescope can do (otherwise one should be working with a smaller telescope), and the interdiction of clouds often takes the matter from difficult to hopeless. Fortunately, there are a few bright stars and galaxies to choose from on this first night.

Red giant stars are comparatively rare (because they live for a much shorter time), but so bright that they contribute most of the light coming from galaxies. This is why a casual look at the night sky will take in many such rubies, like Betelgeuse in the constellation of Orion, or Antares in Scorpio. Like distant lighthouses, they can be seen from far at sea. This highlights the difference between a volume-limited and brightness-limited sample: in a random-volume scoop of space not even one star in a hundred is a red giant, but a sample limited by magnitude includes a much higher percentage because they can be seen even if very far away.

Because this red giant is very similar to the ones producing the light in E019-G013, it can serve as a template in the analysis of the galaxy spectrum. The star's spectrum has the same shape and the same absorption lines as E019-G013—the similarity is stunning—but as soon as it appears on the computer screen two obvious differences show as well. For one, the spectral lines of the star are all shifted to the left (to the blue) compared to their respective positions in the galaxy spectrum. This is evidence of the galaxy's redshift. While this star is barely moving relative to the Earth—only a few kilometers per second—the galaxy is cruising at thousands of kilometers per second. Because of this the Doppler shift has relocated each absorption line in the galaxy's spectrum to a redder color, a shift that can be measured to yield the exact velocity.

Another more subtle difference will play a key role in our study of elliptical galaxies. The lines in the star's spectrum appear sharper, more distinct—the galaxy spectrum by comparison looks fuzzy, almost out-of-focus. This is because the galaxy's absorption lines have been broadened: the degree of broadening is a direct measure of how fast stars are moving inside E019-G013. The light from the galaxy is, of course, a composite of the light from millions of red giant stars, and the speeds of these stars range over hundreds of kilometers per second—these are the orbital velocities that stars have acquired to counteract the pull of gravity in E019-G013. As each star has a slight redshift or blueshift, the absorption lines from all the stars don't line up at precisely the same colors, so the composite line is broader and shallower. The higher the speeds, the greater the spread. The *velocity dispersion* (the total spread in the speeds of stars) in the galaxy will be measurable from the comparison of its spectrum with that of a single star.

We manage to squeeze out one more star before the clouds close us down altogether. Fernando and I retreat downstairs to the kitchen and make coffee and eat our sandwiches. We will wait for the clouds to thin out; perhaps they will go away, but we are not putting any money on it. Good nights are exhilarating—the work carries you through, gallops you through, if it's going well. Really bad nights, with rain and snowstorms, or ice on the dome, aren't so bad either—at least you can turn in early. The worst are nights like this one, not good enough to work, not hopeless enough to quit. On this night we will shoot one more galaxy as it plays hide-and-seek through the clouds. Finally, at 3 a.m., disgruntled and discouraged, Fernando and I stow the telescope and turn off the equipment. As we leave the dome to a cold, gusty darkness, Fernando wishes me good-night and adds "*Posiblemente mañana será mejor.*" But tomorrow night will be no better.

On the third night the weather breaks, and skies remain gloriously clear for the rest of the run. The work goes well and rapidly; fatigue mounts as we work nonstop from dusk till dawn, but we are cheerful now—I because I am getting what I came for, Fernando because he is less bored when he can do his job.

Five days later the run is finished. I have accumulated good spectra of fifty-five ellipticals for our project, and about one hundred coarser spectra of other galaxies, data for another project. This observing run has been a success, but many such explorations will be needed along our journey. I would be returning to Las Campanas in November of the same year to take

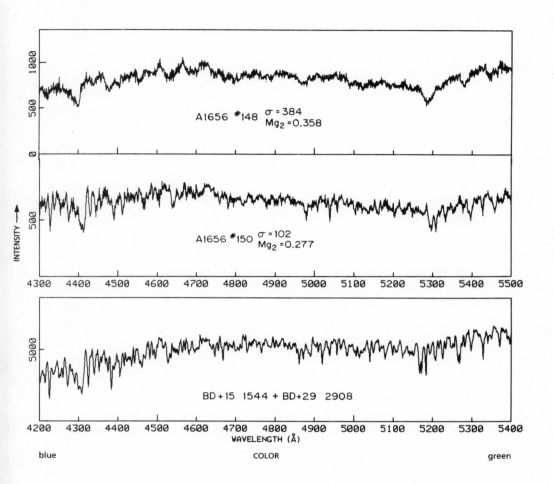

Spectra of elliptical galaxies, showing the broadening of the lines due to increasing *velocity dispersion*. The bottom spectrum is that of a single star, used as a template. The middle spectrum is from an elliptical galaxy of moderate velocity dispersion—the typical speeds of stars are about 100 km/sec. The dispersion is even higher in the top spectrum, which is from a galaxy where the typical speed of stars is much higher, about 380 km/sec. The individual lines become less distinct as the dispersion increases, as is shown by the absorption lines due to magnesium atoms that are indicated by the arrow. Note also that the absorption lines of the galaxies are *redshifted* relative to the star. *(Reference: A. Dressler, 1984* Astrophysical Journal, *vol. 281, p. 512.)*

spectra of ellipticals on the other side of the sky, the part which in February was hidden by the yellow Sun and its blue daylight. Later this year Roger and Roberto will undertake the same kind of spectroscopy with the 158-inch Anglo-Australian telescope at Siding Spring Observatory in northeast Australia, and Sandy and Dave will take spectra of Northern-Hemisphere elliptical galaxies from the Lick Observatory. Before we have finished gathering spectroscopic data in 1984, there will be two more runs at Las Campanas, and more observations by other members of the group at Lick and the Kitt Peak National Observatory in the U.S., Siding Spring in Australia, the Royal Greenwich Observatory in Great Britain, and the South African Astrophysical Observatory.

Dave will coordinate another type of observing that is just as vital to our quest. He and others of the group will use smaller telescopes, like the 1.3-meter at Kitt Peak National Observatory, to measure the brightness of each galaxy in the sample. Most of the data will be taken by directing light onto *a photomultiplier tube*, a device that looks like an old radio vacuum tube and produces an electrical signal whose strength is directly proportional to the amount of light focused on its small collecting screen. For each galaxy the light within circular apertures ranging from a few arcseconds to a hundred will be collected; in this way not only the total light, but how it is distributed, the *light profile*, will be measured. Certain stars, whose brightnesses have been measured many times before and have thus become "standards" to be used in calibration, will also be observed.

In March 1981 Dave and Sandy spent a cloudy night at the Kitt Peak National Observatory near Tucson, Arizona, waiting in vain for the perfectly clear skies that are essential for photometric measurements. There they got to know Gary Wegner, an astronomer from Dartmouth College whose photometry program had also landed him in the cafeteria. For Gary photometry was a well-developed skill; the care and patience required to make a measurement to one percent or better, a task that astronomers all too frequently fail to accomplish, comes naturally to him. This night, and during the disappointing cloudy nights that followed, Sandy and Dave discussed our project with Gary, at great length. Eventually they asked if he was interested in joining up. The gloom of bad weather and the magnitude of the task didn't seem to discourage Gary, who quickly said yes.

With Gary, the group had reached its final complement of seven and was hard at work collecting data. We would continue in bunches of two or three until 1983, when all seven would assemble for the first time at Cambridge University at the invitation of Donald Lynden-Bell. There we would begin to form our first coherent picture from the collage of thousands of measurements for nearly 500 galaxies, a hodgepodge of data acquired with ten different telescopes from observatories on five continents. We had returned from our first incoherent scouting reports. Our true journey into the realm of deep space had barely begun.

Gary Wegner may fall a little short of Superman, but he bears a striking resemblance to Clark Kent. On the surface he is quiet, almost bashful; he dresses conservatively, speaks deliberately and carefully, never swears, and rarely lets his emotions show, but when he laughs it's usually a guffaw.

Gary is a family man. He comes from a large but tight clan. Born in Seattle in 1944, he grew up near the town of Bothell, Washington, surrounded by the kind of rural landscape that is fast disappearing. He developed a fondness and fascination for nature, learned how to fish and hunt and climb. His interests and values seemed to come with the territory. His marriage to Cynthia Kay Goodfellow has tallied twenty-five years and five children. The Wegners live in a new colonial home on some wooded acres outside the Yankee town where they raised their brood, Hanover, New Hampshire. They are American "folks," and proud of it.

But, like Clark, Gary has a few surprises under his street clothes. He has opinions, strong opinions, of people and happenings, though you may have to drag them out of him.

"My mother worked when I was little and different ladies cared for me," he remembers. "One nice lady was Ruth Bruzas, who always read to me from the encyclopedia when I brought it out and turned to the astronomy section—I was fascinated by the idea of there being other worlds besides our own.

"One question that puzzled me greatly when I was very young was who God is and how did the world come to be. I asked everyone to the point of seeming impertinent. By no means do I want to convey the impression that I am an atheist, but what I realized from this is that most of the mumbo jumbo that people tried to feed me was completely unconvincing and that nobody knew the answer to this mystery.

"I would work outside with my parents and often my father took me hunting or fishing with him. From this I began to appreciate and wonder about the forces of nature. I saw trees that were so large that ten or more of us had to join hands in order to reach around them. I was impressed by the power of the sea, seeing the breaking surf and shipwrecks, and riding the great rollers in small boats doused in salt spray—being so close to whales that I could see the barnacles on their hides and their cowlike eyes looking

DONALD LYNDEN-BELL

SANDY FABER

ALAN DRESSLER

ROBERTO TERLEVICH

DAVE BURSTEIN

GARY WEGNER

ROGER DAVIES

fearfully at me. There were also the great mountains with their spires of stones, snow, and ice, and the turbulent rivers churning down from the heights over great waterfalls. Sometimes we saw so many ducks and geese covering the entire sky that it was one vast cackling cloud that went over for what seemed like hours; my father told me to take a good look, for when I was grown, he predicted, all of this would be gone. Unfortunately, he was right.

"I remember at about this time first hearing about atoms—I think this came from talk about the A-bomb. However, no one I knew was a mathematician or physicist, so I had to wait many years before I was to learn more of this and the scientific method."

One year, when there had been atom bomb tests in the Pacific, Gary's family was vacationing on Vancouver Island. They received daily reports of the location of the radioactive cloud that was drifting their way. A Canadian woman came out of her house and screamed at them, "You Americans are responsible for all this!" Gary grew up with the strong sense that war was horrific and senseless, a belief that grew into a near-religious conviction after he visited Nagasaki many years later.

"My mother always loved to read, and we often read books together, taking turns reading out loud. My Grandma Gardner also loved reading and gave me astronomy and space-travel books. I became an insatiable reader, although I was not a particularly outstanding student. I always found school very boring and did not do well in arithmetic."

But Gary was so far ahead in reading skills that he found most of his education insufferably boring: "School seemed so childish—in the second grade I was poring through *National Geographic* and reading everything I could get my hands on about Antarctica, while my teacher was covering 'see Spot run' and pretending to be some little bunny character from a children's book." Later, his almost painful shyness made the social aspects of high school intimidating. Although he had many schoolfriends, he purposefully avoided the clubs and activities that were supposed to make school "fun." What *was* fun, on the other hand, was the big, wide world and the possibility of "figuring it out."

"In 1954 Mars came near; my cousins had a small telescope and we gazed at Mars for hours. We couldn't see anything, but I was determined to have a telescope of my own in 1956, when the red planet was to come even nearer. I worked at odd jobs, mowing lawns, picking berries, and so on, in order to save enough for that telescope. Nobody could tell me what kind to get, but I did get one and saw Mars. However, by Christmas of that year my father and mother had found out more about telescopes and gave me a real astronomical one. Then I was really hooked on astronomy.

"My friends and I would stay out all night with the telescope, star maps, cameras, and binoculars, watching everything wheel overhead. In 1957 we saw Sputnik I and II and tried to make our own rockets. We conducted some of the most dangerous experiments you ever saw—I shudder to think back. This was not only in pyrotechnics, but electricity as well. My parents were less than pleased when the police showed up one day—our 'rockets' bore a disturbing similarity to pipe bombs, and neighbors were complaining about the noise.

"I became interested in spectral analysis and bought war-surplus prisms and lenses. This is not unlike many boys of that era, but what probably set me on a track to become a professional was my later mathematics courses in high school. This taught me rigorous thinking and made me realize that I could now *do* mathematics, which I had already recognized as the real way of thinking about nature."

Of the seven on our team, Gary's drive to be an astronomer seems to have been the most determined. By age sixteen he had presented a paper to a meeting of the Astronomical Society of the Pacific, held in 1960 in Eugene, Oregon. Gary had taken spectra of the Moon which he compared with spectra he took of Earth rocks from his extensive collection. From their colors and reflectivities he was able to make some educated guesses about the Moon's composition. At the meeting Gary heard some of the best astronomers of the day. The discussions were lively and upbeat, and Gary left inspired and fully committed to becoming an astronomer. In 1963, the year of his graduation from high school, Gary traveled to Washington, D.C., where he met legendary astronomer Harlow Shapley and was awarded third place in the Westinghouse Science Talent Search for his project on the origin of the solar system.

He began college at Washington State that year, where he received an excellent grounding in physics and basic mathematics. However, he soon realized that he needed to move to a university with astronomers if he wished to follow that road, so he transferred to the University of Arizona in 1965. He feels that he was received coolly by most of the faculty (unfortunately, professors immersed in research often seem to have little use for undergraduates), but a few were very attentive to his education, and he worked with one on studies of the Sun.

For graduate work Gary moved on to the University of Washington. In a fine education in general astronomy, Gary was particularly influenced by K. H. Böhm, a pioneer in the study of white dwarf stars—the extremely dense remnants of normal stars like the Sun. Not only did Gary choose to do his Ph.D. thesis on white dwarfs, but it became the focus of his research for years to come. He received a Fulbright fellowship to conduct research at Mt. Stromlo Observatory in Australia for two years; there he would make the first studies of white dwarfs in the southern sky. Later he received an appointment in Oxford, and met, on travels to the Radcliffe Observatory in South Africa, the former Astronomer Royal of Great Britain, Sir Richard Woolley. Woolley liked the young man so much that he offered him a staff position at the South African Astrophysical Observatory in Cape Town. Gary moved his family to South Africa and stayed there three years.

Eventually Gary moved to Dartmouth College, where he is now the Margaret Ann and Edward Leeded '49 Distinguished Professor in Physics and Astronomy, and Director of the Michigan-Dartmouth-MIT Observatory. It is said that he is the only member of the faculty to come to Dartmouth from the West Coast without ever having crossed the country. But he brought his world of experience and good heart.

As dawn comes to the final night of my February 1981 observing run at Las Campanas, I pack my papers, charts, and tapes, and then drive up to the high peak of Manqui that overlooks the Manquis ridge, to watch the sunrise. I am waiting for the blue flash of light that marks the Sun's breakout over the razor crests of the Andes. A green flash is sometimes seen as the Sun sets; because the Earth's atmosphere disperses light like a weak prism, the Sun when low on the horizon is actually spread slightly into a spectrum—the blue and green light trailing the red image and setting last in a momentary spot of green. The blue light, not the green, should be the last to go, but when the Sun contacts the horizon, the light path through the air is so very long that the blue light is gone. A large amount of airborne dust intercedes between our eyes and the sunlight. Dust preferentially absorbs more blue light than red (why the setting Sun is distinctly redder than the midday Sun), and at the horizon there is enough dust to swallow *all* of the blue light. The "bluest" light that remains is the green; hence the green flash.

As the Sun rises over the Andes, however, it is already "up" a few degrees in the sky, high enough that much of the blue light survives. While the green flash is unreliable because the true horizon is often obscured by clouds or fog that scramble the fragile event, the blue flash will not disappoint as long as the sky is clear above the mountains. I wait for half an hour, increasingly certain that the Sun will break over the ridge at any moment, convinced by the ever-brightening sky that the time is near and I must not look away. (Unlike the sunset, when one follows the slow disappearance of the Sun's disk, the crucial moment comes abruptly.) Finally, when I am certain that the whole ridge must soon explode with light, my patience is rewarded by an intense, searing spot that has the diamond-blue luster of a welder's torch suddenly ignited. It lasts but a second—which seems long— before burgeoning into an intense golden fire too intense to behold. Soon the ridge with its covey of now-sleeping telescope domes is flooded by a warm-yellow light.

So have begun a million-million days on Earth, but these mountains are young and have seen but a thousand-million ascensions of the Sun. We are such newcomers, infants in Earth's realm; it has been not even one million days since Aristarchus and Aristotle walked the land.

# 5 • ALAN IN WONDERLAND

We like to think that most Europeans living in 1491 didn't know that the world is round. More probably they knew, but just didn't care. Unless you were a merchant, a philosopher, or a sailor, chances are you got along just fine imagining a saucer-shaped Earth that the Sun, Moon, planets, and stars climbed over each day—few people required a more sophisticated cosmology. True, Greek philosophers had surmised the spherical shape of Earth two millennia earlier; Aristarchus of Samos had even come close to estimating the exact size. But such matters were purely academic, and academia was even less a part of pop culture in the fifteenth century than it is today.*

As we approach the twenty-first century, it has become more difficult to ignore the true geometry of Earth. Trips covering a fair fraction of the world's circumference have become common; we are becoming accustomed to the idea that, for example, a flight from California to Japan begins with a course set almost straight north toward Alaska. A fair number of Earth's people now have to deal with time zones—the mistake of picking up the phone at dinnertime and waking Cousin Guido in Sicily out of a dead sleep is a hard-to-ignore consequence of living on a round world. Then, too, there are those pictures of Earth from space, especially ones showing a recognizable piece of terra firma, which make a subtle but indelible impression. Thus, even

---

*We revere Columbus for his courage in setting out to demonstrate once and for all that the world is round, but this machismo was at least partly fired by a stubborn refusal to accept very good scientific evidence regarding the size of the Earth. In his funding proposal to Ferdinand and Isabella, Columbus grossly underestimated the Earth's diameter by a factor of three, which made a westward journey to China competitive with the difficult voyage east around the Cape of Good Hope.

though the Earth isn't so curved that we are reminded every time we look out the window (as would be the case if we lived on a body as small as or smaller than the Moon), in our mind's eye we are becoming accustomed to the fact.

The seven of us on our cosmic voyage had something in common with those sailors of the fifteenth century. We too were stuck with an innate sense that the space in which we live is flat, even though we have come to understand—at least intellectually, through our efforts to map and describe the universe—that this too is an illusion sprung from limited vision. We live in what appears to be a three-dimensional *Euclidean* space, a framework of three axes fixed at right angles, for which one needs only three numbers to specify the position of any object. But what about time? Although the coordinate of time can be ignored in our living rooms—we can aim for the sofa with the assurance that it hasn't moved far in the last hundred-millionth of a second (the time it takes for light bouncing off the sofa to reach us with an updated position)—this sense of simultaneity enjoyed in our perceived surroundings breaks down in the realm of deep space. Galaxies hundreds of millions of light-years away are seen at positions they held hundreds of millions of years ago. Each galaxy is viewed at a different epoch, times spanning millions of years for even the nearby galaxies and billions of years for large chunks of the universe. In the grander world, then, *four* numbers are needed to describe an *event*—both where and when something "is." Space and time weave together in what is actually a four-dimensional world.

Just as the disappearance of a ship below the horizon seems a puzzle if one misconstrues the Earth to be flat, our misconception that the universe is only three-dimensional leads to apparent paradoxes about the big bang and the expanding universe. For example, since all galaxies seem to be rushing outward from the Milky Way, it would appear that we occupy the center of the universe, the place where the big bang occurred. Many people ask, after they are shown that creatures on any other galaxy would make the same observation and come to the same conclusion, "Well then, where exactly in space *did* the big bang happen?" Others focus their thoughts not on the center of the universe but on its "edge." Surely, they say, some galaxies are out leading the way, those that form the outer shell of the expanding sphere unleashed by the big bang. Wouldn't beings on those galaxies look back to see galaxies at their backsides with only emptiness out in front? And, from the most sophisticated, comes a worse puzzle: how can we look farther out into space, and thus back into time, and see a bigger and bigger volume, when we know that the universe was *smaller* in the past?

These and similar conundra are born from basic misconceptions about the nature of space and time, the ones we get from *intuition*, that is, the circuitry evolved in our brains to deal with perceptions of the immediate world. In order even to make sense of these questions we must begin to accept the true nature of space and time, a reality totally outside our experiences. This reality was first recognized at the beginning of this century, at the same time astronomers were arguing about the nature of the nebulae and the scope of the universe and puzzling over the energy source of the stars. Much has been written, and written well, about these profound concepts that rank as some of the great achievements of human thought, and little justice will be done to the beauty and subtlety of the ideas by my abstraction that follows. The aim is only to extract the flavor of the revolution in thought that gained us a foothold in a strange world whose paths we were poorly outfitted to follow.

We quite naturally imagine space as a fixed framework, unbending and static, and time as a ticking clock, steady and precise—absolute standards against which all events can be measured. This was the world that Newton described so brilliantly with his "new math," a world where the trajectories of apples and planets yielded with equal ease to *the calculus*. Newton's "laws"* are an excellent description of the world we perceive with our senses, but his system is only an approximation, albeit a magnificent one, that breaks down when we deal in vast distances and great speeds.

Albert Einstein was the first to grapple successfully with this larger domain, when he was trying to understand how the laws of physics would appear to different observers in relative motion. Newton's laws of *mechanics* describe how objects move in response to external forces and give simple rules to transform from one observer to another. For example, consider a baseball pitcher tossing a lazy 50-miles-per-hour curveball forward from the backseat of a convertible automobile traveling at 60 miles per hour. The pitcher would not be surprised to learn that a batter on the sidewalk will be terrorized by a blazing 110-miles-per-hour fastball. Such simple rules had been found to hold as exactly as could be measured for some 200 years, but by the end of the eighteenth century a few disturbing contradictions had surfaced.

One was the surprising result of an experiment conducted by physicist Albert Michelson and chemist Edward Morley. They discovered that the same

---

*"Laws," in this context, are better thought of as rules or relationships deriving from a more general theory that explains a broader set of phenomena.

value is measured for the speed of light regardless of whether an observer is moving toward or away from a light source. Though the speed of a car is minuscule compared to the speed of light, at least in principle one would expect to measure a small change in the speed of light coming from the convertible's headlights. Our sidewalk batter would probably expect a slightly higher speed for the light emanating from the headlight than for the light trailing the taillight as the car sped past, and neither, one might think, would be the same as the speed of light coming from a parked car. But the experiment performed by Michelson and Morley showed that this is not true: the speed of light is exactly the same for all three. Light reaching Earth from the nearly stationary Sun, from a galaxy receding at 10 percent of the speed of light, even from a quasar moving with 90 percent of the speed of light, is all clocked at exactly the same speed, 186,282 miles per second.

This enigma was a blow to some overconfident physicists of the late nineteenth century who thought that the major discoveries of physics were behind them, and that all that was needed was a little fine-tuning. (Michelson himself had said that "the future of physics lies in the next decimal place.") Einstein was one who puzzled over such issues. He was not so much concerned with the Michelson-Morley experiment, but with a more fundamental failing of the most important physics of the day, Maxwell's equations, to transform properly when using Newton's simple presciption for transforming from one moving frame to another. In 1873 James Clerk Maxwell had published a brilliant synthesis of the "laws" of electricity and magnetism, reducing the motions and forces associated with charged particles to a theory with four elegant mathematical equations that include the velocity of light as a fundamental constant. Maxwell's formulation showed that electric and magnetic "fields"—mathematical descriptions of how force varies with distance and motion from a charged particle—are actually manifestations of a single *electromagnetic* force, the strongest force in the universe. Maxwell's triumph was the greatest achievement in physics since Newton's; its validity was undisputed. Nevertheless, his equations were apparently valid to observers in only one particular frame of reference—any other observer sailing past at a constant speed should find the equations invalid. This didn't seem likely.

It had been proposed that the "correct" frame of reference, the only one where Maxwell's equations were valid, was defined by *the ether*, an infinitely stiff, invisible, massless stuff permeating the universe (at this time generally thought to be only the Milky Way system of stars). The ether, it was supposed, conducted light waves the way water conducts water waves—this was an ad hoc explanation for how light could travel across a vacuum. But experiments involving the speed of light, like the Michelson-Morley experiment, had already raised considerable doubt that there was any such thing as ether, by showing that whether one was traveling "upstream" or "downstream," the speed of light waves is the same.

Einstein proposed a more ingenious solution: he found a way to retain the validity of Maxwell's equations for all observers moving with a constant speed relative to each other. To do this, Einstein had to abandon Newton's principal tenet that space and time are *absolute*, replacing it with the notion that measurements of time and space are *relative*, that is, they depend on the relative motion of the observer and source. Also, he adopted as a proven fact the puzzling observation that the speed of light is constant regardless of the relative motion of source and observer. This set of ideas led to the *special theory of relativity*, a clean mathematical formulation stating that lengths and times are measured differently by observers moving relative to each other.

For example, a football halfback will perceive a football field as shorter (the distance between yard markers will shrink) as he rambles down the field, an infinitesimal effect at a speed of ten yards per second, but a 25 percent effect if he reaches the blinding speed of 150,000,000 meters per second, half the speed of light. Not only will this truly lightning-quick halfback (finally, an accurate sportscaster's metaphor) see a shorter field than the fans in the stands, he and the crowd will disagree on exactly when he crosses each of the yard markers. This is simply because the time read from the stadium clock depends on one's distance from the clock—the light showing the position of the clock hands (or the numbers of a digital readout) travels at only a finite speed. For example, a fan who is equidistant from both the halfback and the clock will record a slightly *later* time than the halfback. This is because as the halfback crosses the yard marker both the light of his image and the clock's image leave on their journeys to the fan, but the halfback himself must see an image of the clock that left earlier so that it would rendezvous with him at the appropriate moment. With some thought it becomes clear that synchronization and simultaneity are concepts that have little meaning if observers are moving relative to each other. Further analysis will reveal that the time interval between crossings of the yard markers will also be different for the runner and crowd; the runner will conclude that the stadium clock is running slower than his Rolex, and, paradoxically, a spectator will conclude that the Rolex is running slower than the stadium clock. It's all a natural consequence of how time is measured, how information is communicated, and the finite speed of light that is measured the same by all observers. Of course, this is only a fanciful example of what is now a widely verified experimental result—that any method of timekeeping, from Big Ben to an atomic clock, will appear to slow down if observed in a frame which is moving rapidly.

Einstein's special theory of relativity is constructed with mathematical rules that squeeze distances and stretch times from one observer to another in just such a way that Maxwell's equations remain the same—*invariant* in any frame that is translating at a constant speed. In an odd way, this new description of reality continued the Copernican tradition of displacing our world from a privileged position to one of indifference. It was, nevertheless, the crucial step away from the notion of space as a vast, rectangular theater metered by a single, eternal clock. Space and time were now known to be relative quantities. But this relativity theory was indeed a "special" one, applicable only to the case where one observer went cruising with constant velocity past another. The more general case of observers changing their relative speeds—accelerating, whether by speeding up in a straight-line path or swinging by on a curve—was not covered by the theory. Furthermore, though the special theory of relativity was a great success in generalizing Maxwell's equations, it virtually wrecked the other great achievement of physics, Newton's laws of gravitation, which did not transform in the same way. Different observers would find Maxwell's equations the same if they were moving uniformly relative to each other, but the laws of gravitation would now be different.

It took Einstein ten additional years after his 1905 paper on special relativity to construct a more general theory that included gravitation—this was the greatest scientific achievement of his legendary career. The key concept addressed exactly that issue which was not covered by the special theory—what different observers would record if they were *accelerating* with respect to one another. Here Einstein came up with a dazzling insight—that gravity is indistinguishable from other accelerations. Newton's famous equation $F = ma$, force equals mass times acceleration, combined with Newton's law of gravitation that quantifies the force between two bodies, leads directly to the notion that gravity is an acceleration applied to a body with mass. Einstein took this a step further, realizing that to perform an experiment in the presence of a gravitational field, like Galileo's experiment of dropping a heavy and a light ball from a tower, is exactly equivalent to performing the experiment in an accelerating frame. For example, at the surface of the Earth we are subjected to a force that accelerates (increases the speed of) free-falling objects thirty-two feet per second for every second that the object is falling. In one of his famous "thought experiments" Einstein imagined himself in free space (away from Earth's gravity), enclosed in an elevator accelerating at the same rate as Earth's gravity—a speed increase of thirty-two feet per second every second. He hypothesized that it would be impossible to tell from any physics experiment whether Earth's gravity or pure acceleration was responsible for forces on objects within the elevator. The experiments would look identical. This hypothesis, which Einstein called the *principle of equivalence*, is technically the same as equating the gravitational mass of an object (the "weight") with its *inertial* mass (the resistance to a change in uniform motion) in Newton's equations.

From this premise a startling and fundamental new property emerged— that space itself must be curved by the presence of gravity, and thus the presence of matter. This is a direct consequence of the principle of equivalence—easy to show, if not to accept and believe. We begin by noting that a light beam is nature's chalkline—it defines the structure of space itself by drawing the straightest possible line through space. Now we ask what happens in Einstein's accelerating elevator if a light beam passes from one side to the other. From the point of view of riders in the elevator, the beam travels a curved path because as it crosses from one wall to the other the elevator is picking up speed. If light beams define space, then it follows that space itself must be curved in an accelerating frame, and since this is exactly equivalent to the presence of a gravitational field, space must be curved by the presence of matter as well. With his general theory of relativity Einstein predicted that light passing close to the Sun should follow a slightly curved space, with the result that the Sun would appear to alter the positions of stars as it passed in front of them, as if the Sun were a huge, weak lens. In an expedition to view the total eclipse of 1919, star positions close to the edge of the Sun were measured by Sir Arthur Eddington, expressly to test Einstein's amazing prediction. These observations confirmed the general theory of relativity as well as could be measured. In the years since, the theory has been confirmed to astonishing accuracy.

Einstein's special theory of relativity had described a four-dimensional world, but it was still a "flat" world, one where the laws of Euclidean geometry, for example, that the angles of a triangle sum to 180 degrees, and the circumference of a circle is equal to pi times its diameter, are still valid. But the generalization of relativity to accelerating frames of reference—the general theory of relativity—brought with it the realization that space-time is curved, warped around objects with mass. If we were able to make extremely precise geometrical measurements, even in the vicinity of the relatively weak gravity of Earth, we would find that the rules of Euclidean geometry are not precisely correct. In the presence of the Sun's stronger gravitational field, the effect is pronounced enough that departures from flat geometries have been measured. One can even imagine a place where mass is so densely

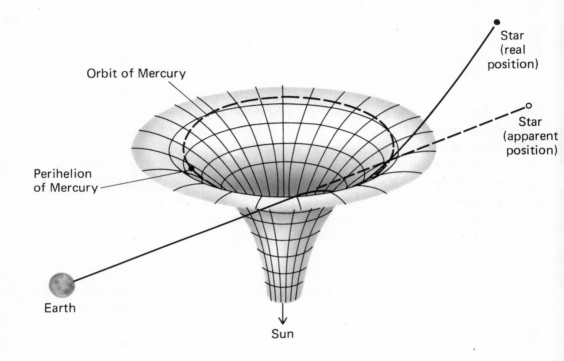

Star
(real
position)

Star
(apparent
position)

Orbit of Mercury

Perihelion
of Mercury

Earth

Sun

The bending of light in space curved by the presence of mass. The drawing is for the light of a star passing near to the Sun on its way to Earth, but space shown is a two-dimensional surface viewed in three dimensions for clarity. The planet Mercury is so close to the Sun that its orbit shows effects due to general relativity. Mercury's closest approach to the Sun during its orbit—*perihelion*—advances an observable amount each year, something Newton's laws of motion would not predict. *(Reference: J. Pasachoff*, Astronomy: From the Earth to the Universe, *Fourth Edition, 1991. Philadelphia: Saunders College Publishing, p. 412.)*

concentrated that space is completely curved around it—such places have been dubbed ''black holes'' because even light is pulled back in its attempt to escape. In the vicinity of a black hole the effects of curved space-time would be very obvious, but no more obvious than the peril to one's life that such observations would entail.

Einstein preferred to use this notion of curved space as a replacement for gravity itself: gravity, he said, is another illusion sprung from an incorrect perception of geometry. When objects are not free to travel their force-free paths, referred to as *geodesics*, they feel accelerations. For example, we are not free to fall in the natural curvature of our space toward the center of the Earth. What we call ''gravity'' is the reacting force of the Earth on our feet that prevents us from falling freely toward the center of the Earth along the natural curvature of our space. In contrast, an astronaut orbiting the Earth, and with it the Sun, floats weightless along the geodesics of curved space-time and feels no forces. Where, in this case, is the gravity of the Earth or Sun? Nowhere, says Einstein, it is an illusion.

Einstein's notion that gravity is not a mysterious force, but rather a manifestation of geometry, is intimately connected to the idea that the world we live in has more true dimensions than the three we perceive as ''space.'' One way to help visualize this is to consider a two-dimensional world where we, as three-dimensional creatures, have a godlike perspective. For example, consider balls rolling on a billiard table—not the usual kind with felt over slate, but a flexible surface like a stiffly stretched sheet of rubber. Such a *membrane* is a two-dimensional surface, so this is a start at imagining a two-dimensional world. As a ball rolls across the pliant surface, we see it stretch the surface, creating a depression where it comes to rest. Imagine some smaller balls, marbles for example, that are rolled toward the billiard ball. They themselves make small impressions in the membrane, of course, but the bigger effect is the way they are drawn into the hollow created by the billiard ball. The paths of the marbles are bent as they pass by—they are following the acceleration-free geodesic lines. Indeed, if one passes too closely it may actually get caught in the hollow of the larger ball, circling until friction robs it of its energy and it spirals into the center.

All this is pretty obvious to us, the three-dimensional supercreatures, but we have been cheating by affording ourselves a three-dimensional view of this two-dimensional world, by seeing the hollows that the balls make in the membrane. Imagine what we might conclude if we had only a strict top view, with no perspective to the side, and saw the balls rolling in their strangely curved paths. Expecting the paths of the marbles to remain straight, unless they are acted upon by a *force* (Newton's first ''law''), we would quite naturally conclude that a strange force is drawing the marbles to the billiard ball. We might even call the force "gravity." What was obvious before, in our three-dimensional view, is now mysterious in the two-dimensional *projection* of the three-dimensional world. In his concept of gravity, Einstein suggested that our perceived three-dimensional world is a projection of a higher-dimensionality universe. Viewed by creatures able to sense the additional dimension(s), the mystery of the gravity—that we are held to the Earth's surface, that the Moon is drawn to orbit the Earth, and the planets to orbit the Sun, and so on—would be no mystery at all, just geometry.

We are now closer to addressing the questions that led off this chapter, the questions of "where did the big bang happen?" and "what's beyond the edge of the universe?" To go further we'll need to replace billiard balls and marbles in our "flat world" with more intelligent two-dimensional beings— the ones who have at last achieved the status of being too thin, if not too rich. In their world the *"flatlanders"* perceive left and right, back and forth, but have no sense of up or down, just as we cannot imagine a *fourth* dimension in our world. Again, we'll situate ourselves with a godlike perspective of the flatlander world, properly realizing that there is an up and down and that their world can curve in it. In fact, this time we'll give the flatlanders a truly curved world—a sphere, remembering that on the surface of this sphere they can sense only left and right. Unable to sense up and down, the visualization of a sphere is beyond the flatlanders.

Like the billiard-table world, this fantasy world can be used to demonstrate the idea that gravity is no more than a manifestation of geometry. With their limited vision, the flatlanders believe that they live on a plane, and that the rules of plane geometry will apply. But it is clear to *us* that the rules of two-dimensional Euclidean geometry will be violated if the flatlanders stray too far. Imagine two of the flatlanders walking away from each other, even in completely opposite directions. They try to walk as straight as they can— veering neither to the left or right (geodesics!)—certain that they will not meet again, according to the rule that straight lines on a plane cross at only one point and diverge beyond. But *we* see that, in reality, the two flatlanders are walking great circles on their unperceived spherical world, and destined to meet on the other "side." When they map their journey they find their paths mysteriously curved—a mysterious "force" has pushed them back together. They call the mysterious force *gra* (well, after all, it's a shorter world).

Finally, we are ready for the questions about the big bang and an ex-

panding universe. In this fantasy world the perceivable universe is the surface
of a sphere, so an *expanding* universe is modeled by having the sphere grow
in size, like an inflating balloon. Galaxies can be little sequins glued to the
balloon's surface. As the balloon expands, what is actually happening is that
space itself is stretching, carrying the galaxies with it as it grows. On a small
enough scale, space will appear flat, since, for the same reason our ancestors
perceived a flat Earth, a sizable chunk can be well approximated by a plane.

Where then, did the big bang occur in this world? Let the air out and
see. Clearly it took place not at particular galaxy "spots" but at the center
of the balloon, and at a time when the balloon was infinitely small. We now
realize that it is quite impossible for the flatlanders to have a complete sense
of where the big bang occurred—strictly speaking, it isn't "on" the two-
dimensional space that they are able to perceive, but in the "down" direction,
the center of the sphere. However, we three-dimensional beings also understand that when the balloon was infinitely small, all of space was concentrated
at this imaginary central point, so that in fact *every* point on the surface was
once also the central point. The big bang happened everywhere.

Who will see an edge to the universe, a view from which there will be
no more galaxies? Clearly, there is no edge on the surface of the sphere—it
wraps around continuously—and everyone sees the same view of galaxies
receding as far as the eye can see. If there is a boundary, in any sense of the
word, it is in the up-down direction, a direction of space that the flatlanders
cannot directly perceive with their two-dimensional senses.

We should note that the Hubble expansion works just the same in this
two-dimensional world. As the skin of the balloon stretches uniformly, the
distances between galaxies will increase in direct proportion to their separations—galaxies twice as far apart will separate twice as fast as those with
half the separation. It's also apparent that galaxies are not streaming *through*
space, just being carried along with it; each galaxy will be at rest within its
space if the Hubble expansion is smooth and uniform.

Finally, the riddle of looking back in time to an apparently larger volume,
one that should be smaller when the universe was younger, is more of the
same. The light arriving from the past, from great distances, has traveled a
long circuitous route on its journey, perhaps halfway around the flatlander's
spherical world as it grew and grew in size. This light from the past, from
a once-small volume, arrives on its geodesic trajectories in curved space-
time in a wider-than-expected fan of directions. As it reaches one of the
present-day "sequin" galaxies it appears to come from a very large volume
of space, and, in this way, the much smaller past universe seems to surround
the larger, present universe. In a certain sense, we see a magnified view of

the past; galaxies seen at great enough distances, far back to the beginning of the expansion, would appear in the sky as very large indeed, but impossibly faint as their light is spread over such a large region of the sky.

It may seem strange, but notions like these are a far better description of the real world than those of our intuition.

Unique among our group of seven researchers was Donald Lynden-Bell. Like the great Einstein, Donald is a theorist: his career has not been based on performing experiments or making observations in order to pry some of nature's secrets from abundant though often baffling clues. Rather, Donald has made profound contributions by examining bits of evidence that empirically minded astronomers have collected. But, more important—and this is what makes him a great theoretician—he has succeeded in uncovering and understanding new phenomena by vigorous application of his superb analytical mind to previously discovered patterns of nature's behavior.

Donald was the elder statesman of our group, not just by age, but by experience and accomplishment. At first he would flounder, the proverbial fish out of water among this group of empiricists who spent most of their time worrying about equipment breakdowns, ''seeing'' and weather, calibrations, computer programs and computer failures—the hundred details that make the difference between good data and bad. But Donald would take this opportunity to choose to immerse himself in these aspects of the project as well. From an adventurous trip to the South African Astrophysical Observatory to try his hand at demanding photometric observations, to the tedious job of editing data files, he would gain an appreciation of how the other half lived. He would learn to put aside the true theorist's tools, pencil and paper, and enter the world of numerical analysis, writing computer programs that processed data and ran simulations of physical systems. The day in our voyage was coming when Donald would not just help with the rigging, but take the helm as we sailed our way through waters of new ideas and challenging implications of our work—as we all knew he would.

Donald's world, in both personal and professional terms, is considerably removed from the day-to-day concerns of most of us. He comes from a family with a strong line to the British military and another to the Church of England, and his passion for science also reflects, in many respects, a quest for honor and truth. While many lives may be best described in terms of events, activities, and encounters, Donald's is defined by his ideas—his lifelong pursuit of meaning in the physical world and spiritual world.

He was born in April 1935. His mother came of an intellectual family;

she studied English and enjoyed music at Oxford. His father, Lachlan Arthur Lynden-Bell, had strong interests in art and painting, but he left school at seventeen to join up for the 1914–1918 war. As a boy, Lachlan Arthur had inherited a fine 3.5-inch telescope from his grandfather, who had been a friend of the astronomer Sir John Herschel. Thus, a generation later, young Donald was shown the sky and the planets; from the age of about thirteen Donald enjoyed surveying stars, planets, and nebulae for himself. He was always interested in how things worked and can remember from an early age being mystified by how a refrigerator can cool things without any input of coldness itself. But he was no prodigy: he learned to read very late (at the age of nine), and at school he excelled only in mathematics.

By college, what Donald read began to shape his future career. He remembers in particular the works of Sir Arthur Eddington, the brilliant physicist and inspirational teacher. "I particularly enjoyed Eddington's popular books *The Nature of the Physical World* and *Space-Time and Gravitation*. The lively debate at the beginning of the latter caught my interest and thereafter the philosophical overtones held my attention. Much of Eddington's work had such overtones and natural philosophy had great appeal. I can remember the struggle he had to understand how time could be relative to an observer and not be an absolute. However, a full explanation of the 'Twin Paradox,' in which one identical twin comes back from his journey younger than his twin brother, only occurred to me much later in life."

The special theory of relativity discussed in this chapter was one of the subjects that helped Donald develop his considerable intellectual potential. The "Twin Paradox" that he mentions was poorly understood at the time when Donald began to think about it. The paradox arose because, in special relativity, two observers moving past each other at speeds comparable to that of light will see each other's clocks moving more slowly. It would seem that a brother who watches his twin speed out into space will see him return younger, but his traveling brother, who sees *his* twin receding at near the speed of light, would reach the same conclusion. Clearly, they cannot both be younger. The resolution of this paradox, Donald later discovered, had to do with the period of *acceleration* when the traveling brother turns around and heads back. This is the time during which the clocks are "set right."

Donald continued his education at Cambridge, where he was supervised by such great minds as Abdus Salam, who coauthored the quantum-electrodynamic theory, at the foundation of modern physics. It was Salam who realized Donald's deep interest in the physical world and advised him to change courses from pure mathematics so that he could learn some of the real facts of science from physicists themselves. Almost immediately, Donald immersed himself in the study of a fundamental, complex issue concerning the nature of absolute and relative motion. Maxwell had discovered that when electric charges are at rest, they interact simply through the electric force, which falls inversely as the square of the distance. But when charges move, there is complementary sideways force—magnetism—that is proportional to the *current*, the product of charge and motion. Donald developed an analogous picture for gravity in which there is an analogous "gravomagnetic force" that gives a sideways motion to "currents of matter." According to Donald, this is the true origin of what we call the Coriolis force, which makes it difficult, for example, to walk a straight path from the inside to the outside of a spinning merry-go-round. This is truly fundamental research that addresses questions as basic as "Is there an absolute frame of reference in the universe?" Donald has applied his work on this subject to a problem posed by the great Austrian physicist and philosopher Ernst Mach. Mach pointed out that when a bucketful of water is hung from a rope and spun, the surface of the water assumes a dishlike (parabolic) shape. Mach asked: would the same thing happen if the bucket was stationary but the whole universe was spinning around it? Mach said yes, but theoreticians have argued about the answer—even the question—ever since. This is a particular example, representative of many more like it, of an issue that challenges, entertains, and obsesses Donald.

The following year, Donald's graduate education turned further toward astrophysics. He decided to work with a professor who was very enthusiastic about the role of magnetism in astronomy. He remembers that "after a happy start things did not go well and I got bogged down in using a very primitive early computer to try to solve a not very important problem. I was saved by a course at the Royal Observatory at Herstmonceux which introduced me to Sir Richard Woolley, the Astronomer Royal, who suggested some interesting problems in stellar dynamics. Thus, I switched subjects and wrote my thesis in 1960 on stellar dynamics of galaxies—the formation of their spiral arms and the evolution of viscous disks. All these subjects were ripe for development as the next two decades showed."

*Stellar dynamics*, how stars move complex multibody systems like galaxies, would become Donald's forte, and it would soon lead to one of his most important contributions. After receiving his Ph.D., Donald was awarded a Harkness fellowship for study in the U.S.; he and his future wife, Ruth Trescott, moved to Caltech, where Ruth finished her Ph.D. in chemistry. Here Donald had his first substantial contact with optical astronomers, including Maarten Schmidt, who was then working on finding the optical counterparts of strong sources of radio waves, work that led to the discovery of the nature of quasars—the strongest light sources in the universe. Later in his career Donald would be the first to propose that these ultraluminous beacons are powered by the release of gravitational energy as huge quantities of mass are pulled into a giant black hole, the densest concentrations of matter we know of.

His years at Caltech established important scientific relationships and exposed Donald to many new areas of astrophysics, but none was more productive than a collaboration with Allan Sandage. "On Woolley's advice," Donald remembers, "I offered my services to Allan Sandage for three months. Sandage and Olin Eggen were by then nearing the completion of their survey of high-velocity stars in the Milky Way. By contributing my knowledge of stellar dynamics to the arguments, we were surprised to discover that we could deduce that the high-velocity star component of our galaxy had to form very rapidly—on the free-fall time scale. We also deduced that the correlation between the chemical abundances of the elements in the stars and the characteristics of their orbits would have been preserved from the formation period. These 'fossil' correlations tell one a little of the history of how the [Milky Way] galaxy formed." The paper Eggen, Lynden-Bell, and Sandage wrote is considered a classic in the field because it was the first to recognize a relationship between the way stars move and the percentage of heavy chemical elements they contain. This in turn is very revealing of the way and the order in which the stars in our galaxy formed.

Donald's prowess in the field of stellar dynamics was to lead to many important contributions, like the mechanism he and Caltech physicist Peter Goldreich proposed for sustaining the spiral pattern common to so many galaxies. However, his best-known contribution is at the very heart of how a galaxy forms. It goes by the odd name "violent relaxation," a turn of phrase that sounds like either an oxymoron or a very unpleasant vacation. Violent relaxation refers to the collapse of a system of masses, for example a young galaxy of stars that is falling together under the influence of gravity. In such a situation collisions or even close encounters between stars are very rare—a galaxy is almost entirely empty space. One might think that there would be no overall change in the shape of the system, since no energy or momentum could be exchanged between the particles. But Donald showed that the overall gravity field generated by all these myriads of stars would vary violently during the collapse event, and that the energy stored in this field would accomplish the transfer of energy from some orbits of stars to others. In this way the galaxy can "relax" (violently) into a new shape that will be stable over time.

When they returned home, Donald spent a few years at the Royal Greenwich Observatory with its headquarters at Herstmonceux Castle on the south coast of England. Then he returned to Cambridge, where he was to become Director of the Institute of Astronomy. Ruth also became a professor at Cambridge, in chemistry. There they bought what Donald calls a "fine old house" in which to raise their children, Marion and Edward. Donald has gone on to much more important work in astrophysics, all the time struggling with even more esoteric subjects like the role of free will in a mechanical universe, and the general tension between science and Christianity. He is a man of ideas, a relentless pursuer of the most challenging problems the universe has to offer, dedicated to the engagement of the human mind to their understanding and solution.

Remarkably, Einstein had developed by 1915 what still appears to be the proper machinery for describing space-time, years *before* Hubble's proof that Andromeda is another galaxy and its implications for the vast scale of the universe. Einstein was able to use a new, powerful type of algebra developed by nineteenth-century German mathematician Bernhard Riemann to express the theory of general relativity in elegant form—this elegance was, to many, evidence of its correctness. But when Einstein began to study the implications of his theory, he found to his disappointment that the universe should have no stable, static solutions—it would have to contract or expand, but it could not stand still for more than an instant.

In those days the universe was considered by most to be synonymous with the Milky Way galaxy. They imagined a vast, perhaps limitless profusion of stars mixing and circulating like guests at a cocktail party, but generally not going anywhere. Einstein's general theory of relativity did not allow such a static situation. To restore the preferred notion of a static universe, Einstein was led to include an arbitrary, repulsive force to counteract gravity; he did this with great reluctance because, unlike other facets of the theory, it was totally ad hoc. Ironically, about fifteen years later Hubble's data showed that the universe is not stationary, but expanding. For many, this was an even more stunning success of the general theory of relativity: it had, in fact, predicted the most salient feature of the universe-at-large—its dynamic motion. It seemed that the theory might be further applied to predict the complete history of space-time; the past, present, and future of the universe seemed within reach.

Physicists responded by working in earnest to apply the theory to the observable universe. The solutions of Einstein's equations admitted two dramatically different futures for the universe—eternal expansion or eventual recollapse. Bold enough to ascribe a global geometry to the whole universe, theorists mused over the properties of space implied by these two specific alternatives. In doing so, they were ignoring the small-scale curvature of space, for example, around massive objects like the Sun or a black hole. Instead, they considered the curvature of space averaged over huge scales, a regime so large that only the average matter density in the universe, not its specific collection into galaxies or stars, was important. In a sense they were replacing the rather sudden bending of a light beam as it traveled in the curved space of a single galaxy with the smooth, continuous curvature that would result if the same amount of matter were dispersed evenly through the cosmos. We do much the same thing when we ignore the peaks and valleys on the surface of the Earth and describe it as a sphere.

What Einstein's equations revealed is a connection between the geometry of space and its dynamical state. Consider two cases of an expanding universe like our own. In one case, the energy of expansion is not large enough to overcome the pull of gravity of all the matter in the universe; as a result, the expansion will eventually be halted by gravity, and the universe will recollapse. This state is referred to as a *closed* universe. On the other hand, if the expansion energy is great enough, the universe will overcome gravity and expand forever—this is an *open* universe. Which kind of universe we live in depends on the density of matter; if it is high enough, gravity closes the universe and it eventually recollapses, otherwise, it is open and expands forever.

The connection to geometry that Einstein found is this: in a closed universe space is *positively curved*—like the spherical world of the flatlanders described above. In this world, what appears to be a flat disk (in two dimensions) is really a "dome" whose surface area is less than the Euclidean value of pi times the radius squared.* This closed universe has a finite volume but it has no edges—a light beam sent out in any direction will eventually find its way back to its source. The universe is destined to reach a maximum size and reverse—the idea became popular that after falling back together to a "big crunch," the universe might repeat the big bang over and over. (However, more recent work has suggested that a "big flop" is more likely and that no new universe would issue forth.) A closed universe is actually the ultimate black hole—nothing can escape it. From this sprang the pop-science-fiction notion that one universe might secretly nest within another.

---

*In the four-dimensional case, closer to our own universe, this would be equivalent to saying that the volume of a sphere would be less than ⅓ times pi times the radius cubed.

On the other hand, in an open universe, one whose mass density is too small for gravity to stop the expansion, the curvature of space is negative. Returning again to the world of the flatlanders, this is like a saddle shape—it curves up in one direction and down in the other, i.e., space never "closes." What the flatlanders call a disk is, in an open universe, really a potato chip shape whose area is greater than pi times the radius squared. In contrast to the finite volume of the closed universe, the volume of an open universe would be infinite from the moment it was created from the singularity of the big bang, a disturbing concept to say the least.

With such a direct connection between the fate of the universe and its geometry, the lure was irresistible to see if the universe conformed to such simple descriptions. Observational cosmologists began in the 1960s to push telescopes and detectors to the limits, literally. Allan Sandage wrote a seminal paper outlining the range of world models and proposed observations that might measure the geometry of space and so specify the universe's destiny. The 200-inch Palomar telescope was the instrument of the two competing groups: Sandage, Jerome Kristian, and James A. Westphal competed with J. Beverly Oke and James E. Gunn. Both teams hoped that by studying the size and brightness of extremely distant galaxies one could learn whether the space of the universe was positively or negatively curved, thus whether the universe would continue to expand forever or be brought to a halt and recollapse. Another way to look at these observations, less rigorous but more easily grasped, is to think of them as attempts to measure the expansion rate of the universe as it was in the distant past and compare it to today's rate, in order to see if gravity is applying the brakes hard enough to stop the universe's expansion.*

Despite prodigious efforts and the talents of many dedicated people, it appeared by the end of the 1970s that these attempts to find the geometry of the universe had failed. The methods relied on the concept of standard candles and/or yardsticks—the assumption that some kind of object or collection of objects had the same brightness or size over a large volume of space—but it slowly became evident that galaxies had evolved sufficiently over the time span in question to invalidate these assumptions. In other words, it wasn't enough to have a standard candle that was constant over a large volume of space—one needed a candle that glowed at the same brightness for all time as well. Even the direction of the needed corrections could not be accurately

---

*As was mentioned earlier, the Hubble diagram of redshift versus distance was the primary diagnostic in this attempt, since the straight-line relationship originally discovered by Hubble should, at great distances, curve up or down, depending on whether the universe is "open" or "closed."

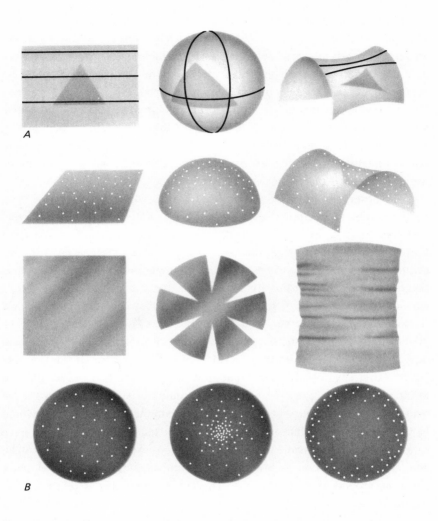

Two-dimensional analogues to three-dimensional space. The plane corresponds to a flat universe, the sphere to a positively curved ("closed") universe, and the saddle to a negatively curved ("open") universe. Heavy lines in the top figures show what parallel lines look like in these spaces, and the distortion of a simple triangle is also shown. The second row shows toy universes of these spaces with dots representing galaxies, and the bottom row shows what a map of galaxy positions made by a "flatlander" would look like if no account were taken of the curvature of space. The row in between demonstrates why these distortions occur: The dome with a distance R from its pole to the equator will have an area less than $\pi R^2$ (Euclidean geometry)—the shortage of "material" is shown by the triangular gaps in the disk. If a saddle is squashed, the flat area produced will have more surface than the product of its width times length—this extra material is shown by the "folds." *(Reference: J. Pasachoff, Astronomy: From the Earth to the Universe, Fourth Edition, 1991. Philadelphia: Saunders College Publishing, p. 603.)*

predicted: were galaxies bigger in the past or smaller, brighter or fainter, and by how much?

But had this quest actually failed? Many today might say that, in retrospect, it succeeded brilliantly. What the investigations did show is that the universe is nearly balanced in kinetic and gravitational energy, that is, very near the dividing line of open and closed. Here "nearly" is used in its astronomical context, meaning to within a factor of ten. In common life, a factor of ten is considered horribly inaccurate, akin to no success at all—one would not expect the directors of General Motors to survive long if they could balance costs and revenues only to this tolerance. (Well, actually, I'm not so sure.) But in the cosmological context, this turns out to be a genuine triumph, one that plants the seed of enigma.

The observation that the universe is, as far as we can tell, not far from the borderline separating "open" from "closed" can be seen as a great mystery, for it is an extremely unlikely result. The universe could have started with any ratio of gravitational energy and kinetic energy, from 10 trillion to one *billionth* and beyond, in both directions. With this enormous range, it seems extraordinarily improbable that the universe is near to a balance between the two, implying a universe that is just "open" or just "closed," but not extremely one way or the other, as the odds would have predicted. Some scientists, swerving perilously close to philosophy and ontology, regard this as a mere manifestation of our being here. If the universe were "extremely closed," they argue, it would have swelled and recollapsed in little more than an instant; we humans would not even be here to ask the question "why this universe?" Likewise an extremely open universe would have dispersed to a dilute, thin gas before any stars or galaxies could have formed, destined to swell to formless oblivion. Again, we would not be here. Their explanation, in brief, is that the universe must be this way if we are to be around to contemplate it. This line of reasoning is referred to as *the anthropic principle*.

Others have followed the more traditional scientific approach, though they too have been led to a radical explanation. The universe is not negatively curved or positively curved in any global sense, they argue, it is locally *flat*. (You mean that after all this rigmarole about curved space it turns out it's flat? Actually, it is only the space part that turns out to be flat; space-time is still curved.) Near balance—to within a factor of ten—cannot possibly be a coincidence, they maintain, because to be so close to balance today, the universe would have to have been balanced to phenomenal precision when it was very young, i.e., its deviation from perfect balance in the distant past would have been infinitesimal.* The universe, conclude these theorists, must actually be in *perfect* balance (though we are not yet able to verify this experimentally), and some process in the early universe must have made it just so.

It was just this problem, the apparent ''flatness'' of the universe we see around us—the highly improbable fact that space is not strongly curved in either a negative or positive sense—that propelled Alan Guth of MIT in his development of the *inflationary universe* model. There will be more to come about this, but the extraction of a key concept will nicely round out this discussion of the shape of our universe. Guth suggested that what we call the universe is not a global reality, but just a small piece of the fabric of space-time that underwent a period of unimaginably rapid expansion during its extreme youth, enlarging what was a microscopic, smooth patch into a vast realm. The time and scope of this supposed event would have been $10^{-35}$ seconds after the big bang (an unimaginably tiny fraction of a second after creation) and an expansion factor of some $10^{50}$ (about 100 doublings in size, also unfathomable). The idea here is that, regardless of the curvature of the initial patch from which the universe grew, the period of inflationary growth would have flattened it out completely. The simplest analogy is with a balloon, again. The curvature of the tiny balloon is very steep, but as it is blown up, its surface becomes a better and better approximation of a flat plane. Imagine doubling the size of a balloon 100 times in succession and you get a good idea of why the universe would turn out very flat indeed, no matter how tightly curved it was when inflation began.

The inflationary universe model has offered an explanation for this perplexing observation that space around us is at least approximately flat, by claiming that it is *exactly* so. The model is also helpful in explaining other remarkable properties of the universe, not the least of which is why it is expanding at all. Presumably, the universe is now just coasting, a legacy of the big push of inflation. But did inflation actually occur, and is space truly

---

*Consider, as an analogue, the accuracy in speed required to pitch a baseball from Earth so that it comes to a stop at Jupiter. One needs to pitch the ball at a speed close to 30 miles per second (108,000 miles per hour) to do the job. The exact number is a detail of the specific circumstances, but the point is that an error of one percent would be enough to have the ball go sailing past Jupiter by some 40 million miles. To have the ball come to rest close enough to Jupiter to be considered ''a strike'' requires a precision of 1/1000th of a percent (one part in a hundred thousand) in the speed. Since the ball is being pitched at about 100,000 miles per hour, an error of only one mile per hour, the speed with which the pitcher strides forward as he releases this 108,000-miles-per-hour fastball, will be enough to make him miss.

that flat around us? These questions remain open and the subjects of intense study. What is certain, though, is that the simple notion of a global universe containing all there is, coherently constructed from a single solution of the equations of general relativity, has passed into history. The new order has tried to restore a degree of perpetuality to the universe, identifying our observable universe as but one small part that received some rather drastic re-engineering. One might even wonder if the issues collected under the category "anthropic principle" will boil down to "creatures like us could have evolved only in such a flat space," for reasons of time and homogeneity.

It is unfortunate that we have not developed minds that allow a more intuitive approach to the intricacies of space-time, for it would make our task here simpler and our appreciation greater. Humans doubtless would have evolved such a facility if we were able to resolve time at the level of a billionth of a second, for the light-travel time even in our small world would then have been an important factor. We would be able to think and act in space-time with ease. Biology did not permit such development; our nervous systems with their chemical pathways cannot begin to approach such speed of processing and communication, and furthermore there seems to be no use for such a lightning-fast reactive ability in this world. Perhaps if we were creatures of a much larger scale, we would have found such abilities useful or even essential, but this too is awfully hard to contemplate.

If there were a need, however, there is little doubt that such talents could have been developed. For demanding tasks like hitting a baseball or working on the high trapeze, brains have a remarkable ability to set up networks of neurons that approximate the behavior of complex mathematical equations, in lieu of actually solving them analytically. Through experience, our brains are able to wire up these remarkable networks that allow us and other animals to execute amazingly complex maneuvers that would tax whole companies of mathematicians and engineers. We could undoubtedly develop a *sense* of space-time.

Lest this seem idle speculation, however, consider that the situation is likely to change radically in the next century. From ancient times our species has endeavored to boost its innate capabilities. Machines are nothing more than extensions of our bodies, increasing our strength and endurance, honing our powers of vision and precise action. Likewise have we multiplied one facet of our minds as well, that of memory, by compiling vast records of experience, ideas, and information into books and pictures. The development of computers is amplifying this last activity enormously, but the true impact of computers is likely to be in increasing our cognitive powers, the ultimate expansion of our abilities. It is both an alluring and terrifying prospect, but it seems inevitable that the human brain will one day be linked to the central processor of a computer infinitely more subtle and complex than anything under development today. If and when such a synthesis occurs, it will certainly help us overcome the limitations imposed by our biology, launching us to a direct, sensation-guided perception and investigation of the world beyond our world.

# 6 • MAPMAKER, MAPMAKER . . .

Marc Davis and I staked out a small patch of Santa Monica beach real estate one afternoon in late summer, 1977, escaping the gritty smog that too often blanketed the Observatories' offices in Pasadena. The cool sea breezes quickly cleared away thoughts of work and astronomy, and after the obligatory gossip about our astronomer friends, and the inevitable talk of women and sports that men in their late twenties have been known to have, we fell into a silent appreciation of the sun, sea, and sand. These were tangible results of nuclear fusion and supernova explosions that had curiously led to we two humans pursuing a fundamental cosmic question: where did we come from? And for Marc, then an assistant professor at Harvard, an institution famous for denying tenure to assistant professors, and for me, a two-year postdoctoral fellow, the question "Where are we going?" loomed every bit as large.

At least for the course of the summer, Marc knew exactly where he was going. He had come to the Observatories with a mission: his single-minded purpose was to build an *instrument*—astronomer shorthand for any piece of equipment that attaches on the end of a telescope and collects the data. In particular, Marc was going to construct an electronic camera to record light as it emerged from the spectrograph of the Harvard-Smithsonian sixty-inch telescope at Mount Hopkins, Arizona. He seemed immune to the vanity that often afflicts instrument builders—wanting to do it *their* way—and was quite happy to copy a highly successful design that Carnegie astronomer Stephen Shectman had built for the Palomar 200-inch telescope. Now Shectman was assembling an improved version for the new du Pont telescope at Las Campanas, and Marc was duplicating it, down to the last wire, integrated circuit, and resistor. Shectman's pursuit of novel, inexpensive, and efficient instruments was his scientific passion. Marc's interest in detectors was, in contrast, only a matter of expediency: he would take this or any other shortcut to get to the *science*, his passion.

Why couldn't Marc find a commercial company that made the high-sensitivity camera he needed and simply buy it? Why did he instead have to invest months of his own effort and tens of thousands of dollars for parts and circuitry? The answer lies in the faintness of astronomical objects: for example, the distant galaxies that were Marc's targets, which are thousands to tens of thousands of times fainter than the faintest stars one can see with the naked eye. Even with the light-gathering power of a sixty-inch mirror in the telescope, the collected light would be so feeble that, when split into finely divided colors by the spectrograph, the energy would dribble in packet by packet. Each of these *photons* would have to be individually detected and recorded. In comparison with the faint glow of distant galaxies, a nighttime sports event recorded by a commercial TV camera was a fireball.

Much of Marc's task would revolve around electronic circuitry, a sort of specialized computer Shectman had designed to detect and record these incoming photons. The simplified description of this instrument I gave earlier glossed over a number of subtleties by suggesting that detecting light this way is as simple as clicking off beads on an abacus. Unfortunately, the extreme faintness of the light prevents such direct detection. The row of electronic junctions called *diodes* that Shectman was using were designed to produce a tiny electric current when struck by light, but they had no hope of doing so when the signal was just a single photon. So it was necessary to first pass the light that made up the spectrum through a series of *image intensifiers* that amplified the light, turning a single, feeble photon entering one end into a brilliant flash at the other, a flash bright enough for the diode to "see." But this chain of intensifiers, each one like a miniature television screen, also provided the first complication. Just as a television glows in the dark after it is turned off, each little bright spot lingered for a while, even when the signal that triggered it was long gone. Shectman knew it was crucial to count each photon once and only once, so he had designed a primitive brain into the electronics, one that would register a photon only if the signal went from dark to light, but would ignore a persistent spot as it faded away. Hundreds of times each second these electronic bean-counters would check the signal in each of the 1,000 diodes, comparing voltages to those recorded on the previous sweep. When the circuitry decided it had indeed "seen" a photon, it flipped a switch in the microcircuits that said, "I have a signal. Ignore subsequent signals in this location for the next 100 clock periods."

Then the specialized computer would compute the average signal from adjacent diodes (each spot of light stretched over more than just one) in order to find a more precise position, corresponding to the color of light in this bit of the spectrum. With this embellishment, one could determine not only which diode was at the center of the light spot, but where along the diode, to one part in four, the center was. In this way the row of 1,000 diodes actually functioned as 4,000 tiny collecting bins. And, better by far than any real abacus, the circuits could make these decisions and calculations thousands of times every second.

Integrated circuits, the little silicon chips that look like multilegged bugs, made it all possible. Each had dozens of transistors—tiny electronic switches—etched into a paper-clip-sized package. Like a microscopic railroad switchyard, this network of hundreds of little switches directed tiny trains of electricity with blinding speed, controlled by wee conductors who performed, with uncanny accuracy, "If this, then do that, but if this *and* that, then do this . . ." Such is the role of *logic circuits*, the miniature decision-makers that have changed the conveniences of modern living and unleashed a flood of information/data, simply by doing the simple operations humans have taught them, but near perfectly, and millions of times faster.

By early fall, Marc had completed his copy of Shectman's "photon counter" and taken it back to Harvard. By winter, fellow Harvard astronomer Dave Latham would mate it to a chain of image-intensifiers; Marc's graduate student John Tonry was well on his way to finishing the computer programs that would collect and process the data. The team, rounded out by John Huchra, one of astronomy's most prolific observers, would gather the hardware and software for the trek to Arizona and, after a few months' struggle, get the whole package together and working at the telescope. By spring they were recording spectra of galaxies, on their way to obtaining more redshifts than had been done by all the observers since Vesto Slipher, the astronomer who had started it all in 1912 by recording the spectrum of Andromeda and a baker's dozen of other nearby galaxies.

It is uncommon in science for such a fundamental measurement as galaxy redshifts, understood by 1930 to be of crucial importance, to be so little exploited for so long. When, in 1950, Vera Rubin chose as her master's thesis the question of whether the universe as a whole is rotating, the first real attempt to look for deviations from a perfectly smooth Hubble expansion, she found scarcely more than 100 published redshifts for the attempt. Later that decade, legendary astronomers Milton Humason, Nicholas Mayall, and the young Allan Sandage published a generation's work that still amounted to only 600 redshifts.

Redshifts would be vital to understanding how the universe was arranged (perhaps *constructed* if you were of the mind to think in those terms) because they provided the simplest reliable way to make a three-dimensional map of the positions of galaxies. Galaxies cover the two-dimensional map of the sky like wallpaper flowers, but they are not all at the same distance, far from it. Hubble demonstrated that a galaxy's redshift, which yields its velocity away from us, is in direct proportion to that galaxy's distance from us. This was what one expected for a smoothly expanding universe, which, by the 1930s, was recognized as a basic description of the one we live in. In addition to being a fundamental property, redshifts are also a very simple quantity to measure: the shift of the spectral lines, like the Doppler shift in pitch of a train whistle as the train approaches and then recedes, gave a direct measurement of a galaxy's speed, regardless of how far it was from us.

Why, then, with such crucial information to be had on the *three*-dimensional distribution of galaxies, had so few redshift measurements been made and published in the forty years since Slipher? The paucity of such essential data was largely due to the fact that, decade after decade, making these measurements had remained a slow process—most galaxies are very faint, and the photographic plates used to record their spectra were sadly inefficient. There were few telescopes with large enough mirrors for the work to go reasonably rapidly. Although astronomers with access to those telescopes would from time to time obtain spectra of certain galaxies of particular interest to them, none committed to the decades of effort it would have taken to deliver many thousands of redshifts.

But it may well be that the "expense" of making these measurements was not the sole roadblock. An expectation that the results would be unexciting also seems to have discouraged any of the major players from devoting their astronomical careers to measuring redshifts. One might have thought the lure of producing the first true three-dimensional map of part of the universe would be great. However, the near-dogmatic belief that the universe was both isotropic, the same in all directions, and homogeneous, smooth in its density distribution, apparently worked to persuade astronomers that such a map would be uninteresting, just a random spray of galaxies here and there with little of the structure that a mapmaker was looking for. Probably it was just not worth the huge effort.

This prejudice arose from the commonly held image that the big bang had spread matter very evenly in all directions. In a way, this had been one of the most important discoveries of early cosmology—that the sky was covered fairly uniformly with galaxies. But this was a conclusion based on the *two*-dimensional distribution, that is, Hubble had shown that galaxies were to be found in every *direction* in the sky. But practically nothing was known about the three-dimensional space distribution, which might be hard to untangle when looking at a map of the sky in which galaxy distances are not at all obvious (because of the great range in true size and brightness that galaxies have).

Still, there were clues, even in the two-dimensional distribution, that the cosmic landscape might not be a smooth sea, but have islands and continents to delight a mapmaker. Caltech's irascible astrophysicist Fritz Zwicky had noted conspicuous groupings ranging from tens to hundreds of galaxies in the galaxy catalog he assembled in the 1930s, something that was confirmed and underscored in the 1950s by a huge compilation of thousands of very rich clusters, many of them very distant, by graduate student George Abell's careful inspection of the then-new Palomar Sky Survey. But, from the point of view of the large-scale structure of the universe, these aggregations seemed less important, like the globular star clusters that encircle our galaxy, which are an intriguing and instructive feature but nonetheless a sideshow compared to the Milky Way itself. To many astronomers, these rich clusters were no more than statistical accidents, those rare places where an unusually high density of matter in the early universe had led to numerous galaxies bound to each other by the stronger gravity. Rich clusters might be interesting, perhaps even useful, but they were not regarded as a fundamental feature of the landscape. The picture, then, was one of a countryside, with farms completely covering the hillsides and the valleys, now and again a village, and, rarely, a great city.

There were two outspoken critics of this view, Fritz Zwicky and Gerard de Vaucouleurs from the University of Texas at Austin. Nothing shows better the importance of redshifts than the *inability* of these two men to convince their colleagues that the distribution of galaxies in space was far from smooth. Each had a slightly different view of what that distribution looked like, and they quarreled as much with each other as with their other colleagues. Both were way out ahead in realizing that the large-scale structure of the universe was complex and interesting, not the dull, smoothly expanding splatter others imagined. But because they had few redshifts and thus had to rely on two-dimensional maps of the sky, they couldn't make their cases convincingly.

Swiss-born Fritz Zwicky had shown his brilliance by predicting the existence of neutron stars, an exotic end-state in which some stars collapsed to the density of an atomic nucleus. He was, throughout his career, an iconoclast of the first rank; this and an arrogant tone made him persona non grata with the more traditional astronomers who controlled the Pasadena empire of great optical telescopes. The maverick view that concerns us here was Zwicky's insistence that *all* galaxies belonged to clusters. Zwicky had been the first to produce a large catalog of faint galaxies, charting their positions on a flat map of the sky. As he went along he identified clusters, thousands of them, wherever he thought the density was anomalously high. Then, around strong concentrations Zwicky drew what most astronomers thought to be fanciful boundaries—farfetched, meandering borders that reached to the frontiers of neighboring clusters. In Zwicky's world all the farms were affiliated with one or another town or city, no matter how remote they seemed. There were no "loners."

De Vaucouleurs also believed that the universe was organized into units bigger than individual galaxies. He had been impressed by Vera Rubin's willingness to look for rotation in the local distribution of galaxies, her search to see if the whole universe, like the Milky Way galaxy, is turning. Emboldened, de Vaucouleurs proposed that our galaxy is part of a flat *supercluster*—a galaxy of galaxies. Though the observational roots of this notion went all the way back to Humboldt and Herschel in the eighteenth century, the idea had never really taken hold. De Vaucouleurs's suggestion was broadly attacked, especially from Pasadena, the Mecca of cosmological research since Hubble's time, where the vision of a smooth universe prevailed. Even Zwicky, who believed in superclusters, vigorously opposed de Vaucouleurs's notion of a *Local Supercluster*, because de Vaucouleurs saw the supercluster as the basic unit rather than the cluster. For de Vaucouleurs, the presence of a dense knot of hundreds of galaxies, the unit that Zwicky was focusing on, was a minor matter. The Local Supercluster did in fact contain such a rich knot, the Virgo cluster, but to de Vaucouleurs this was just a matter of happenstance—the supercluster itself was the important organizational structure. In de Vaucouleurs's world, the vast counties of farms were the primary structure. The presence of a town or city was incidental.

This clash of concepts seems almost trivial in hindsight; in perspective it is clear that these two stood much closer to each other (and to reality) than they did to the rest of the astronomical world. Without galaxy redshifts to form a three-dimensional space-map, Zwicky and de Vaucouleurs had little success convincing their colleagues that the superclustering they described was genuine. Instead, most astronomers believed they were seeing only chance alignments of galaxies at vastly different distances, much like the stellar constellations—random collections of stars at a wide range of distances, whose apparent association into larger units is only an illusion that vanishes if viewed from another place within the Milky Way.

The frustration of this debate as to whether superclusters were real or an illusion peaked in the 1970s, when Jim Peebles assembled a map of the positions of a *million* galaxies, as counted by astronomers at Lick Observatory who had surveyed the sky using photographic plates. In Peebles's sky map an individual magnificent galaxy, with its enormous power and complexity, didn't even rate a puny dot—such is the oppression of cosmology. Instead the sky was divided into hundreds of thousands of tiny squares, and each square was made bright or faint depending on the number of galaxies counted in that tiny zone. This picture, the first to display a vast area, the majority of the sky in fact, revealed the most striking image yet of a cobweb of galaxy chains, but again few believed these to be real features. Even Peebles, whose work supported the notion that clustering of galaxies on larger scales was natural and expected, doubted the reality of the sharp features on the map he himself had helped make. It was time for new measurements of redshifts of galaxies to bring this map into full relief and begin to disclose the true structure of the universe.

∞

By the mid-1970s the possibility of adding vast numbers of new redshifts was at hand. Light detectors with much greater sensitivity than photographic plates, and a new generation of optical telescopes, made it easier to propose and justify more ambitious projects like mapping the distribution of galaxies. As we look back at this era we see a classic example of an important recurring pattern in scientific work: the approach of a specific, pointed search in contrast to the general statistical survey. Both approaches can, and in this case did, lead to the same discovery, but the differences in method and program can be striking.

In both attempts the goal was to get many new redshifts for galaxies and produce a three-dimensional map of some region of the sky to test for the reality of the apparent structure in the flat sky maps. Some researchers concentrated their attention on specific instances, choosing to test whether "filaments" in Peebles's Lick galaxy-count map were real associations— clumps in three dimensions as well. Others followed the prescription that it is better to study an unbiased sample—a large, randomly selected region— than to base general conclusions on what might be special cases. Only by finding the general degree of clustering, they reasoned, would the significance of the extreme examples be understood. It is a common division in science: some focus on the pathologies in order to define the general rule, others prefer to carefully constrain the common phenomena in order to provide a framework for the extreme.

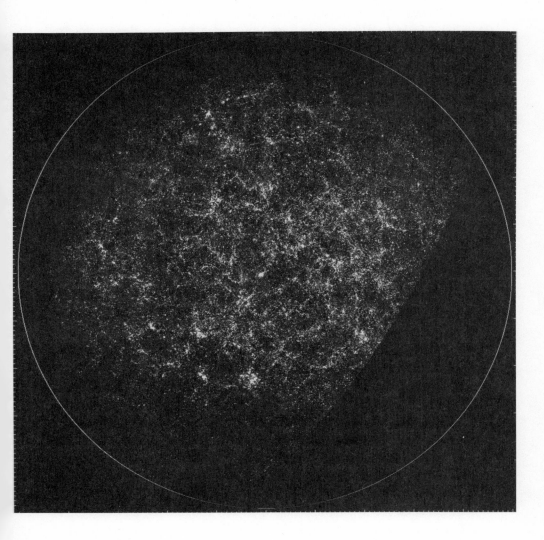

The map of the positions of a million galaxies made by Jim Peebles from the Lick Sky Survey. The filamentary appearance suggests a rich, subtle structure. The dark area to the right shows the limit of the survey, which did not reach far into the southern sky.

The approach of "the specific" was the first to bear fruit. Some astronomers began to focus their attention on the dense knots and filaments that
looked like they might be superclusters. By 1976 many of the 2,700 redshifts
that now had been measured were concentrated in these "interesting" regions. Several research groups realized that by filling in the gaps with new
redshifts for only a few dozen carefully chosen galaxies, they might be
able to show if these high-contrast regions were chance alignments or real
associations. The most striking result came in 1978 from Stephen Gregory
of State University of New York and Laird Thompson at Kitt Peak Observatory, who had concentrated on measuring redshifts for galaxies in the region
surrounding the extraordinarily rich cluster of galaxies in the direction of the
constellation of stars called Coma Berenices. They found a continuous chain
of galaxies from the densely populated Coma cluster to another Abell cluster
a full twenty degrees away in the sky. Although it had long been known that
this second cluster lies at roughly the same distance as the Coma cluster (the
average redshift of galaxies is the same for both clusters), it had been assumed
that the two clusters were much too far apart on the sky to be physically
associated. The connecting bridge of galaxies showed otherwise. Building
on de Vaucouleurs's notion of a supercluster, Gregory and Thompson put
forward a new picture: that the Coma cluster and its lassoed companion are
just high-visibility jewels embedded in a broad necklace that stretches over
a large region of the sky—a supercluster. And they went further by demon-
strating the obvious corollary of their notion of superclustering: just as there
are vast regions where galaxies have been collected, so too there are huge
voids, regions of space where few, if any, galaxies are found.

The first demonstrations that certain superclusters were real physical
associations, and that they were often found adjacent to large voids, made a
crack in the façade of a very smooth universe, but the astronomical community as a whole was not convinced. After all, these were special regions,
chosen because they showed up so vividly in the two-dimensional galaxy
maps of the sky. Might not the average volume of space have a much more
even galaxy distribution, thereby saving the notion of a smooth universe?
After all, a survey of the domiciles of midtown Manhattan residents might
give a very slanted view of average American homes. Were these studies
really providing a fair picture of the habitat of a typical galaxy?

The answer to this question could come only from large, unbiased redshift
surveys. In 1981 Swiss astronomer Gustav Tammann and Allan Sandage
published a major new resource, a catalog with redshifts for 1,246 nearby
galaxies spread around the sky. This catalog, when used to make a space
map, removed all doubt that de Vaucouleurs was right: the Local Supercluster

is a reality, not just a phantom projection but a coherent collection of galaxies into a fat pancake of colossal proportions.

In the same year Robert Kirshner from the University of Michigan, Augustus Oemler, Jr., from Yale, Paul Schechter from Kitt Peak, and Stephen Shectman from Carnegie reported the results of another unbiased search. Using Shectman's photon-counting detector, they measured redshifts for 133 galaxies in three corners of a vast triangular patch of the sky behind the constellation Boötes and found a huge, gaping hole where galaxies were not to be found. If these three sample regions were typical, they were enclosing an enormous volume empty of galaxies. Certainly, the team found galaxies in this direction of the sky, but their redshifts always tended to be above or below a certain range, leaving a yawning gap in between. But the survey was small. This meant that the chance for a fluke "hole" was great, so in succeeding years the group went on to survey the entire area, not completely (there were too many targets) but sparsely, by choosing a representative sample of a few galaxies in each of hundreds of patches that covered the vast region. When finished, they concluded that although the region was not completely empty of galaxies, it was remarkably so, with about one-tenth the average density of the universe. Of course, other voids had been located by this time, but this *Great Void*, as it came to be known, had been found by accident, in a "blind search," and it was ten times larger in volume than the largest empty region found by Gregory and Thompson. The Great Void stretched for the extraordinary distance of 200 million light-years.

By now the long-cherished paradigm of a smooth universe was teetering. What followed was an episode that is wonderfully typical of science, which evolves from model to model, finding it difficult though necessary to abandon previously revered ideas. Many were now acknowledging that the universe was not smooth, but how best to describe the emerging picture, and what implications did it have for how the universe came to be?

As the new, surprising observations were reported in colloquia and published in the professional journals, the level of rhetoric and controversy rose, and those engrossed in the subject began to take sides. Curiously, even cold-war politics played a minor role. Cosmologists in the Soviet Union had for many years formed a solid front behind the ideology of pre-eminent Russian physicist Yacov B. Zel'dovich, whose model for the early universe predicted that matter had first condensed into huge sheets and filaments that much later broke up into individual galaxies. These astrophysicists had long believed that the universe would be found to be lumpy, not smooth. Their picture became known as the "top-down" model because it proceeded in the order of big structures breaking up into smaller ones. This was opposite to the

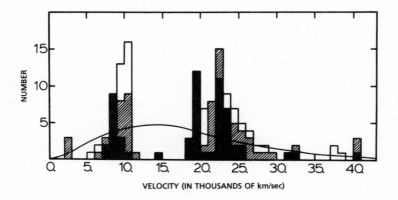

The discovery of Robert Kirshner and his collaborators of a "great void" in the distribution of galaxies, seen here as a gap in the histogram of redshifts for their survey. *(Reference: 1981 Astrophysical Journal Letters, vol. 248, p. 47.)*

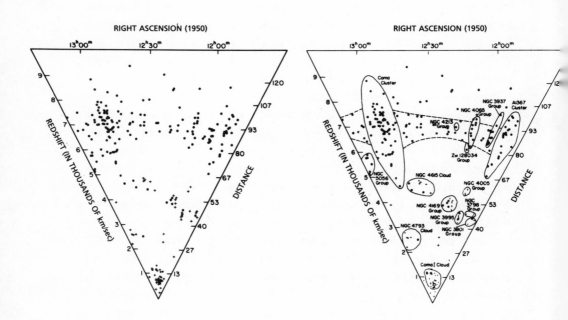

The map of galaxy positions made by Stephen Gregory and Laird Thompson, which showed voids and a chain of galaxies connecting the Coma and A1367 clusters. The view from our galaxy is from the bottom point of each triangle. Looking out to a "wedge" of space, each galaxy's redshift (approximate distance) and celestial "longitude" (position on the sky) is shown by a dot. *(Reference: 1978 Astrophysical Journal, vol. 222, p. 784.)*

order of structure building that was popular in the West, represented by Peebles's "bottom-up" scheme, where a hierarchy of structure developed from galaxies joining into groups, and then groups combining into clusters and finally superclusters. Behind the Iron Curtain, superclusters were seen as the building blocks, the first pieces from which everything else followed, but the Western picture called for galaxies to form first and aggregate to form the bigger units. Israeli astrophysicist Avishai Dekel attempted to promote a kind of détente, a compromise "east-west" cosmology incorporating the best features of both, but, as was typical of the period, he was coolly received by both sides.

Those who preferred the top-down, large-structures-first model, or otherwise had a fancy for seeing apple carts upset, delighted in the new evidence for large-scale features in the galaxy distribution. This, they contended, was *the* decisive evidence that the universe is organized into large structures, huge sheets and chains of galaxies weaving through a space that is otherwise totally empty. They argued that the emerging picture of a universe with a high contrast in density between galaxy habitats and voids supported the interpretation that large structures, rather than the galaxies themselves, were the building blocks of the universe. Could distinctive features of such vivid contrast have grown from a casual gravitational clumping of randomly strewn galaxies? To many it seemed more likely that, following Zel'dovich's idea, superclusters appeared first, as sheets of gas from which galaxies were sliced like a cookie-cutter pressed into rolled-out dough.

The supporters of Peebles's bottom-up model were nowhere near ready to concede that the Western picture had to be abandoned in light of the new data. Their view was that the new evidence was based on samples either too small, or biased in a way that gave an incorrect reading of the situation, much as a political poll might fail if it asked too few voters or focused on a special group. This reasoning formed the basis for a counteroffensive that was marshaled on two fronts. One assault was directed at the statistical certainty of the evidence. By this time everyone agreed that there was some higher level of organization than galaxies: the issue was now one of degree. Certainly the Great Void in Boötes was impressive, but just how unusual was it? In a universe where galaxies are spread out in a completely random fashion, such a large, empty region would be extraordinarily unlikely, it's true, so finding even just one might easily invalidate this working hypothesis. On the other hand, by this time everyone also recognized that galaxies are clumped by gravity to some extent, and the likelihood of a big hole in a clumpy distribution is much greater. For an example of this effect you need look no further than the page of print before your eyes. If you force your eyes far out of focus you will see white tracks snaking their way down this page—rivers of breaks between words. Continuous, long trails would be very rare in a page full of randomly spaced *letters*, but the association of letters into words makes their probability much higher. To evaluate properly how extraordinary is the Boötes void required a very good understanding of the strength of galaxy clustering on smaller scales. Also, the particular and somewhat peculiar sparse sampling pattern Kirshner, Oemler, Schechter, and Shectman had used in their survey made it difficult to model what was to be expected for different patterns of large-scale clustering. The result was a persistent, unresolved debate of the significance of the Great Void.

The second assault came in the form of more observations, designed to show that these large structures were just lucky shots. This is where the effort by Marc Davis to build a better detector for measuring galaxy redshifts was to play a vital role. Marc was one of those unwilling to abandon the bottom-up clustering model; he himself had worked on it with Peebles. By 1982, five years after our visit to the beach, Marc and his collaborators Huchra, Latham, and Tonry had finished their redshift survey, which became known as the CfA (Harvard-Smithsonian Center for Astrophysics) Survey. The CfA survey showed some large-scale organization, to be sure, but a three-dimensional map showing the locations of these 2,400 "representative sample" galaxies revealed no extraordinary clumping. It appeared that a little gravity pulling galaxies together from an otherwise random distribution might work just fine. Now at the University of California, Berkeley, Davis was working to make this argument quantitative, by developing a computer program that would simulate the bottom-up clustering model and make comparisons with the CfA survey.

Later in the decade his team would add, through a collaboration with Brazilian astronomer Luiz da Costa, a second CfA survey done from the Southern Hemisphere. Together these two samples, compared with the model universes Marc made in a computer, reinforced his belief that the case for a very lumpy universe had been greatly exaggerated. By this time, however, a bombshell had been dropped in the war; ironically, Marc Davis himself had lit the fuse.

Our group of seven, now gaining momentum in our expedition into the world of elliptical galaxies, paid close attention to the battles over the shape of the cosmic terrain. The consequences of a lumpy as opposed to a smooth universe would have important implications for what we were trying to do, and, although we hadn't planned it that way, it would become the very heart of our effort. We knew that the elliptical galaxies we had chosen to study would be good markers of the high-density knots, filaments, and sheets of galaxies our colleagues were finding—of all galaxy types ellipticals cluster most strongly. A 1980 letter by Dave Burstein to the rest of us, written as he was helping to choose the elliptical galaxy sample, made special note of his impression that the ellipticals tended to be arranged in great sheets that crisscrossed the sky. But our interest would go well beyond mapping the *positions* of these galaxies, what the other groups were so vigorously pursuing. We were about to measure the *motions* of these galaxies through space, and this would be a different way to judge the lumpiness of the universe, one that would offer additional, crucial information.

Here is the crux of the matter. A perfectly smooth universe, if such existed, would expand completely evenly, each galaxy receding with the space as it stretches to a larger and larger volume. But a universe with any lumpiness will depart from this uniform expansion. The reason is simple. Regions in which the density of matter is higher than average, for example a supercluster, will contain greater gravitational force—the combined pull of all the galaxies on each other—than an empty region like a void. This means that in denser-than-average regions gravity will work *against* the overall expansion and as a result the region will expand more slowly than a region of average density. Conversely, galaxies lying on the boundaries of near-empty regions will appear to separate from each other more rapidly than would galaxies in a region of average density. As a result, if the motions of galaxies in a region of space can be measured and compared with what is expected for a smooth expansion, the presence of an extra-large concentration of matter, or a dearth of matter, will be disclosed in the pattern of velocities. Galaxies in the denser region will have "extra velocity" *toward* the center of the extra mass, and galaxies near voids will appear to have extra velocity *away* from their centers. Astronomers have coined the term *peculiar velocity* for these extra motions, differentiating between the "normal" velocities galaxies have as a result of the expansion of space—what Hubble discovered—and the extra, peculiar velocities that arise if the universe is lumpy. When you see the words *peculiar velocity* from here on out, as you often will, you should think of the *extra velocity*, in addition to the normal Hubble expansion, induced by a lumpy distribution of matter.

When we began, the emphasis was on learning about the intrinsic properties of elliptical galaxies, and what this might teach us about the way they, and other galaxies, formed. Over the next three years things would change. We would learn how to apply what we were learning about ellipticals to finding their motions through space, and soon we would be fully embroiled in measuring these peculiar velocities. In this way we would enter into this burgeoning enterprise of surveying the universe and eventually find ourselves fully embroiled in its controversies. We were about to wander far from our original destination.

Investigating the lumpiness of the universe by looking for these peculiar velocities, the extra motion each galaxy might have in addition to what a smoothly expanding universe would give to it, is one of those cases where it is simple to decide what to do, but far from simple to actually do it. When an astronomer measures the velocity of a galaxy, by taking a spectrum that shows the Doppler shift of the spectral lines, she is finding the *total* motion, the sum of the velocity that galaxy gets from the overall expansion of space and the extra, peculiar velocity. The problem is how to separate the "normal" expansion velocity from the extra "peculiar" motion, especially when, as is true, the peculiar velocity accounts for only a small fraction of the total. The key piece of information needed is the galaxy's distance from us. With this we can predict what the expansion velocity alone should be for this galaxy, by using Hubble's relation that this velocity is directly proportional to the distance to the galaxy. Then this expected expansion velocity, which would be correct for a smoothly expanding universe, can be subtracted from the velocity actually measured for the galaxy. What is left over is the extra motion, the peculiar velocity, that the galaxy might have as a result of the gravitational pull of an uneven distribution of matter in its vicinity.

Such measurements of galaxy distance that are made independent of the redshift of the galaxy (which would give the correct distance only in a uniformly expanding universe) had been problematical. As a result, only a little had been learned about the peculiar velocities of galaxies as the seven of us began our journey. Even though measuring the distances to elliptical galaxies had not been our focus, we knew from the start that if we could make a breakthrough in this direction we could open a new trail through this largely unexplored territory. That's exactly what happened.

Early attempts to look for these subtle departures from a smooth Hubble expansion—the peculiar motions a lumpy mass distribution would generate—were difficult, intriguing, and controversial. In 1975 Vera Rubin and Kent Ford had been the first to go boldly where only Hubble had gone before. Hubble, remember, had looked to see if the merry-go-round motion of our Sun around the Milky Way galaxy is reflected in the speeds of neighboring galaxies, but instead he had discovered a bigger effect—that all galaxies are moving away. He had discovered the expansion of the universe. Rubin and Ford picked a more distant sample of galaxies than the ones Hubble had first looked at, a sample constructed to include galaxies that were approximately all at the same distance from our galaxy, that is, they populated a shell centered on our galaxy.* If the Hubble expansion were perfectly smooth, they reasoned, each galaxy would have about the same recession velocity. What they found, however, was that galaxies on one side of the sky seemed to be receding faster than on the other side—a lopsided Hubble expansion. Assuming that Rubin and Ford had succeeded in selecting an equidistant sample, their result could mean only one thing—the Hubble expansion was not perfectly smooth. This could be a reflection of a large peculiar velocity of our galaxy—the Milky Way cruising along at about 500 km/sec in a particular direction—or peculiar velocities of a similar amount, and over vast regions of the sky, for the subject galaxies themselves. Either way, the conclusion would be the same: the Hubble expansion was not smooth.

This was a radical result that flew in the face of accepted wisdom of the time, in particular, a clever argument by Sandage. He had reasoned that peculiar motions must be less than about 100 km/sec, or else many nearby galaxies would, by chance, be headed *toward* us and would show blueshifts.[†] They don't. With this counterevidence, and the potential pitfalls in the Rubin-Ford approach that others pointed out, few researchers took the result seriously. The common criticism was that Rubin and Ford had not succeeded in selecting a sample that lay at a common distance in all directions. If galaxies chosen in one part of the sky were on-average just a bit farther away, then this side of the sample would have a significantly larger velocity away from us than the other. But this anisotropy (not being the same in all directions) would in this case be due to a failure to sample fairly the smooth Hubble expansion, rather than genuine peculiar velocities of galaxies.

The real shakeup came just a year later, when research groups at Berkeley and Princeton measured an anisotropy in the intensity of radio waves that

---

*Their method of doing this was to use only a certain variety of spiral galaxy which, they believed, always came in approximately the same true size. By measuring the apparent size on the sky to estimate how far away each one was, and keeping only those that appeared to have roughly the same size, they were selecting a sample for which all the galaxies lay at approximately the same distance. Their assumption about galaxy size was a subject of later debate.

[†]There are only a handful of galaxies that are so nearby that their velocities away from us should be less than 100 km/sec in a smoothly expanding universe, and two of these do show blueshifts. If galaxies had random peculiar motions of, say, 500 km/sec, as the Rubin and Ford result implied, dozens of the galaxies whose Hubble expansion velocities were less than 500 km/sec would have these canceled by a peculiar velocity in our direction, and would also show blueshifts. This is not observed, so the implication is that peculiar velocities are not that large. But, as we will see, there is a hidden, incorrect assumption in this reasoning.

bathe the universe from all directions. This "cosmic microwave background" (CMB) light had been predicted in the late 1940s as a fading glow of the big bang, and discovered serendipitously in 1965 by researchers at Bell Laboratories. What are commonly called radio waves are just another form of light, differing from visible light only by their slower frequency of oscillation and correspondingly longer wavelength. The Bell scientists were looking in what is called the microwave part of the radio spectrum in order to find a "channel" sufficiently free of natural "noise" for good transmission of telephone calls. In the process they discovered an unexpected diffuse sea of radio photons arriving from all sky directions—it turned out to be a stunning confirmation of the big bang theory and the first recognized signal from the early universe.

The CMB, coming as it does from the cosmic fireball, is representative of the universe-at-large. As such, this sea of radiation is perfectly suited as a reference for measuring the peculiar velocities of galaxies. Those in perfect Hubble flow are at rest, merely floating with respect to the CMB bath, but those with peculiar velocities are swimming in the sea of photons. In 1976 physicists at Berkeley and Princeton discovered that our own galaxy was in fact swimming through this microwave photon sea. The sensitive instruments they built, carried aloft by high-flying balloons, showed that the CMB radiation is slightly hotter in one direction of the sky and slightly cooler in the opposite direction. The natural explanation: our galaxy is moving *through* space (not just carried along with the expansion of space) at about 600 km/sec with respect to this cosmic backdrop. Here was indisputable evidence of a peculiar velocity much more rapid than had been thought possible. Peculiar velocities were a reality, and they were large.

The Copernican principle, named after the Polish astronomer who put the center of the solar system at the Sun, where it belonged, expresses that we should always expect to occupy a typical rather than a special place in the universe. The first indisputable evidence for a departure from smooth Hubble expansion, the large peculiar velocity of our Milky Way galaxy, should have awakened the astronomical community to the likely possibility that such large "extra" velocities were common, i.e., that the Hubble expansion was not all that smooth, with all its implications for large-scale lumpiness of the distribution of matter. It is revealing that few astronomers were bold enough to go that far at the time.

Instead, rather cautiously, the interest in peculiar velocities now moved to a single question: where are the lumps of matter that have accelerated our galaxy to this large speed? Facing the direction in the sky where the Milky Way is headed, the most conspicuous collection of galaxies is the populous Virgo cluster and the high-density environs of the Local Supercluster that surrounds it, the region that Gerard de Vaucouleurs had made such a fuss about. This rich area of galaxies is hardly on a line with the peculiar velocity of the Milky Way, however—it is the difference between left and right field on a baseball diamond. Nevertheless, Virgo and its entourage galaxies are in *roughly* the right direction (well, at least it is a fair ball), so it made some sense to look and see if this concentration of matter was partly responsible for the big pull.

Just as we were beginning our study of elliptical galaxies, the first attempts to measure the peculiar velocities of large numbers of galaxies were beginning to make real headway. In 1982, with a new method of estimating the distances to spiral galaxies, University of Arizona astronomer Marc Aaronson and his collaborators published distances they had derived for 306 spiral galaxies, all of which had Hubble expansion velocities of less than 3,000 kilometers per second. With these data they made the first accurate map of the local Hubble flow, placing each galaxy at the distance they had determined and tagging on its velocity measured from the Doppler shift. When the team compared this map to the smooth velocity map that would have accompanied a completely *even* expansion of space, they found extra motions, the peculiar velocities that indicated a lumpy distribution of matter over large volumes of space. Their map showed a systematic pattern of peculiar velocities: not only our Milky Way galaxy, but all the other galaxies around it, were streaming toward the higher-density region around Virgo. This demonstrated that at least part of our galaxy's motion through space can be credited to the pull of this overdense region. As expected, the greater-than-average density of the Local Supercluster leads to greater-than-average gravity, which, in effect, fights the expansion of the universe, generating departures from smooth Hubble flow.

The Aaronson et al. study was a breakthrough, the first time a map had been made that showed how the matter in the universe is distributed, not from counting galaxies, as others had done, but by using the motions of those galaxies to trace the matter distribution. One might wonder why such a roundabout procedure is necessary, when it is much more direct to simply look and see where the galaxies are. The answer to this question, and why the measurement of peculiar velocities is absolutely fundamental, is that the motions of galaxies are due to the combined gravitational effect of *all matter* in the region, not just what can be seen in the glowing starlight of galaxies. As we shall consider in a later chapter, it had been thought possible (and is now regarded by many astronomers as certain) that space contains more matter than what is visible as galaxies. For this reason a test of the smoothness of the Hubble expansion would be crucially important because it would provide unique information about the lumpiness of the *mass* distribution in the universe, including any and all matter, whether or not it emitted light, as do the stars and gas in galaxies.

Besides opening the door to this kind of fundamental information, Aaronson et al. provided a remarkable clue that the lumpiness of the matter distribution went far beyond the Local Supercluster. Their work showed that not only was the direction of our galaxy's journey through space off line with the Local Supercluster, but the pull attributable to this sizable overdensity was less than half of the 600 km/sec speed that the Milky Way had acquired. Both the direction and speed were way off, so much of the peculiar motion of our galaxy had to come from elsewhere, and this meant farther out in space, which further meant, something *bigger*.

Our study of elliptical galaxies would become important in this search for other lumps of matter that might be tugging the Milky Way around, because our sample was evenly spread around the sky, and it extended far beyond the Local Supercluster. We would eventually find a method of determining the distances to our elliptical galaxies and this would allow us to make a map of the Hubble expansion, as Aaronson and collaborators had done, but on a grander scale. This was still years off and unknown to us in 1982, when we began to look at the early data we had collected and what they might tell us.

The initial stages of our work had gone well; by the end of 1981 we had brought back, from our first round of observing at the Kitt Peak, Anglo-Australian, Lick, Las Campanas, and South African Astronomical observatories, spectra and photometry of about 300 of the 400 destinations on our galactic itinerary. With such a wealth of information it seemed that we were not far from our original goal of checking the remarkable relations between velocity dispersion, luminosity, and metal abundance found by Terlevich, Davies, Faber, and Burstein. Recall that this group had found mutual correlations of three quantities: the typical speed of the stars in an elliptical galaxy, the galaxy's brightness in absolute terms, and the enrichment of "heavy" chemical elements—metals—by generations of stars. If these correlations held up, measurements of these quantities could be used to determine distances to our sample galaxies, and this would allow us to make the most extensive map of the Hubble expansion yet attempted. But most of the data that we had collected were *raw* data, which is to say, sundry numbers recorded on computer disks and tapes that would require processing and calibration before they yielded the measurements we needed.

Such *data reduction*, as astronomers call it, would be our first hurdle, but a bigger problem would be cross-comparing all these measurements and convincing ourselves that data gathered from so many different telescopes and instruments, processed with various algorithms at each of our home institutions, could be combined into a uniform set. In our previous projects most of us had taken a smaller number of measurements, usually from a single telescope. Even when several sources of data were available, there were rarely enough duplications to check if the measurements were in good agreement. So, wary of the difficulty of combining different data sets, we arranged for each of our observing teams to observe many of the same galaxies in order to see how well the measurements agreed.

Although such knowledge leads to confidence, it invariably leads to disappointment as well. Scientists are fond of chiding each other with the cynical adage: "To know a quantity accurately, measure it just once." Our evolving data set, with its hundreds of measurements and dozens of cross-checks, afforded us new "opportunities": the opportunity to learn that different data sets were not easily combined; the opportunity to find out that systematic errors were larger than we had hoped; the opportunity to be "realistic" about how well such measurements could be made. These were important lessons, but painful, a little like looking at yourself in one of those magnifying shaving mirrors.

Another substantial problem was something we had never considered. None of us had ever worked within such a large group, and we were about to learn how complex it was to bring everyone together to accomplish these tasks. And this was only the beginning, because after processing of the data was complete, we would have to work as a group to analyze and interpret the results. Assuming we could all agree on what we had found, we would then have to write up the results for publication and present them at astronomical meetings and at invited talks at academic institutions. Spread out as we were from England to California, each of us with other scientific projects and diverse responsibilities (like teaching), these tasks were to prove much more difficult than we had imagined. They would take years.

Still, the pot was beginning to simmer. In 1981 John Tonry and Marc Davis published a paper that analyzed the velocity dispersions (the typical speed of the stars) and metal abundance (the amount of heavy chemical elements) for hundreds of elliptical and their cousin S0 galaxies from the CfA redshift survey. A principal conclusion of their paper was that the correlations found by Terlevich, Davies, Faber, and Burstein were *not* observed for a new, larger sample of galaxies. This was exactly what Terlevich et al. had worried about happening, because of their tiny sample of twenty-four galaxies. But, as usual, deciding who was right was not simple. Tonry and Davis's measurements of velocity dispersion and metal abundance were only a by-product of the primary goal of the CfA survey to map the galaxy distribution, for which spectra of comparatively low quality were needed. Measuring redshifts required only that the dark absorption lines in a spectrum be *detected*, that is, finding the strong dips in light intensity that occur at certain colors. However, to determine the typical speed of the stars in a galaxy (the velocity dispersion) or extract the depths of the absorption lines to measure the amount of heavy chemical elements (metal abundance) was the next step up in detail. Accurate measurements of these quantities required the collection of much more light. By analogy, it may be easy to see a person walking in the fading light of dusk, but identifying the person may not be possible without enough light to see important details.

What Tonry and Davis had to offer, therefore, were measurements of many more galaxies than the twenty-four Terlevich and collaborators had studied, but each measurement was significantly less accurate than those of the earlier study. They reported that they saw little evidence for the "delta-delta" relation of Terlevich et al., the strong tendency for a given galaxy with an unusually high or low velocity dispersion (for its luminosity) to have a correspondingly high or low metal abundance as well. It was the discovery of this correlation that had fired the hope that there was a "second parameter" for the correlations of these three quantities. That is, in addition to the first parameter, luminosity, that drove the velocity dispersion and metal abundance up and down, a second parameter like the shape of the galaxy might explain why the velocity dispersion and metal abundance were both too high or too low. Just as height and waistline could be used together to predict the weight of a person much more accurately than just one of these alone, the first and second parameters could be used to predict a galaxy's luminosity, its true brightness. With this, its distance could be determined by comparing its true brightness with how bright it appeared in the sky; distances of sufficient accuracy could be used to map the Hubble expansion. No wonder members of our group were worried. If Tonry and Davis were correct, the basis of our project, correlations we hoped to use to learn about the formation of elliptical galaxies and to measure their distances, had been undercut. This massive effort of the seven of us would be largely a waste of time.

This concern dominated conversations among members of the seven when two or three of us would meet at the odd conference, or converged with the specific intent of moving the work along. For example, in July 1982 I attended a conference at Cambridge University on the subject of galaxy formation. Here our colleagues presented the latest ideas in the theory of how galaxies might have condensed out of the hot gas left by the big bang, and the observations that were or might be useful to test these ideas. At this conference Roger Davies and I spent many hours discussing the progress and difficulties in our large elliptical galaxy program, and specifically the challenge from Tonry and Davis. Both of us felt that our group had enough data to test the delta-delta correlation again. Depending on the outcome, we could either counter Tonry and Davis, or, if we came to the same conclusion, quickly seek out a face-saving fallback and a new direction for our project. We recognized that if the fundamental motivation for our huge program was fatally flawed, it would be better for us to come out with our own paper detailing this. If, on the other hand, the delta-delta correlation survived, it would be best to dispute the claim by Tonry and Davis before it became common lore.

It appeared that we had enough data to make this test, even though we were well shy of completing the observations for our all-sky sample. But much of the data was unreduced, and no cross-comparisons had been made. Roger and I decided that a judiciously chosen subset of the data could be used. For example, I had already processed all the spectroscopic data from my two 1981 observing runs at Las Campanas; these were coherent data in that they all came from a single telescope, spectrograph, and detector combination. Roger was well on the way to reducing an equally large sample that he and Roberto Terlevich had observed with the 150-inch Anglo-Australian telescope. We reckoned that photometric measurements of overall galaxy brightness were available for about seventy-four of these galaxies, and that this sample, as accurate as the original Terlevich et al. data set but three times larger, would be enough to make the delta-delta test. Roger and I felt a great sense of urgency that these data, or some similar sample, be used to check the delta-delta relation, and that a report of our group's progress and findings, in response to Tonry and Davis's paper, be given at the upcoming meeting of the American Astronomical Society.

Our conversations also encouraged Roger to take an important step in the organization of our project. He wrote a letter to Dave Burstein in early August congratulating Dave on his new appointment as assistant professor at Arizona State University. In the letter Roger bemoaned the incoherent state of the data samples and volunteered to coordinate the compilation and combination of spectroscopic data from the Las Campanas and Anglo-Australian observatories, and, when they became available, the Lick data as well. By undertaking one of the major parts of the project, making sense of all of the spectroscopic observations, Roger was in effect suggesting that the collection and organization of all the data, the job nominally assigned to Dave, was too big a task for one person. In the letter, Roger suggested that Dave, who was almost a fanatic when it came to structure, order, and detail, concentrate his efforts as "arbitrager" of photometry, the largest and most complex of the data sets.

It was a small structural change, but an important one. Roger and I were particularly impatient to move things along, and this first stage of parceling out responsibilities seemed to help. Dave started to put a great deal of time into the task of compiling previous photometric measurements and incorporating new ones, and when Roger returned to his new staff position at the Kitt Peak National Observatory, he dove into his assignment by comparing the spectra I had obtained with those done by him and Roberto. Donald Lynden-Bell had earlier in the year applied for a grant from NATO (which sponsored exchange meetings of scientists from NATO countries), and by the end of the year we learned that we had been given $5,000 to cover expenses for all seven of us to meet in Cambridge the following summer. Things were indeed beginning to move, at long last.

In the early fall of 1982 I visited Sandy Faber in Santa Cruz. I found out that Sandy appreciated and shared the concern Roger and I had about the Tonry-Davis challenge to the delta-delta relation, and we determined to try to assemble a sample immediately that could test the matter. Sandy had a large number of spectra taken with the three-meter Shane telescope at the Lick Observatory from which she had measured the strength of absorption lines in the yellow-green region of the spectrum due to the chemical element magnesium—these "magnesium line-strengths" would be our measures of the metal abundance in each galaxy. She hadn't yet succeeded in measuring accurate velocity dispersions—the typical speed of the stars as indicated by the broadening of the spectral lines—but by this time such velocity dispersions for many of these galaxies, and photometry as well, had been measured and published by other astronomers.

We assembled a list of slightly more than one hundred galaxies and made two plots, one showing luminosity against the velocity dispersion and the other showing luminosity against the strength of the magnesium absorption lines. Each showed a correlation, which we modeled as a straight line running down the center of the distribution of points. This would describe the average relation in each diagram. Then we measured the *residuals* from the mean relation, i.e., how much each galaxy lay above or below the best-fitting straight line. Finally, we plotted the delta-delta relation—the residuals in velocity dispersion versus the residuals in magnesium line-strength. The good news was that the correlation seen by Terlevich et al. was there in the new, much larger sample. The not-so-good news was that the correlation was weaker—a wider scatter from a straight line than what had been seen in the earlier study.

At least we now knew that the premise of our project was not demonstrably false, even though it appeared that the correlation of the residuals was not as remarkable as the first twenty points of Terlevich et al. had suggested. We recalled that in that study the other four points lay far from the excellent relationship, and were discounted as "deviant." Terlevich and his three collaborators, including Sandy, knew that this was a risky move, and, indeed, it now appeared that these four galaxies had not been deviant, just a bit worse-than-average. Excluding them had painted a too-rosy picture of the delta-delta diagram. Such are the perils of the statistics of small numbers.

Though relieved about the survival of the delta-delta relation (whatever it meant, the relation indicated something interesting that required investigation with our larger sample), Sandy was very disappointed when she saw that, with the new data, the identification of the "second-parameter" as the shape of the galaxy image didn't hold up at all. A plot was made of the residuals, in either velocity dispersion or line strength, against the elongation of the galaxy—a continuous scale from round to very oblong. With the earlier sample the correlation with the elongation of the galaxy had seemed real, and this suggested that, together, luminosity and elongation determined what the measurements of velocity dispersion and magnesium line strength might be. For the new data the plot could best be described as a "scatter diagram"— a shotgun spray of points indicating little or no relation between the two quantities. (For example, for a group of 1,000 randomly selected children, a plot of age versus weight will show a clear, though far-from-perfect, correlation, while a plot of either age or weight versus day of birth, running from 1 to 365, would be a scatter diagram.) The hope of linking the delta-delta correlation with the details of the collapse of an elliptical galaxy to a certain shape was almost certainly lost, and with that loss the identification of the second parameter was again problematic. Small-number statistics had indeed struck again. We would have to look for another second parameter in the new data if we were to succeed in improving the accuracy of distance measurements to the point where we could use our data to make a good map of the Hubble expansion.

Before we finished our report to the other five members of the group, we made one more set of plots that was a harbinger of things to come, puzzling and important things at that. We made the same plots relating luminosity, velocity dispersion (typical speed of the stars), magnesium line-strengths (abundance of "metals"), but this time we made the plot for just the forty-odd galaxies that are members of the Virgo and Coma clusters. We had long wanted to do this, because such a sample would be relatively free of errors in distance—all the ellipticals in a rich cluster are at nearly the same distance from the Milky Way. This meant that we could reliably use the relative brightness of these galaxies as a measure of their *absolute* brightness (luminosity), which was the principal parameter in the correlations with velocity dispersion and metal abundance. Previously we had very few members of these clusters, but Sandy had gathered data on a fair number of Virgo galaxies, and I had spent several nights at Las Campanas in February 1982 collecting data for twenty-six galaxies in the elliptical-rich Coma cluster.

We were in for a surprise when we plotted these data. The correlations between luminosity and velocity dispersion, and between luminosity and line-strength, were markedly better than for the galaxies in our large sample, most of which were in small groups of galaxies or fairly isolated—we called them *field* galaxies to distinguish them from galaxies in rich clusters. This smaller scatter was in itself encouraging: it indicated that by measuring either velocity dispersion or magnesium line-strength, the luminosity of an elliptical galaxy in a cluster could be fairly well predicted. Finding a second parameter would be icing on the cake, of course, but for the cluster sample it seemed that just the first parameter, luminosity, was very well correlated with the properties one could measure from a single good spectrum. Furthermore, and this was the kicker, the variations from the good correlations—the residuals—were no longer themselves correlated. In other words, the delta-delta relation had all but vanished for this first well-studied cluster sample, while we knew that it had survived the Tonry-Davis challenge for our more complete sample that included galaxies outside of clusters.

The better correlation of velocity dispersion and line-strength with luminosity, together with the disappearance of the delta-delta relation, could have been telling us that cluster elliptical galaxies are somewhat different from field ellipticals. Perhaps the cluster galaxies formed in a more coherent way and thus coupled better to the primary character of a galaxy—its mass—which we measured as the total luminosity of its stars. Perhaps, like psychologists studying learning achievement, we had found a sample of children who were all raised in the same environment, allowing us to correlate with schooling and teachers, while ignoring the troublesome secondary variables like nutrition, home education, and domestic tranquility that muddled any simple interpretation of why some learned more easily than others.

We clearly liked this interpretation; we came back to it again and again.

It catered to our hope of sorting out how galaxies formed and what influenced their development. Again we rejected the most obvious interpretation: that the delta-delta relation arose because of errors in luminosity that were the direct result of errors in the *distance* assigned to the galaxies. For field galaxies, distances were traditionally determined from redshifts through the application of the Hubble relation for uniform expansion. In contrast to this, galaxies in a given cluster were all at roughly the *same* distance, so the Hubble relation was not needed. To explain what we had found, we needed only to conclude that the Hubble relation didn't work all that well because, in fact, the expansion was *not* very smooth (as it shouldn't be if the universe was very lumpy). The disappearance of the delta-delta relation in the cluster sample seemed to say just that. It made perfect sense and was the most appealing explanation, but we couldn't, and we wouldn't, believe it. Not yet.

I realized after my time with Roger in Cambridge and my work session with Sandy in Santa Cruz that, with the difficulty in getting all seven of us together, the many professional commitments each of us had, and the complex task that stood before us of assembling one of the largest, most homogenous data sets ever attempted in this field, it would be a long time before we reached our goals. I had never worked on a project like this—no end was in sight. My anxiety level was rising, as in one of those dreams where you struggle against the wind but never get anywhere.

Seven years into my career I knew that I was very much a "brick-by-brick" person. I divided my projects into chunks that could be done in a couple of years at most, even if they were part of long-term studies. This had become my chief frustration with the elliptical-galaxy project. I had often argued with the group that a smaller sample, say 100 galaxies, should have been chosen to test the conclusions of Terlevich et al.—the delta-delta relation and the identification of the second parameter—before the much larger and more difficult all-sky sample was attempted. But this was not my project: I had not started it, was not around for the initial planning, and, for the first time in my career, was not the major contributor. It was becoming obvious to me that in this program there would be no "bricks," no steps along the way to set and tamp, to test with one's weight and then to stand on and admire the new view. This was more like building an airplane: you couldn't fly until all the pieces were assembled.

With the group of seven concentrating so heavily on the all-sky sample, I decided that my salvation might be to home in on the ellipticals in rich clusters, where Sandy and I had seen noticeably better correlations and the remarkable disappearance of the delta-delta relation. Clusters, after all, had been my specialty ever since graduate school; I knew a lot about them and particularly liked the way each provided a well-defined, coherent sample of galaxies all at the *same distance* from our galaxy. Furthermore, these cluster ellipticals were challenging because they were fainter than most of the galaxies in the all-sky sample. Taking spectra of them required large amounts of observing time; it was the kind of project astronomers at the Carnegie Observatories traditionally pursued. Now, I hoped, I would be free to gallop off from the group without recrimination, maybe even make an important contribution that the others would appreciate. Best of all, if it went well, I could write a paper. For me, writing a paper for one of the professional journals brought with it a sense of accomplishment, satisfaction, and closure. I would then relax a bit, taking great pleasure in having made some progress in understanding this task we were about. Another brick.

I went back to Las Campanas in March 1983, and reobserved all the Coma ellipticals for which I had taken spectra in 1982, adding four more to bring the total to thirty. With so many repeat measurements, I could not only improve the accuracy of the data but assess the measurement errors as well. I also took spectra of twenty-three Virgo ellipticals; because the Virgo cluster is closer than the Coma cluster, these were brighter and therefore took less observing time.

I returned from Chile enthused and went at these data with a passion. This was another great benefit of my job with Carnegie; I could fully immerse myself in a research project because we Carnegie scientists had more control over our time than our colleagues in universities or industry with their many responsibilities. In less than a month I had the spectrographic observations processed and analyzed and had located photometric measurements made previously by other astronomers. Now I had measures of magnitude (brightness), velocity dispersion, and magnesium line-strength for fifty-four galaxies in the two clusters.

The results Sandy and I had found were reinforced. Both the velocity dispersion and line-strength correlated well with galaxy brightness. (Again, it was because these were *cluster* galaxies that ordering galaxies by their relative brightnesses was trivial—the usual uncertainty in calculating the true brightness of each galaxy, which requires an estimate of its distance, was avoided.) Most important, as it had seemed the previous year, the delta-delta correlation was reduced to near nothing.

Again, the disappearance of the delta-delta diagram for the elliptical galaxies in clusters suggested a straightforward but powerful interpretation:

that the distances derived from redshifts for ellipticals *not* in clusters were subject to significant errors. In their paper, Terlevich et al. had in fact pointed out that the simplest explanation of why they found the delta-delta relation for their sample of ellipticals, almost all of which were not in clusters, was that distances to these galaxies were not accurately given by the Hubble relation between redshift and distance. Inaccurate predictions of distance would result in errors in the luminosity calculated for each galaxy, and errors in this primary parameter would give rise to a correlation between all quantities that depended on it. For example, assigning a mistakenly too-high luminosity to a galaxy meant that both its velocity dispersion and magnesium line-strength would appear too small because both would be the correct values for a lower-luminosity elliptical galaxy. This "correlation of mistakes" could be the delta-delta diagram.

Terlevich et al. had rejected this explanation, but the disappearance of the delta-delta relation in clusters, where all galaxies were at a common distance, cried out that this was the correct conclusion. However, just as they had, I too found it difficult to believe that the Hubble relation could be so grossly in error. Galaxies would need to have peculiar velocities—motions in addition to smooth Hubble expansion—upwards of 1,000 km/sec to explain what had been seen. Hard to believe, I concluded. More probably, elliptical galaxies in clusters were just different—better behaved—and this was an important clue to the way galaxies formed in and out of clusters. Time and time again, I and the other members of our group would come to this conclusion, rejecting the simple explanation in the hope that we were learning about galaxy formation. It was a mistake, but one that would help us in the end.

If I missed the mark on that one, on another important matter I was right on target. At issue was the peculiar velocity of our own Milky Way galaxy, in particular, the hot question: how far away were the concentrations of matter that had accelerated our own galaxy to its speed of 600 km/sec (the peculiar velocity that had been inferred from the hot and cool spots in the cosmic microwave background light)? I had a new and accurate measurement of the distance of the Virgo cluster relative to the Coma cluster, a cluster in the same direction but more distant: the Hubble expansion velocity for Virgo was about 1,000 km/sec and that of Coma about 7,000 km/sec. But my new measurements indicated that the Coma cluster was only about six times as far away, not seven, as the ratio of their velocities demanded, *if the Hubble expansion was uniform.*

I had made these new determinations of the relative distance of the Coma and Virgo clusters by comparing the relationships of velocity dispersion and

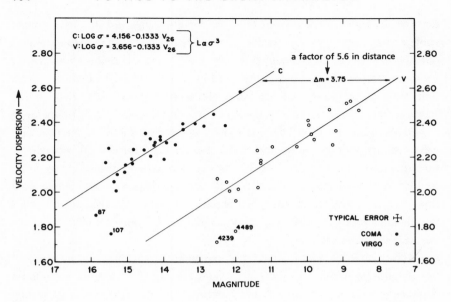

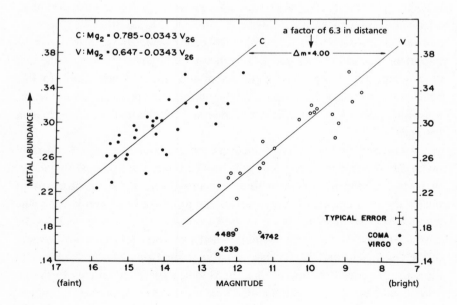

Plots of galaxy brightness versus velocity dispersion (the typical speed of the stars), and galaxy brightness versus metal abundance, for galaxies in the Coma and Virgo clusters. The same relationships appear to hold for both sets of galaxies, and they are remarkably tight, but there is an offset in brightness that corresponds to a factor of six in distance. *(Reference: 1984 Astrophysical Journal, vol. 281, p. 512.)*

magnesium line-strength with galaxy brightness.* For example, the correlation of galaxy brightness with velocity dispersion (the Faber-Jackson relation) was a straight line with modest scatter for each cluster. For both clusters the relation had the same slope, but the two lines were offset—galaxies in the Coma cluster were thirty-six times as faint at the same values of velocity dispersion and line-strength. If the ellipticals in Virgo and Coma belonged to the same breed, a reasonable assumption, then the difference should be entirely due to the greater distance of the Coma cluster. According to the way light intensity falls, as the square of the distance, the Coma cluster would then be six times as distant as Virgo in order to account for the fact that the galaxies were thirty-six times fainter, that is, $6^2$. Could this be reconciled with the ratio of seven for the Hubble expansion velocities, which implied that the ratio of distance was seven rather than six? Easily, if our own galaxy were moving 200 km/sec in the direction of the two clusters. This peculiar motion of our own galaxy would be adding to the proper speeds of both clusters in a smooth Hubble expansion—the Milky Way was running along after them, so to speak. To correct for our "point of view," we would have to add 200 km/sec to each, which would mean that their true velocities would be $1,000 + 200 = 1,200$ km/sec, and $7,000 + 200 = 7,200$ km/sec. These corrected Hubble expansion velocities, now in the ratio of six to one, would agree with the relative distance I had derived from comparing other properties of the galaxies, the apparent brightness, velocity dispersion, and magnesium line-strength.

Like the Aaronson et al. investigation using spiral galaxies, then, my study of ellipticals had deduced a peculiar motion of our galaxy toward the Virgo cluster. And as in that study, the value I found was relatively small— about 200 km/sec. This was small in the sense that it was much less than the 600 km/sec peculiar velocity implied by the variation in intensity around the sky of the cosmic microwave background. Back in 1980, Paul Schechter had also gotten a similar result when he applied the Faber-Jackson relation to

---

*The essence of this technique, which will reappear again and again, is that one quantity, velocity dispersion, can be measured independent of the galaxy's distance, while the other quantity, in this case galaxy brightness, does depend on distance. Thus, the scheme is to derive a distance by using a correlation between a quantity that doesn't depend on distance to predict the value of one that does.

A possible point of confusion is the disentanglement of the gross redshift of the whole galaxy, caused by its motion in the Hubble expansion, from the small spread in redshifts due to the different speeds of the stars *inside* the galaxy. One results in a shift of the entire spectrum to longer wavelengths, the other, a broadening of the individual spectral lines (in the case of velocity dispersion) or, if the galaxy is rotating, a shift of the spectral lines from one side of the galaxy to the other.

elliptical galaxies spread around the Local Supercluster. Paul, who had one foot in each camp of observation and theory, had exceptional skills in analytical analysis that allowed him to pioneer the technique that we were now using to measure velocity dispersions from a galaxy's spectrum. For theoretical reasons, Paul expected a higher value for the motion of our galaxy toward Virgo; he was disappointed with his observational result. In his paper he called for a larger galaxy sample and improvement of the technique (more accurate distances to yield better peculiar velocities) along the lines of the paper by Terlevich et al., the very activity in which our group of seven was now engaged.

Such measurements of the Milky Way's peculiar velocity within the Local Supercluster, for example, its "infall" toward the Virgo cluster, were an important first step in mapping the amplitude and variations from a smooth Hubble expansion. Newer measurements had confirmed, with high accuracy, our galaxy's motion of 600 km/sec with respect to the cosmic microwave background, an extremely distant frame of reference. As stressed by Marc Davis and Jim Peebles in their key 1983 review paper, a comparison of this number to measurements of our galaxy's peculiar velocity with respect to *closer* reference frames would test the lumpiness of the mass distribution in the local universe.

Their argument went something like this. Imagine many boats floating in a river. Suppose the pilot of one boat measures a speed of 10 knots relative to the shore, but virtually no motion with respect to other nearby boats. From this observation she might well conclude that her boat was part of a flotilla drifting downstream with the current. On the other hand, if she measured 10 knots with respect to the shore, but 8 or 9 knots with respect to the other boats, she would reasonably deduce that her motion was a more local effect, that is, *her* boat was doing most of the moving and the others were nearly at rest. In this analogy, the measurement relative to the shore is like measuring the Milky Way's speed relative to the distant cosmic microwave background, and the measurement with respect to other boats is like measuring the speed relative to nearby galaxies. If your speeds relative to the nearby and distant frames of reference agree, it is likely that your motion arises on a very local scale, but if the speed relative to the distant frame is large, but relative to the local frame is small, this implies that you and others around you are moving together, so the motion of the whole arises on a larger scale. In the case of galaxy motions, this tells us how large are the irregularities in the mass distribution that are perturbing the smooth Hubble expansion.

To make their point, Davis and Peebles looked over the various published studies that had attempted to measure our galaxy's peculiar velocity in the direction of the Virgo cluster with respect to local frames of reference. They concluded that the most reliable measurements were ones that gave values of about 400 km/sec. This amount was a fair fraction of the peculiar velocity of our galaxy measured with respect to the distant "shore," the cosmic microwave background. (In fact, it was every bit of it if you considered how much of the motion was in this direction alone.) In the same way as with the boat analogy, Davis and Peebles concluded that the peculiar velocity of our galaxy arose *locally*, within the Local Supercluster, because the velocities with respect to very near and very distant frames of reference were comparable in size. In other words, they concluded that most or all of the total peculiar velocity of 600 km/sec arose from the gravitational pull of mass within the Local Supercluster.

Perhaps Davis and Peebles carried a bias from their theoretical expectations when they chose the higher values as the best ones. Perhaps because I had too great an attachment to my own work, I was prejudiced toward the lower values of about 200 km/sec for our galaxy's peculiar velocity with respect to Virgo, a value Schechter, and Aaronson and collaborators, had also found. Fair or not, and uncertain as it was, the choice had important implications. A value of 200 km/sec for the "locally referenced" motion was so much smaller than the 600 km/sec of the "cosmically referenced" motion that just the opposite conclusion seemed in order: that the peculiar velocity of our galaxy arose from lumpiness in the mass distribution significantly larger than the Local Supercluster. With the formality that regrettably accompanies scientific writing, I challenged Davis and Peebles's conclusion in the paper I wrote:

> When this possibility is considered along with the large component of [the peculiar velocity of our galaxy] in a direction perpendicular to the centroid of the Local Supercluster, it seems likely that the pull of this nearby density enhancement does not account for a large fraction of the [CMB] component. This, in turn, would imply that the density fluctuations that are accelerating [the Milky Way] and the Local Supercluster are farther away and of much larger scale.

In other words, we were moving more slowly with respect to nearby boats, and more rapidly with respect to the shore, so we and the other boats were moving together toward something more distant. I probably should have put it like that in the *Astrophysical Journal*.

I had gone out on a limb by suggesting that the lumpiness of the universe, being measured for the first time in the distribution of *all* mass, seen and unseen, is over very large distances. Davis and Peebles had gone out on a limb, too, in coming to the other conclusion. Fortunately for me, I would turn out to be right, and it would be the labor of the seven of us that would eventually prove it to be so. Of course, this was years away, but at least I went to our first group meeting in Cambridge with a different attitude, one I found shared by Donald Lynden-Bell, that mapping the Hubble expansion was what our project was really about.

I was pleased, self-satisfied, with the contribution I had made. Writing the paper had tranquilized me considerably. I had become less concerned with the slow pace of our large project. However, despite my newfound enthusiasm to map departures from smooth Hubble expansion, I had become less confident about our ability to do it accurately enough—to me, the strong delta-delta relation for field ellipticals indicated that they were more poorly behaved than cluster galaxies. The connection I should have made, but didn't, was that the strong delta-delta correlation itself betrayed the presence of large peculiar velocities that riddled the calculation of distances from the Hubble relation, the very effect that I had forecast in my paper but couldn't see before my very eyes.

Margaret Geller and John Huchra had inherited from Marc Davis the so-called "Z-machine," the sixty-inch Mt. Hopkins telescope mated to the Shectman photon-counting detector that had measured more redshifts—"Z's," as they were called—than all the previous observations put together. They had decided to target the Z-machine on galaxies in Zwicky's voluminous catalog, in this way probing more deeply into space than Marc had done in the first CfA survey. They knew exactly why they were doing this: they were going to pour cold water, once and for all, on those exorbitant claims of fabulous lumpiness in the distribution of galaxies, like the Great Void in Boötes, where galaxies were sparse across a hundred million light-years, or the Coma supercluster, where a dense bridge of galaxies stretched as far. Geller and Huchra didn't buy all that "large-scale inhomogeneity" stuff. The universe as it had been studied from *Haaarvard* was tidier, better behaved. Geller and Huchra were armed with the big gun; they were determined to restore law and order to the territory.

Looking for structure as large as the Great Void meant casting the net farther from our galaxy, in order to map out a region large enough to contain such an extensive variation in the distribution of galaxies. To go farther into space, they would have to select fainter galaxies. But the number of target galaxies rose rapidly as one went fainter, so an all-sky survey was out of the question.* In order to keep the program tractable they cut down the covered area to a number of *slices* six degrees wide, stretching one-third of the way around the sky. Since Huchra and Geller were certain that the result of their deeper mapping would be basically uninteresting—merely refuting the outlandish claims of other studies—they chose as the first region a strip in which many galaxy redshifts had already been measured, the one containing the very rich, well-studied Coma cluster. They would need only to add about 600 new redshifts to raise the total number that had been observed to more than 1,100, thereby completing the deepest complete sample of any region of space. They hoped that the presence of the Coma cluster might help them salvage *some* useful science from an otherwise dull project. With the velocities of many galaxies in the vicinity of the Coma cluster, they should be able to discern the pattern of galaxies falling toward each other—the way a dense cluster formed from the smooth background of field galaxies. It was an important phenomenon to investigate in the context of Peebles's "hierarchical" universe, the model that built its structure from the bottom up, and one in which Huchra and Geller had become increasingly interested.

This was a collaboration born of and nurtured by great friendship. Margaret's growing involvement with observational astronomy, a departure from her training in astrophysical theory, was a direct result of the fact that she and John got on so well. Furthermore, it was a symbiotic relationship. Some theorists (the ones observers seem to prefer) thrive on a diet of fresh data, and an observer like John, who usually spends more than 100 nights a year at telescopes, used up too much of his time and energy taking and processing data to be able to give a lot of attention to what it all meant.

In 1985 Valerie de Lapparent was a graduate student at Harvard. Her interest in pursuing an observational rather than theoretical Ph.D. thesis had prompted her adviser at École Supérieure J.F. in Paris to pack her off to the Center for Astrophysics to finish up. Margaret and John had been casting about for something sensible to give Valerie to work on, and the analysis of the redshift distribution in the "slice" seemed like a reasonable choice. She would do a little of the observing, although John would do most of this himself, and she would become familiar with the data reduction, although there were regular CfA support staff who would do most of the analysis of the spectra and measure the galaxy redshifts as a matter of routine. Valerie

---

*The number of objects increases roughly as the volume, which grows as the distance raised to the third power. So to go three times as far as the first CfA survey, which had about 2,500 galaxies, would have required about twenty-seven times as many redshifts, over 60,000. With the "one-at-a-time" approach Geller and Huchra were taking, this would have taken many decades to complete.

would have the luxury of worrying only about the results—she would write the computer programs that would plot and analyze the distribution of this large sample of galaxies, to find out what could be learned about large-scale structure from such a map. Margaret and John were fairly certain that Valerie's analysis would merely confirm how boring the distribution of galaxies really was. Worried that the project might not make a very exciting doctoral thesis, they encouraged her to work on other projects while the data were being gathered.

So it was that no maps were made of the first slice until all of the 600 new redshifts were available. One day in the fall of 1985, Valerie's computer programs were ready to plot out the results, in the by-now traditional form of a "pie diagram" that showed the galaxy's direction along an arc that swept over 120 degrees of the sky, and the galaxy distance—in this case simply the redshift—given as each galaxy's velocity in the Hubble expansion. As the diagram crept out of the plotter, John felt an awful sickness; something was terribly wrong with the redshifts. The slice had big empty zones with no galaxies, embraced by thin, graceful curves dense with galaxies, as if a bath of soapsuds had been cleaved by a sharp blade. This could not be; it was too incredible. But what could have gone wrong?

The three began to go over the data again and again, checking, rechecking, and cross-checking, and looking for strange correlations that would show that mistakes had been made at the telescope, in the processing, or plotting. By the next day they knew the answer: Nothing had gone wrong. This is how the universe is put together. They were perhaps the first humans to see the pattern so clearly, so undeniably. There would be no going back to the smooth universe. What they had set out to restore, they had destroyed, forever.

It certainly isn't fair to say that de Lapparent, Geller, and Huchra *discovered* that galaxies are arranged in a highly nonrandom way, weaving their way across space in very large sheets and chains, opening and closing huge voids. The work of many astronomers, particularly studies in the previous ten years, had repeatedly pointed the way. But the study by de Lapparent, Geller, and Huchra turned a conceptual corner by providing a clear image that none could ignore—the first "slice of the universe" gave the fabled picture worth a thousand words. Now the quest had been taken to a new plateau, and questions were recast and refocused: Just how strong were these patterns, and did other, equally large regions of space retain the same character? Would even larger slices reveal still larger features? Could the influence of gravity alone shape such stark contrasts in density? How *had* a presumably smooth big bang evolved into such a lumpy distribution?

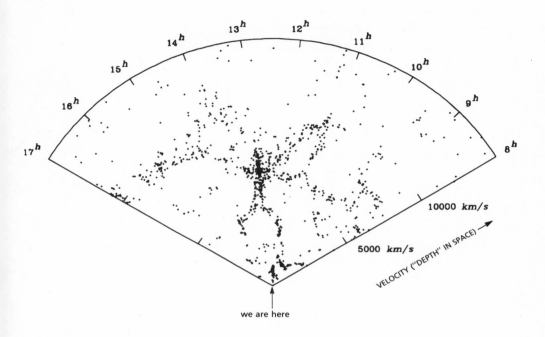

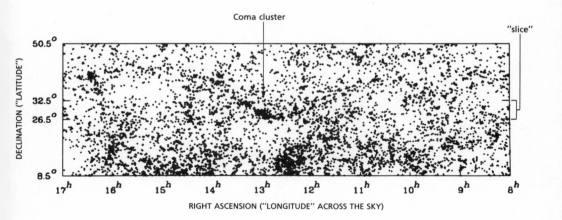

The first CfA "slice of the universe." The slice of the section of sky at the bottom, with each galaxy shown as a point, corresponds to the fan-shaped diagram above, with the distance of each galaxy determined by its velocity (redshift). The distinct structure, far from smooth, is only hinted at in the bottom picture, but comes through clearly in the pie diagram. *(Reference: de Lapparent, Geller, and Huchra, 1986* Astrophysical Journal, *vol. 302, L1.)*

There would be more slices, more analysis, and more arguments. As trite as it may sound, the end of this debate was just another beginning. Each of us working to understand the large-scale structure of the universe was now sure that this was the place to be, and that even more exciting days of discovery lay ahead.

# 7 • SEVEN SAMURAI

On August 7, 1983, at Cambridge University, the seven of us met together for the first time. Here, from this once sleepy English village on the river Cam, Isaac Newton had followed his mind to the sky and returned with a universe not only describable but comprehensible. He captured the great cosmic force of gravity in a few symbols and a simple mathematical expression of their relation. Once-mysterious paths of moving objects became sudden slaves to Newton's ''laws''—from pebbles to planets, they could move only along the courses he would predict. Newton changed the relationship of humans to the universe. His brilliant formulation of the rules of nature was a bridge from Earth to the heavens that opened the way for humans to step beyond our world, to trespass a realm once reserved for the deity.

We seven would add our efforts to those of generations of scientists who had tested the comprehensibility of the universe on larger and larger scales. Newton had taken his law of universal gravitation to the Moon and beyond, finding that he could embrace in a single stroke what Galileo, Copernicus, and Kepler had observed: precise orbits for planets around the Sun, and for the many moons that encircle Jupiter and Saturn. Centuries later these same rules would be tested on ever grander systems and found to describe accurately the motions of stars as they orbit within our Milky Way galaxy, and the trajectories of entire galaxies as they circulate in giant clusters. In our time Newtonian gravity had been applied across the gigantic realm of the Local Supercluster, and before our own journey was over, we seven would carry it to a distance ten times larger still—half-a-billion light-years. What Newton had discovered in the seventeenth century would work perfectly in a realm far larger than he could imagine. His was a true triumph of the mind.

But, as Confucius never said, "The journey to Andromeda begins with but a single light-year." And it seemed that we were at least that far from our goal as we began in 1983 to chart our own course through the superclusters, the habitat of elliptical galaxies. We had made many measurements, to be sure, but not all were processed into a form that would tell us what we wanted to know. As we learned more about these measurements, we would find that there were many details, and some associated problems, that we had not yet considered. And we were about to find out that there would be personal matters that would both aid and frustrate our desire to be on our way.

Donald Lynden-Bell, our host in Cambridge, was, at forty-nine, the usually merry but sometimes cantankerous director of the Observatories at the Institute of Astronomy. Skinny and angular, bald—with a ring of fine white hair hanging on at brow level—Donald could have easily played the part of a country vicar in some Dickensesque tale; indeed, his family had a long and serious tradition within the Anglican Church. Now, through fourteen-year-old Edward, he was brought into contact with modern social mores, about which he displayed frequent exasperation. In fact, to be plain about it, Donald was a prude and not the slightest bit ashamed of it. To his credit, though, he was an unusually tolerant prude; his disapproval of rough talk and undignified behavior was communicated mostly by uncomfortable scowls.

But Donald was all smiles the crisp, sunny day he greeted us, as we looked forward to two productive weeks in the land of lush green lawns and cathedral-like colleges. Donald welcomed Gary, Roger, and me into his home, where we four and Edward partook of what I considered to be a rather spartan lifestyle—especially at mealtime—but I was soon to learn that these meetings would be long on labor and short on amenities. Nonetheless, the first of the "in the lab at eight, home at midnight" days began with a certain excitement. Perhaps some of us thought that in two weeks, with enough tenacity, we would have the first results of our two-year-old project. We were wrong. On that very first day we learned of a large hole in the data set: the spectra taken by Sandy Faber at the Lick 120-inch telescope had not been analyzed. Sandy had suffered serious back problems that year—she had spent March 1 to June 15 of 1983 in bed, and even now she could move only slowly and with considerable pain. What energy she had was diverted into what would become the Keck ten-meter telescope on Mauna Kea, the extinct volcano on the big island of Hawaii. Our elliptical galaxy program had been pushed to the back burner and little had been done since my visit to Santa Cruz almost a year earlier. With one-third of the crucial measurements of velocity dispersion missing, any grand plans to whip all the data into shape were dashed on the spot.

And it soon became obvious that Sandy's back problems had taken a toll on Dave Burstein as well. Dave and Gary had gone to Santa Cruz for the previous two months to work with Sandy. Unfortunately, Sandy was so incapacitated that all she could do was help Gary measure the diameters of galaxies on little 3 x 5-inch Polaroid prints—even that had been excruciating. Taking on what should have been a two-person job, Dave had embarked on a marathon computer session, trying to patch the thousands of incongruous measurements of galaxy brightness—the *photometry*—into a coherent sample. Learning how to use a new computer system to boot, Dave drove himself to exhaustion by working twelve-hour days seven days a week, trying desperately to have all of this photometry data ready for our group to analyze in Cambridge. He needed a breather when he arrived in Cambridge, but instead of a second wind he caught a stiff zephyr of new responsibilities. Donald, Roger, Roberto, and I were unaware of the degree of Dave's effort and how it had driven him close to the edge. The stage was set for a blowup.

Certain challenges became apparent immediately. The list of photometry data that Dave had labored so long and hard on in Santa Cruz would become the focal point of our discussions of how we would proceed. Dave had brought all this information on magnetic tapes and would now spend several long days trying to transfer the data and the programs he had written to a computer off in another building of the Observatories where we were working. The very fact that this could be done was something new for us, and a minor miracle, but it would take a good deal of work to accomplish for the first time. Dave disappeared.

Meanwhile, Sandy and Gary continued to worry about the uniformity of the selection of elliptical galaxies, something they had been investigating with those little 3 x 5-inch Polaroids back in Santa Cruz. They and Roberto would now discover some disturbing trends about the sample of ellipticals chosen from the southern-sky ellipticals, and they would set about to devise a remedy. Roger and I took up where we had left off comparing the measurements of velocity dispersions—the typical speeds of stars—for galaxies that had processed, analyzed spectra. The further we got into it, the more there seemed to do, and the further we seemed from deriving a final set of values. We all ran off to do our assorted tasks, leaving Donald alone to think about the "big issues" and become noticeably impatient. He was accustomed to starting his work with finished, published data. For the first time he was working on a project with observers busy evaluating, combining, and validating data sets. Donald was eager to get on with it, and had many thoughts on how to proceed. We just weren't ready.

It is hard to make clear the aggravation of these diverting, yet crucial, tasks without giving some examples. Roger and I were trying to compare the measurements of velocity dispersion that had come from the spectra. Recall that the typical speed of the stars in an elliptical galaxy—what has been called *velocity dispersion*—can be measured by how much the dark absorption lines in the spectrum are broadened by the Doppler shift: higher speeds result in broader spectral lines. We had deliberately measured some of the same galaxies with more than one telescope, each of which had its own spectrograph and detector system. Spectra from these different sources had been processed by different computer programs as well. Now we found that while some of the velocity dispersion measurements agreed very well, some did not. Were these discrepancies simply because some of the measurements were less accurate (because less light had been collected, for example) or were there real differences in the instruments that took the data, or in the way our computer program extracted the width of the spectral lines? By cross-comparing observations of the same galaxies taken by different observers at different telescopes, Roger and I were trying to find out.

We also worked on another problem that might seem truly piddling, a pesky little detail that we expected to have a small but non-negligible effect. We knew that we couldn't let gremlins like this run free in the data set or sooner or later they would come back and bite us. So minor a matter is this that its description may bore you, but we couldn't let it bore us. To take a spectrum, one first passed the light from a galaxy through a defining aperture. As it happened, the spectrographs at different telescopes had different-size apertures. So, as a result, different fractions of the central light of each galaxy went into each spectrum. In his Ph.D. thesis Roger had shown that the typical speed of stars in an elliptical galaxy, as measured from the blended light of millions of its stars, falls slowly but steadily as one looks farther from the center of the galaxy. This was not really a surprise—the planets far from the Sun circle more slowly than those closer in—but unlike the case of the solar system, where almost all the mass is concentrated at the Sun, the amount of slowing-down with increasing distance cannot be simply predicted for elliptical galaxies. We needed to know how big this effect is, first because we were trying to compare different observations of the same galaxy taken with different-size apertures, but more important, because our sample included galaxies over a very wide range of distance. This meant that the same-size aperture admitted light from a much larger region in a distant galaxy compared to a nearer galaxy. (Recall how when you framed the family portrait in the viewfinder of your camera, you had to move farther and farther away until you could include little Nibbles tugging at Uncle Mike's cuffs.) Unless we made allowances for this effect, two identical galaxies observed at very different distances would be incorrectly described as having significantly different velocity dispersions.

None of this would have been a problem if we could have gone to the telescope and dialed in the size of the aperture we wanted, so that we would have used an aperture four times smaller for a galaxy four times farther away, say, in order to be sure to cover the same area. However, spectrographs came with only a few sizes of apertures, much like the ready-made frames at the art store: if you want 8 x 10 or 11 x 14, great, but if you bring in a 7 x 17 masterpiece, you're out of luck. Likewise, we had to observe with the aperture that came closest to what we wanted, and sometimes this wasn't all that close.

Roger and I looked carefully at some data that would provide what we called an "aperture correction" to apply after the fact to all the data we had gathered. For about twenty galaxies Roger and Roberto had obtained spectra covering a range of positions moving out from the center of the galaxy, and I had taken spectra for several dozen elliptical and S0 galaxies through both small and large square apertures. After a day's work we had a rough prescription for increasing or decreasing each measured velocity dispersion by a certain amount, typically 5 to 10 percent, to correct the value to what we would have gotten if we could have used a properly sized aperture.

Much of our time was involved in tracking down these little maladies that, if left untreated, would have doomed our attempt to compile a truly uniform set of data. But while Roger and I were concocting remedies for little ills of the data, more serious ailments were being diagnosed by Sandy, Gary, and Roberto. Nothing was more basic to our project than the attempt to sample elliptical galaxies all over the sky to the same limit of brightness. We had made all our observations in the belief that candidates had been ruled in and out properly. Now, at Cambridge, the three discovered that the attempt to assemble a "magnitude limited sample"—all galaxies brighter than a certain limit—hadn't succeeded very well. For one thing, we hadn't come to grips with the fact that a wide band on the sky was left completely unsampled. The northern and southern catalogs from which we selected galaxies didn't overlap, in fact, they left a fifteen-degree swath in celestial latitude (what astronomers call *declination*) unsurveyed. It was as if a map of the world had no record of continents or islands in a round-the-world swath reaching roughly from the equator to some 1,000 miles south. We were already losing another band around the sky where the dust in the Milky Way snuffed out the glow from distant galaxies. There was little we could do about that, but it was becoming clear that we could ill afford to lose another band of the sky. So, using a third catalog as a source list, Sandy and

Roberto had to spend several days with photographs from the sky surveys, finding more ellipticals and verifying that they were within the brightness (magnitude) limit of our sample. Soon we knew that even our trips to the telescopes weren't over—there were more galaxies to observe.

This was depressing, but at least the remedy was clear. A more subtle and potentially more damning problem found by Sandy and Gary was that even for the cataloged regions, applying a brightness limit uniformly across the sky had not been as successful as we had hoped. The catalog of the northern sky, the Uppsala General Catalog, had magnitude estimates made by Fritz Zwicky, the prolific, ill-tempered Caltech astrophysicist. There was some question about the accuracy of his "by-eye" estimates; systematic variations around the sky could be a serious source of selection bias. But there was a worse problem: the European Southern Observatory (ESO) catalog didn't even record magnitudes, listing instead only the *diameters* of galaxies as measured from photographic plates. Diameters could be used to estimate galaxy brightness: with a modest scatter, galaxy brightness scales as the square of the diameter, so, for example, a galaxy twice as big is roughly four times as bright. However, because the two catalogs didn't overlap there were no galaxies in common, which meant there was no direct way to calibrate one catalog to the other. Not surprisingly, then, Sandy and Gary discovered that back in 1980 Roger and Roberto had missed the mark slightly when they used the ESO catalog to choose the southern ellipticals: the southern sample included galaxies that were a little smaller, and hence fainter, than the faintest ones in the north. Inexplicably, they had also missed some of the bigger galaxies in the south altogether.

Sandy knew how potentially deadly this problem was. She recognized that, eventually, when we used our data to map the universe, and compared the distribution and motions of galaxies to "model universes" made by computer simulations, it would be crucial to know exactly how the sample galaxies were selected and how uniform that selection was over the sky. She remembered well the harsh criticism of the study by her mentor Vera Rubin, and by Kent Ford, criticism on similar grounds that had left their remarkable report of large peculiar velocities in scientific limbo. It was for this reason that Sandy and Gary had, back in Santa Cruz, scrutinized photographs of every northern galaxy in our survey, using Polaroid copies made by her graduate student Jesus Gonzales (students invariably get the grunt work) from the Palomar Sky Survey. Independently, Sandy and Gary had carefully measured "by hand" the diameter of each galaxy, establishing the reliability of these estimates through the cross-comparison of their separate measures.

With questions raised about the southern sample, they determined to repeat the process for each of the galaxies in the south, asking Roberto to help in the laborious procedure by making a uniform set of Polaroid copies from the ESO Southern Sky survey. Now, for the first time they could tie the northern and southern samples together by repeating the measurements of some northern galaxies as photographed from the Australian Sky Survey telescope (the northern and southern photographs *do* overlap, even though the catalogs that were made from them don't).

Gary, Sandy, and Roberto were hardly to be seen for the next several days. Much later Sandy told us that the pain of bending over and inspecting charts "just about killed me." But when it was all done, they had produced a consistent set of diameters for all ellipticals, north and south. Finally a diameter limit could be chosen for the southern sample that selected galaxies as faint as but no fainter than, the elliptical galaxies in the north. Since fainter galaxies are, on average, more distant, this meant that at last we had succeeded in probing space to a uniform depth in every direction.

As we continued with our tasks, Dave had his hands full with the photometry data set, a potpourri in which half of the measurements had been made by our own group but the other half came from years of work by dozens of other astronomers. Some of the older data were poorly calibrated and unreliable; including the time at Santa Cruz, Dave had spent weeks trying to sort the good from the bad, and he wasn't finished yet.

This is how it went, day after day: pairs or trios of us working through the increasing list of issues that would have to be dealt with before we could move on to further analysis of the data. We now knew that we would be fortunate just to settle these issues of data selection, quality, and correction before we left Cambridge. But, as we worked, a more substantive goal emerged, one that became the focus of our second week. It was apparent that we needed to do more than simply merge together all the different measurements with whatever corrections were necessary. We had to decide as soon as possible what measurements would be most important in studying the properties of elliptical galaxies. This became the topic of conversation of daily bull sessions in Donald's office. We had already decided to extract from each spectrum the velocity dispersion and the magnesium line-strength, the latter an indicator of abundance of "metals" in the stars of each galaxy. There was little more to do here than make sure these measurements were reliable and consistent. But the photometry data, the measurements of the brightness of each galaxy observed through a range of different apertures, offered a wide range of possibilities, and we were not yet sure how to describe such complex data in a way that would reveal how elliptical galaxies are put together. More and more, the discussion centered on these photometric measures that Dave was working on.

When he arrived at Cambridge, Dave brought with him magnetic data tapes that stored the data and programs he had written in Santa Cruz. Up until this time computer systems were so different from institution to institution that one rarely tried to import what was done somewhere else, but merely brought the data on sheets of paper. If a computer was needed, new programs were written on whatever system was at hand. But now VAX computers from Digital Corporation were spreading throughout the astronomy world. Dave would try to save weeks of work by transferring what he had done on the VAX computer at the astronomy department at Santa Cruz to the VAX system here in Cambridge. He buried himself at the computer, frequently calling on systems managers or other astronomers to help him make this Cambridge computer read the data written at Santa Cruz, and run the programs developed there. Even though it took several days to accomplish this, we all thought it a marvel that it could even be done, and shared Dave's pleasure at the way the world of computers was evolving for our benefit.

But when Dave emerged from his monklike seclusion at the computer, he found that Donald Lynden-Bell offered no congratulations for his effort. On the contrary, he was dissatisfied with what Dave had done and still annoyed that all the data weren't ready for final analysis *immediately*, from our first day in Cambridge. Dave became furious. He had worked so hard to reach this point, stretching himself to near his limit. There was little enough gratitude coming from the group for his effort, Dave thought, and here Donald was being downright critical.

In particular, what Dave had spent most of his time doing was writing computer programs that combined measurements of a galaxy's brightness and made plots of the light recorded versus the aperture size of each observation. Since galaxies are not point sources like stars, the light collected depends on the size of the aperture: unless an aperture is much larger than the galaxy, the light collected increases substantially as larger and larger apertures are used, that is, as more and more of the galaxy's total light is included. A plot of this increase in measured brightness with aperture size is called a "curve of growth." Dave had written a set of instructions for the computer that enabled it to find the parameters of the curve that best fit these data points, using a mathematical form suggested years earlier by Gerard de Vaucouleurs. So, after weeks of taxing work in Santa Cruz and here in Cambridge, Dave was particularly proud of the way he had stitched together the data, understood its reliability and foibles, and could make the best estimate of the total radius and luminosity of each galaxy.

Ironically, these "total" parameters were the nubbin of Donald's dissatisfaction with this approach. The fault, for a change, lay in our stars and not in ourselves. We could measure with great accuracy the light from a single star, but the light from one of these confounded elliptical galaxies dribbled away so slowly toward the outside that one was never quite sure if one "got it all." Far from the center the glow became so feeble that it became fainter than the glow coming from the night sky itself, making accurate measurement more and more difficult. Dave's solution had been to use de Vaucouleurs's "curve of growth" to estimate the parameters of total light and full size, values that went beyond the actual measurements. By fitting the growth curve in the well-measured inner regions of each galaxy, he *extrapolated* to predict the total light and size, assuming that the curve accurately predicted falling intensity further out.

It was a matter of religion. Donald believed that extrapolation was positively evil—the likely source of much of the scatter in the fundamental relations of galaxy brightness with velocity dispersion and magnesium line-strength. He insisted that Dave deliver measurements of size and brightness that were *interpolations*, that is, averages between measured data points. Donald led the team on the search for new parameters, ones that didn't require extrapolations. There in his office, for an hour or so each day, we brainstormed on what interpolated quantities might prove interesting for elliptical galaxies. Our ideas were translated into tasks for Dave: we would ask him to write computer programs that could extract certain quantities, compare them, combine them, reorder them for the hundreds of galaxies in the sample. Our continual revisions and suggestions were adding to Dave's already heavy workload.

The last parameter that came out of these sessions would turn out to be the most important. It was Donald's invention, a particular diameter that was peculiarly defined as "the diameter within which the galaxy has an average surface brightness [light per unit area of surface] of a prespecified value." The average surface brightness of a galaxy image drops as more of the galaxy is included, of course, since the galaxy is rapidly becoming fainter at the outside. We were asking the question: for each galaxy what is the radius at which a certain average brightness is reached? The particular value we chose was 20.75 magnitudes per square arcsecond, about twice as bright as the glow of the night sky itself. Defined in this peculiar way, Donald's new measure of diameter actually combined information about both size and brightness for the galaxy, with the additional virtue that it could be interpolated from the measurements—it required none of the extrapolation that Donald dreaded.

We had been leaning hard on Dave: he alone had mastered the computer and, rather than trying ourselves, we found it easier to ask him to write a little program or make a plot or two to help us in our tasks. Each day Dave looked a little more drawn and tense, but none of us seemed to notice. Donald's last request, for the special diameter defined by surface brightness, was the proverbial last straw. Dave exploded, screaming at us that he was unappreciated and tired of being "used." Why couldn't we get off our behinds and learn how to use the computer? In a rage he reminded us that he was no slave, but a partner in our undertaking.

We were shocked. Considering the physical and mental strain of these two-week assaults on the project, this kind of flareup should have been expected. But it was the first time any of us had seen emotions enter so overtly and seriously into a scientific project. Sandy was suffering too, desperately depressed by her back problems, embarrassed about arriving in Cambridge without her data, now confined most of the time to a cot in Donald's office. Donald was edgy, impatient to get some results, as Roger and I had been. We were learning that this project was going to be trying and difficult. There would be other "blowups" and others would lose control. Personal relations would assume an importance that we had never encountered in our work. Without doubt it was going to complicate the task that lay ahead, but also would give us the opportunity to know each other as people as well as colleagues.

Tempers cooled and we returned to our work. It seemed a good idea to take some time off, to ease the strain and enjoy the charm of this ancestral shrine of study and thought. We set aside an afternoon for "punting." Renting a couple of small flat-bottomed boats, we nudged them up the Cam by pushing twenty-foot poles into the muddy bottom, toward the neighboring village of Grantschester, where we would partake of a picnic lunch. Each of us took a crack at this awkward though strangely satisfying means of propulsion. I remember feeling particularly pleased with myself for mastering the art, when suddenly the muck of the river bottom reached up and seized my pole. The boat drifted on unsympathetically and I stumbled to the back transom, stubbornly refusing to let go. In another instant I would have been treated to a muddy bath, but someone reached out and pulled me back from the brink. The pole departed rapidly, its fixed frame of reference showing us clearly the surprising speed we had attained. With considerable effort, our two crews joined forces to circle back and retrieve

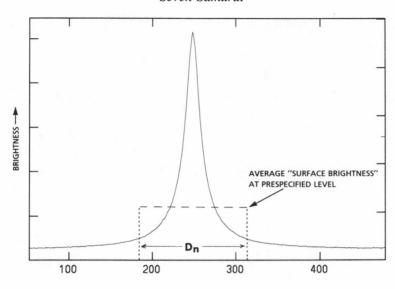

AVERAGE "SURFACE BRIGHTNESS"
AT PRESPECIFIED LEVEL

BRIGHTNESS →

$D_n$

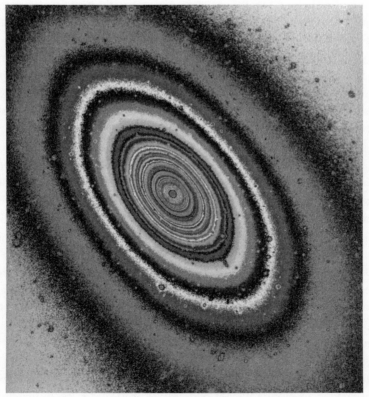

The brightness contours of an elliptical galaxy (bottom). On a crosscut of the "intensity profile" (top), the "donald diameter" $D_n$ is shown schematically.

the pole, and we continued on our way. On many levels it was a metaphor for our continuing journey.

∞

We left Cambridge feeling that we were further from our goal than the day we arrived, but in fact we had accomplished much. We had arrived carrying our collected evidence. Naively, we were expecting that our detective work might "crack the case" quickly, that is, find at least a bit of the how and why of elliptical galaxies. Instead, we left with a more complete "rap sheet" on each of our suspects. This was not just a driver's license description of height, weight, and eye and hair color, but also a tailor's description—dress or coat size, sleeve length, hem length, waistline and inseam, maybe even hat and shoe size.

We were trying to find out how elliptical galaxies are put together, but we didn't yet have a clear idea of which were the important pieces. Like detectives we were assembling details, everything we could think of, hoping as they do that with enough facts some pieces would begin to fall together and a picture emerge. To appreciate what we were doing in 1983, it is helpful to carry a bit further this analogy with the anatomy of a human being. For starters, let's limit our study of anatomy to men—by singling out elliptical galaxies we had done something analogous to selection by gender. Suppose we are interested in understanding variation in size for male humans, and that we take as the primary parameter height—a pretty basic measure of "how much person" we're dealing with. It's obvious that height is fairly well correlated with body weight—taller people generally weigh more. But it will also come as no surprise that this correlation is far from perfect: some people are skinny and some are not. So knowing height alone is insufficient to predict accurately a human male's weight. But suppose we add information about whether the person has "bulk," to use a polite term. From our "tailor's description" of each subject, we discover that one of our measurements, waistline, is a good candidate for this *second* parameter. In other words, the "residuals" from the height-weight relation (whether a male is heavier or lighter than average for his height) are correlated with waistline: a heavier-than-average weight for a certain height goes along with a larger waistline. Using both the first parameter, height, and the second parameter, waistline, as input data, we can make a much more accurate prediction of weight than if height alone is used. Undoubtedly, for a creature as complicated as a human being we would not be surprised to find that even two parameters are not enough to predict weight perfectly. For example, body *proportions* like leg length as a fraction of total height might be another tailor measurement that, if added into our equation, might result in an even better prediction.

Information like this could be used for its own sake just to understand how people are put together, or it might be applied to any number of practical matters. For example, a medical researcher would go beyond the "bulk" parameter and discover that waistline is well correlated to the number of fat cells in the body, and further that this depends on both the genetics and behavior of the human being. This would be a case of realizing that waistline is merely a manifestation of a more fundamental parameter that has to do with biology, and ultimately, biochemistry. In this way, starting from a list of basic measurements for a group of subjects, this simple scientific approach could lead to important conclusions about human development, even cell growth. Or it might be that this information had purely utilitarian value, for example, to a furniture designer who wants to relate height, weight, bulk, and other parameters in order to build a more comfortable chair, or his evil twin who devises an economy-class airline seat that provides the minimum in human needs and requires the maximum in human tolerance.

To help understand the benefit of additional parameters in measuring galaxy distances we can revive an analogy used earlier when we imagined finding the distances to neighboring houses by comparing the apparent heights of people inside. With the additional information we might do a better job determining how far away each one is, by realizing, for example, that people whose leg length is a larger fraction of their body length are taller, on average. The situation was much the same in our study. The primary parameter for elliptical galaxies was already known from Sandy Faber's Ph.D. thesis. This "first parameter" was the total luminosity of the galaxy, closely related to a galaxy's mass. Just as with the people, the characteristics of elliptical galaxies, such things as size, velocity dispersion, and metal abundance, change most noticeably with this simple measure of how much galaxy there is: more luminous galaxies are bigger, have higher velocity dispersions, and have greater metal abundances. But, again, one was clearly dealing with more than a one-parameter family (as we called it), because neither the velocity dispersion nor the magnesium line-strength that measured the metal abundance correlated *exactly* with luminosity. As Terlevich, Davies, Faber, and Burstein had found in their study, at any given luminosity, there was a fair range in both quantities: this implied at least one more parameter. They had nominated the shape of the galaxy—how elongated it was—as a candidate for the second parameter, in the way that waistline served in the above example as a manifestation of a second parameter (though not necessarily the parameter itself, the underlying cause). However, we already knew by 1983 that our new data did not bear this out.

If we failed to identify a new second parameter, we would make little progress in understanding how elliptical galaxies were put together. On the other hand, if we could accurately predict the true brightness (luminosity) of each elliptical galaxy from a measure of its velocity dispersion and some "second parameter," the way weight could be predicted from height and waistline, we would know much more about how ellipticals were born and how they developed. Furthermore, with a more accurate prediction of the galaxy's luminosity, we could better place this galaxy at its proper distance and thereby, galaxy by galaxy, make a map of the Hubble expansion. Driven by the necessity of finding a second and maybe even a third parameter, we worked hard to include every measurable parameter we could think of.

This approach would pay off. Over the next year most of us worked sporadically on the project, but, in the winter of 1984, Donald Lynden-Bell made the breakthrough we needed. While playing around with the tables of parameters Dave had so laboriously compiled, Donald discovered an excellent correlation between velocity dispersion and one of the new parameters. It was the one Donald had invented, the last parameter, the one that had driven Dave over the edge. Dave had facetiously christened it "the donald diameter." (In the year ahead it was often just called "donald," which led to much confusion and some very amusing sentences.)

Donald discovered that the correlation between velocity dispersion and "his" diameter, the diameter within which the average brightness per unit area reached some prespecified level, was far and away the tightest relationship yet found for these galaxies. For the ellipticals in the Virgo and Coma clusters, the data formed a tighter line than even the relation of luminosity and velocity dispersion that had been the cornerstone of our study. By replacing luminosity with the "donald diameter" as the first parameter, he had found a way to use velocity dispersion—a measurement that was independent of a galaxy's distance—to predict with good accuracy a galaxy's size, something that depended *directly* on the distance. This was just what we needed to make a map of the Hubble flow—a distance for each galaxy in our sample.

After learning about this in a letter from Donald, Dave made a confirming report of the surprising tightness of this relation between the donald diameter and the velocity dispersion, in an 1984 April Fool's Day letter to the group. The complex discussion in Dave's letter, including further correlations with several other parameters, demonstrated that we were far from understanding why this particular correlation was such a good one. It was clear, however, that this was the best correlation we had found; unquestionably, this was progress toward our goal.

It wasn't long before we did begin to understand why the donald diameter worked so well. Donald had fortuitously defined a quantity that combined the first parameter, galaxy luminosity, with what turned out to be the true second parameter, surface brightness (the brightness per unit area). Australian astronomer Ken Freeman, always a free spirit and creative provocateur in matters of galaxy formation, had identified surface brightness as the best candidate for a second parameter years earlier when he had served as referee of the Terlevich et al. paper.* In his report Freeman explained why he doubted the paper's claim that galaxy *elongation* was the second parameter, suggesting instead that surface brightness, the average brightness per area within a specified radius, made much more sense. To back up his position he included a plot of surface brightness versus the departures from the relation between luminosity and velocity dispersion; it showed a strong trend, in other words, for most cases, when a galaxy had an unusually high velocity dispersion for its luminosity it also had a high surface brightness, or a too-low velocity dispersion and low surface brightness. The case for surface brightness being the second parameter would have been very strong indeed except for four very discrepant points. Freeman commented, tongue in cheek, that it was okay to ignore the four misbehaving galaxies. He was poking fun at the paper itself, where the authors had chosen to dismiss four discrepant points that were counter to the delta-delta relation the paper was presenting. Roberto Terlevich, indignant in response, wrote to coauthors Faber, Burstein, and Davies that he had investigated surface brightness as the second parameter from day one—it was so obvious—but had rejected it when he saw how poor the correlation was. He wondered what gave Freeman the right to ignore these four ruinous galaxies. Roberto had missed Freeman's barb completely.

However, in the years that followed, the importance of surface brightness as a parameter had become clear. The basic correlation of luminosity and velocity dispersion made sense, of course: more mass implied a greater gravitational force and, as a result, higher velocities for the stars in the galaxy. But the force of gravity depends even more strongly on the distance between the attracting bodies than it does on the total mass. Therefore, it also made sense that, when comparing two ellipticals with the same luminosity

---

*Most professional journals send submitted papers to one or more referees, who anonymously, if they prefer, evaluate and advise on the paper's worth, correctness, and suitability for publication.

(presumably the same mass in stars) the smaller, more compact system would have a higher velocity dispersion because the stars would be closer together and, feeling a greater pull of gravity, would move faster. At a given luminosity, then, a more compact elliptical galaxy (same luminosity in a smaller size means higher surface brightness) should also have a higher velocity dispersion. In fact, this correlation of all three parameters is well known in physics—it is called the *virial theorem*.

We had accidentally discovered a one-step measurement that combined two of the properties of elliptical galaxies, size and brightness, into one— the donald diameter. This simplified things by reducing the three-parameter correlation to only two. Sandy Faber would make this all explicit a year later, when she would plot the properties of elliptical galaxies in a three-dimensional "space" of luminosity, size, and velocity dispersion. She found that the points clustered strongly in a flat distribution that she called "the fundamental plane."* However, if one has the proper viewing angle, that is, along the plane itself, the plane looks like a straight line (consider sighting along the edge of a table). By making a combination of luminosity and size with the donald diameter, we had effectively sighted along this plane, making this relationship among three variables appear as a correlation of only two— a straight line from our line of sight.

Dave began to play with Donald's new toy. He tested to see if the correlation between size and velocity dispersion was as tight for the rest of our sample as it was for the ellipticals in the Virgo and Coma clusters Donald had investigated. Of course, the relation was between *true* size and velocity dispersion, and this had been easy to investigate for galaxies in a cluster since, as before, with all galaxies at approximately the same distance, true size was in direct proportion to apparent (relative) size. This wasn't true for galaxies in the "field," which are found over a wide range of distance. For the field sample, Dave had to take the additional step of estimating the distance to each galaxy so that he could convert the apparent donald diameter into a true size, from arcseconds on the sky to a certain number of light-years in diameter. He did this by placing each elliptical galaxy where it would be if the Hubble expansion were perfectly smooth, that is, at a distance simply proportional to its redshift (velocity). Then he watched the terminal screen as the computer rhythmically placed a point for each galaxy, plotting along the horizontal axis its donald diameter and along the vertical axis its velocity

---

*As often happens when an idea's time has come, this one was discovered independently by S. "George" Djorgovski, working on his Ph.D. thesis under Marc Davis's supervision at U.C. Berkeley.

dispersion. To his disappointment, Dave found that the relation was not nearly as tight as it was for the cluster sample. Again it seemed that the field ellipticals were not as well behaved as their clustered cousins.

Yet there was something else, something very odd. Several times Dave had stared at the screen as the computer slowly worked its way through the galaxy list, a list ordered by "longitude" (astronomers call it *right ascension*) across the sky. Soon, he began to notice that the greater scatter of the points didn't develop randomly, but systematically. Over the minute it took to make the plot he noticed periods of about ten seconds when the points were all coming in too high or too low compared to the scatter around a straight line. This suggested that in some directions in the sky the correlation was very tight, like the cluster sample, but shifted above or below the line, so that when plotted together, the relation for the whole sky looked like a spread-out mess.

Dave isolated one such zone in the sky, the area containing the flat plane of galaxies known as the Local Supercluster, set with its jewel, the Virgo cluster. Dave saw such a shift here, and he thought he understood why. In his letter to the rest of us, he explained his reasoning and included plots that supported his interpretation. Galaxies on the near side of the Virgo cluster, he said, were known to be "infalling"—they had extra velocity away from us, in addition to Hubble flow—toward the cluster, while those on the far side were "backfalling." This was what Marc Aaronson and his collaborators had found in their pioneering study: *peculiar velocities*—velocities in addition to Hubble expansion—for spiral galaxies in the Local Supercluster.

If these large peculiar velocities were present, Dave's use of each galaxy's redshift to estimate distance would be systematically in error, because velocities of the infalling galaxies would include both the smooth Hubble expansion and an additional, substantial peculiar velocity. For example, he would have overestimated the distance of galaxies falling toward Virgo by mistakenly including their peculiar velocities as part of their Hubble expansion velocities. By the simple use of Hubble's relation of distance and velocity, these galaxies would appear more distant than they actually were. With too great a distance assigned to each of them, Dave would have calculated a too-large donald diameter and, as a group, they would lie systematically to the right, below the line that described the average correlation. Galaxies on the far side of Virgo would show the opposite: underestimated Hubble velocities and distances, underestimated sizes. By this line of reasoning, Dave had come to understand that, at least in the Virgo domain, peculiar velocity could account for the systematic behavior he had seen when the points were plotted. In summary, if there were large peculiar velocities for galaxies in a region of space, and this was ignored when using the Hubble relation to estimate their distances, the scatter in the relation between donald diameter and velocity dispersion would look worse than it actually was.

Dave was on the verge of a discovery, but he joined a growing list of those who fell back from the brink. It would have been logical for Dave to postulate that *all* of the systematic behavior that he saw as his plotting program proceeded around the sky was due to peculiar velocities: galaxies in one region moving faster than a smooth Hubble flow, hence their relation shifted to the right; in another region, galaxies moving too slow, shifted to the left. One region more than any other prevented Dave from making the big leap, an enormous volume of the southern sky that showed the greatest discrepancy. To bring it into agreement with the mean relation, Dave would have had to change the distances to these galaxies by more than 30 percent, which would imply huge peculiar velocities of about 1,000 km/sec. This was three or four times larger than the biggest peculiar velocities known, the infall to Virgo.

Alas, Dave was no more likely to have proposed this outrageous explanation in 1984 than I had been in 1983 in my Coma-Virgo cluster paper. Extraordinary claims require extraordinary proof, and in this case Dave knew that there were at least two other, far less radical explanations for what he called the "EGALSOUTH problem" (Dave's nomenclature could get a bit obscure and cumbersome). For one thing, the data from our different observers had not yet been put on a standard system. Spectra of the southern galaxies were obtained by Roberto and Roger at the Anglo-Australian telescope and by me at Las Campanas, while the northern ellipticals were mainly observed from Lick. It was likely that cross-comparisons of north and south would reveal systematic measurement problems that required correction (they did). Even if these corrections were not large enough to explain the entire difference in the relation between size and velocity dispersion, north to south, Dave, like the rest of us, was quite prepared to believe in genuine *intrinsic* differences among galaxies in different regions of space. After all, this was the kind of variation we had been looking for so that we could learn more about the evolutionary histories of elliptical galaxies.

So it was that the correction of all data to a common system became the focus of the next general meeting, three weeks in Santa Cruz starting at the end of July 1984. Sandy was our host for a meeting that she describes as the lowpoint of the project for her personally. Her lower-back problem had grown even worse, and a myelogram in June revealed that little would be gained from surgery. Her back constrained by a plaster cast from armpits to hips, Sandy returned to work in constant pain, and for the first time in her brilliant scientific career depression ruled. The fact that Sandy could not sit at a

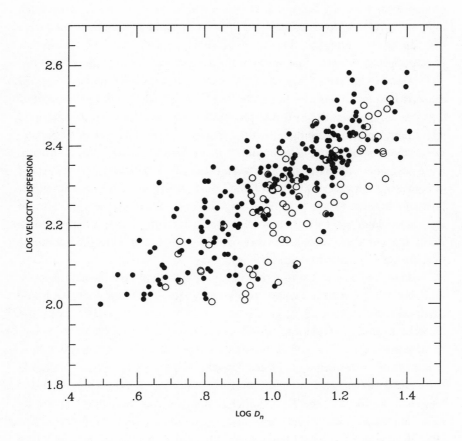

A re-creation, using the actual data, of what Dave Burstein saw when he plotted donald diameter $D_n$ versus velocity dispersion for galaxies around the sky. A particular group of elliptical galaxies in a southern zone of the sky, shown as the open circles, fell systematically to the right of the mean relation for most elliptical galaxies, shown as solid dots. Dave called this the "EGALSOUTH Problem."

computer terminal for any length of time became a severe handicap to our work, because, as the project had progressed, computers had assumed a nearly essential role in data processing and analysis. Very little progress would have been made without the help of Dave Burstein and Sandy's graduate students Jesus Gonzales and Christina Dalle Ore. Even so, analysis of the Lick spectra was still incomplete: the spectra had been processed and prepared for analysis, but no measurements of velocity dispersion had been made. Sandy dreaded what she knew would be a harsh judgment by the group that she was holding up the show. None of us realized how serious her physical problem was—she was not the kind of person to make excuses.

Summers in Santa Cruz were about as close to heaven as anybody had reason to expect in this life, but again, as in Cambridge, "reveling in nature" was low on the agenda. Still, the process of getting to know each other that started at Cambridge continued with a memorable party at Sandy's house to watch the stirring opening ceremonies of the Summer Olympics. And there were occasional diversions: walks to and fro each day through still forests of sky-aspiring redwoods, lunches on the sun-drenched lawn, and the couple of hours I set aside to see if I could drag my carcass up and down Seabright Beach as rapidly as I did as a graduate student. But, for the most part, this Eden was about as remote as the original. The interiors of unfinished-concrete buildings are remarkably similar worldwide.

It was my turn to try to irritate everybody. I had arrived with a lot of new data for galaxies in clusters and a really smug attitude. As I had found earlier for the Coma and Virgo clusters, the elliptical galaxies in two more clusters showed tight relations between galaxy luminosity (brightness), velocity dispersion (typical speed of the stars), and magnesium line-strength (abundance of heavy elements—*metals*). Donald diameters hadn't been measured for these galaxies, so I was still working with these original variables that had been the focus of the Terlevich et al. study. I was particularly struck by how much better the correlation between luminosity and magnesium line-strength was for the data I had taken compared to our larger sample. Even my southern field galaxies, those not in clusters, had a better correlation, I thought.

The first afternoon, I taped up these plots on the blackboard in Sandy's cubbyhole office, and, with the seven of us jammed in among stacks of books and papers, proceeded to preach to the assembled on the benefits of working with homogeneous data sets (especially mine). I expressed serious doubt that the EGALSOUTH problem was real, as opposed to the result of combining data sets incorrectly, or, I intimated, adding in bad data (not as good as mine). For substance, there was much of what I said that was useful and needed saying, but for style, it was a fairly obnoxious performance.

I took a lesson in humility that afternoon. In about an hour I had been returned to Earth or even slightly below. Diagrams were tacked up that showed other subsets of the data, particularly many Lick measurements of magnesium line-strength, and point-by-point comparisons were made of those galaxies common to my data set and the others. It soon became apparent that my data for field ellipticals were no better than anyone else's, although, as was feared, we could also see systematic differences among the data sets that needed correction. As for the cluster sample, everyone agreed that these tighter relationships really were encouraging—perhaps they would become even better when the donald diameter was used in place of luminosity. We resolved to give greater attention to both rich clusters and to the smaller groups of galaxies that could be found in the field sample.

The merging of data sets was now the top priority. Sandy, Roger, and I evolved procedures and computer programs to juggle all the comparisons and tell us what corrections to make. After the first go at these corrections, the shift in the relation of the donald diameter and velocity dispersion from north to south persisted. This meant that the peculiar velocity explanation for the EGALSOUTH problem was still alive. Dave began to look at other parameters in his files to see if he could find any evidence that these galaxies in the south were somehow "different" from the ones in the north. He did, primarily in plots of velocity dispersion versus magnesium line-strength, two measurements that were *independent* of distance and so unaffected by departures from smooth Hubble flow. Finding an offset between the data for north and south reinforced Dave's belief that we had found genuine intrinsic differences, diversity he thought could be attributed to variations in the types and numbers of stars—the stellar population. With intrinsic differences still a possible explanation, the case for large peculiar velocities was still in doubt.

However, all this scrutinizing and reworking of the data was building confidence within our group. The data sets were coming together; we could begin to think about final tables of numbers and analysis of what it all meant. Sandy had settled into the role of group coordinator: the group had its own momentum and plenty to do, but someone had to shepherd its activities from time to time. She had intended to be the group leader, but two factors had worked against it. First, she felt uneasy about asserting herself over Donald Lynden-Bell, because, as she later recalled, he was "so famous, so senior, and so smart." And there was another problem.

"My usual style would have been to take center stage with chalk at the blackboard and control the process, by presiding over key decisions. My back problem was at its worst back then [in Cambridge] and instead I found myself lying flat on a portable cot in Donald's office. It is very hard to lead a group of people from a prone position—have you tried it?—and my energies were at a very low ebb anyway. I found it very comfortable to lie back and avoid taking central responsibility for anything.

"It was the best thing that could have happened to us. The resultant power vacuum allowed each of us to quietly find our own best way to contribute. This lesson has stood me in good stead since. I now think that in small groups of able and motivated individuals, giving orders or setting up a well-defined hierarchy may generate more friction than it is designed to cure. If a good spirit of teamwork prevails, team leadership can be quite diffuse."

So Sandy found this other role—she was, as she describes it, "the glue." In keeping with this style, that day in Santa Cruz she went to the blackboard, but rather than doling out assignments, she opened a discussion of what we needed to do next and who would volunteer to do it. It was agreed that Dave would continue to collect and collate the photometry data and begin writing a paper describing how the data were taken, reduced, and analyzed. Roger would do the same for the spectroscopic measurements. We decided that, in addition to publishing data tables, we would make it easier for other researchers to access and analyze our measurements by placing all data in fixed-format computer files that could be sent on tape or electronically mailed to any researcher around the world.

The calming effect of my 1983 paper on ellipticals in the Coma and Virgo clusters had all but worn off by the Santa Cruz meeting, and my renewed nagging that we publish some results was partly responsible for the new momentum for writing papers. Sandy encouraged my proposal to work on the data for ellipticals in rich clusters, most of which I had obtained myself, and concentrate on the correlations of velocity dispersion and magnesium line-strength with the new donald diameter, which we expected to be even better than the good correlations with luminosity I had presented the first day. I would write a paper that would demonstrate the utility of the new parameter and show how this correlation could be used to find more accurate distances to the clusters. (We were still reluctant to apply the method to individual field galaxies because we thought they might harbor additional intrinsic differences like the kind Dave seemed to be finding.) Donald would be my chief collaborator on this paper and, as with the other papers to come, all of us would share authorship. We left Santa Cruz buoyed, again with no results per se, but with a clear idea of where we were going and enthusiasm for the tasks each of us had to do.

In February 1985, Roger, Sandy, and I visited Dave at Arizona State in Tempe to put the final touches on the merging of data sets. Finally, all the data were put on the same system, and the smaller corrections for aperture size and geometrical effects of Einstein's relativistic space-time were applied as well. Sandy had arrived a week before us and used the time to cull through the photometric data, looking for and eliminating obviously bad data points. Sometimes these were due to transcription errors. Occasionally, bad observations at the telescopes were to blame, for example, a detector that drifted in sensitivity, or a night that seemed clear but actually had hard-to-see cirrus clouds absorbing some of the light. But other times the errors were more subtle and no one's fault, for example, when another galaxy or foreground star fell close to the line-of-sight to one of our sample galaxies, the measurement of galaxy brightness through apertures would be contaminated and thus in error. This final pass through the data cleared up some previously puzzling cases.

Now, using the corrected, cleaned data set, Dave looked again at the "intrinsic differences" he had spotted between the northern and southern ellipticals. *Voilà.* The differences had disappeared. There was no longer any evidence that the ellipticals in the north and the south were intrinsically different. We began to accept, finally, that the most likely explanation for the EGALSOUTH problem was that using redshifts and the smooth Hubble relation to determine distances to these galaxies had resulted in substantial overestimates because they had, in addition to their Hubble expansion velocities, large *peculiar* velocities. The four of us saw glimmerings that we were onto something big. As Sandy puts it, "When I left Tempe I was pretty sure we were in for *some* kind of discovery."

In May 1985 I was back in Pasadena and working on the data for the "cluster paper." I didn't know it, but I was about to convince myself that large peculiar velocities were real. Additional observations I had made with the 200-inch Hale telescope at Palomar Observatory in late 1984 had raised to six the number of clusters with measurements of velocity dispersion and the donald diameter. By and large, these galaxies were more distant and fainter than our main sample; even after considerable effort we had data for only about 100 galaxies in rich clusters—a small number compared to our total sample. The greater distance meant that the errors in predicting distance were proportionally larger. However, for each cluster I could average the data for a dozen or more galaxies, so the final distance prediction was actually very good. Just as with individual galaxies, I could compare the predicted distance of the cluster to what I got when I used its redshift to calculate its distance for smooth Hubble expansion. Any disagreement in these distances could be attributed to a peculiar velocity for the entire cluster. This is what I had done a year earlier with the Coma and Virgo clusters.

Finally, one morning, I had worked through all the details and had reached the point where I was able to compute the distances predicted by the correlation of velocity dispersion and donald diameter. I was pleased to find that, within the uncertainties, for each cluster the distance implied by the smooth Hubble expansion was in good agreement. In other words, there was *no* evidence for large peculiar velocities. Despite the tide of rising excitement in Tempe about large peculiar velocities, it was quite in character for me to hope for a return to this more conventional picture—in science, I'm pretty conservative.

I remember wondering, through lunch, if the results for the field sample, notably the famed EGALSOUTH problem, was just a lot of hooey. Is this what happens when you got involved in a big group? I wondered. Had all the difficulty and confusion led to a big mistake? In fact, it was I who had made the mistake, and though it was a small one, it was deadly. As I rechecked the calculations later that afternoon, I found that for one of the clusters I had made a "sign" error—added when I should have subtracted. Instead of the near-zero peculiar velocity I obtained when I subtracted two numbers of approximately 500 km/sec, the correct arithmetic resulted in a peculiar velocity of about 1,000 km/sec. I was stunned. Disbelieving, I repeated all the calculations again and again, as if I had found a large error in my favor while balancing my checkbook. Finally, I had no doubt that the Centaurus cluster, smack dab in the middle of the EGALSOUTH region, had a huge peculiar velocity. This was not just a single galaxy, but a cluster of hundreds of galaxies moving at high speed. I had become a believer, and I quickly called Dave, Sandy, and Roger to tell them so.

Our momentum continued to build. As we met again in July 1985, in Hanover, New Hampshire, the course was clear: we would calculate the distances to all galaxies in the sample, determine peculiar velocities, and figure out where, how much, and, if possible, *why* they were moving. Gary Wegner, our answer to mild-mannered Clark Kent, hosted the meeting. A lot of what Gary was all about became obvious the day I stepped into his house—a hundred-year-old New England classic with no-nonsense, elegant Yankee decor. This was the kind of guy the pollsters were always looking for: solid citizen with a kind, responsible wife, bright, rambunctious kids—a *functional* family, for God's sake. No wonder they held the first presidential primary in New Hampshire, I thought.

What was surely the most pleasant meeting of all took place at Dartmouth College, a seat of learning as old and as charming as one can find in this country. With the emerging realization that something was about to be born from our long endeavor, spirits were high and relations among the group seemed never better. On a sweltering Fourth of July we barbecued at Lake Mascoma, playing Frisbee until we lost it in the lake. Donald, of all people, led a cooperative effort—we were good at this now—to retrieve the Frisbee from the murky, gently lapping waves, then threw it into the water again where it was lost for good. It was a revelation seeing Donald loosen up, playing like the kind of child he must have been, briefly. The crowning moment came at one of the memorable dinners Gary's wife, Kay, hosted, when Donald, two beers downed, became unusually giddy and launched into a five-minute tongue-twisting poem that began

> I had a duck-billed platypus when I was up at Trinity
>    with whom I soon discovered a remarkable affinity.
> He used to live in lodgings with myself and Arthur Purvis
> and we all went up together for the Diplomatic service. . . .

The rest of us were amazed and delighted. Who was this guy?

It was a hectic time as well. Gary was late in coming home from an observing trip; Roger had to leave early; Roberto, for lack of travel funds, was stuck at the Royal Greenwich Observatory and couldn't come. But, perhaps for the first time since we had set out to gather the data, there was a single direction, a genuine camaraderie, and a gathering energy propelling us toward what had become our goal: to make a convincing case to the astronomical community that galaxies in vast regions of sky were departing from smooth Hubble expansion, at speeds of up to 1,000 km/sec. This would require well-documented evidence and consistency checks to show that other explanations could be ruled out or were very unlikely. We knew that, despite the growing evidence from galaxy maps that the universe was surprisingly lumpy, our claim of a bumpy Hubble expansion would be hard for our colleagues to swallow.

By now we had become confident of our ability to predict the distance to an elliptical galaxy, using the correlation of velocity dispersion and the donald diameter, which had recently been rechristened $D_n$.* The method would be similar to that using the *Tully-Fisher* relation between the rotation speed of spiral galaxies and their luminosities, one that Aaronson and his collaborators had employed in mapping distances and deriving peculiar velocities in the Local Supercluster. Like velocity dispersion, rotation velocity could be measured independently of distance to a spiral, and used to predict the brightness, which is dependent on distance.[†]

It was simple to use $D_n$ to predict distances. We began with the thirty-three elliptical galaxies in the Coma cluster: they defined a tight, straight-line relationship between velocity dispersion and $D_n$. Now, for any other galaxy the relative distance (how much nearer or farther it was than the Coma cluster) could be determined by comparing its measured $D_n$ diameter to the value given by the standard relation for Coma *for that value of the velocity dispersion*. This is simpler in math than it is in English. For example, if a galaxy had the same velocity dispersion as one in the Coma cluster, but had a $D_n$ diameter some 2.5 times larger, it would be assigned a distance 2.5 times closer than Coma. All distances computed this way would be known relative to this one cluster, of course, but this would be sufficient to diagram the Hubble expansion. The method had an average error of 22 percent for each galaxy—not very good on the face of it, but far better than we had done before and nearly as good as the best technique, the Tully-Fisher relation.

We recognized that a major cause for skepticism would be whether ellipticals in different environments—rich clusters, small groups, and isolated ellipticals—obeyed different relationships between velocity dispersion and $D_n$; we ourselves had been expecting to find something of this sort. Therefore we determined distances and peculiar velocities for three subsets, representing different environments, separately, then compared the results. For this we had to write computer programs that would isolate these particular subsets of the sample in different parts of the sky, then apply the correlation of $D_n$ with velocity dispersion and magnesium line-strength (a second, independent way) to find distances to the galaxies.

Donald and Sandy had put a good deal of effort into identifying those

---

*The new nomenclature came after Scott Tremaine, at the Princeton Conference on Dark Matter, overheard a lengthy, animated conversation between Donald and Dave, in which Donald kept referring to measuring and correlating himself. Scott expressed concern that one or both of them might require "professional help."

[†] As for elliptical galaxies, it makes sense that brighter, more massive spiral galaxies rotate faster. To be precise, their stars follow basically circular orbits with speeds that scale as the square root of the mass enclosed by the orbit.

ellipticals in our sample that were grouped together; a catalog prepared especially for us by Margaret Geller and John Huchra that included other types of galaxies in addition to ellipticals was particularly useful. The goal here was to associate those galaxies that were both close to each other in the sky and had approximately the same redshift. Ellipticals in groups were important, not only because they defined a different environment, but also because, as with the richer clusters, a more accurate measure of distance (and peculiar velocity) could be made by averaging individual measurements for several galaxies. Sandy and Dave undertook the job of determining distances and peculiar velocities for these groups that included about one-third of the more than 500 ellipticals in our sample. Roger and I concentrated on the cluster sample, also containing about one-third, and Gary and Donald found distances to the isolated ellipticals.

It was about midnight in the otherwise deserted physics department at Dartmouth when the three teams gathered and, to our great relief, concluded that each of their samples showed very similar results. In all three we saw the very large peculiar velocities in one part of the southern sky, the EGAL-SOUTH region, fairly far away—about 100 million light-years. Before our study, no one had made measurements of peculiar motion in this part of the sky. In other regions we saw evidence for departures from smooth Hubble expansion as well, perhaps not as large or coherent, but the results from the three analyses were consistent. This showed that the effects we were seeing were not the result of intrinsic differences in the properties of elliptical galaxies that depended on their environments. The excitement was overwhelming. We had made a true discovery: not all galaxies were at the distances their Hubble velocities implied, and many were moving in unison with their neighbors at remarkable speeds.

It is an amazing thing, this feeling of discovery. At some time or other in our childhoods we all experience it, as we set out on the daunting endeavor to understand the world we have mysteriously landed on. It is enough that we discover something that is only new to ourselves—a new path through the woods, a hideout, a strange insect, a beautiful rock—we are not looking for recognition, just satisfaction. You can imagine how much greater is the pleasure of discovering something no one has ever seen before.

Even as a five-year-old I think I knew that the world of Earth had been totally ''discovered''—there were no more hidden places no one had ever seen. That is probably what had excited me the night I first saw Saturn through a telescope. Surely a world this vast and far away was wide open for discovery. A scientist or artist has a chance to hold on to this wonder of childhood; the potential for discovery can last a lifetime. Unfortunately, for most adults the opportunities disappear, and the joy of discovery goes with it. We must learn again how to share discovery with our fellows, not just with those whose lives are dedicated to learning and knowledge, but to those who watch from a distance. In our schools we must teach that learning is a lifetime process and that the discoveries of our time belong to us all.

Our new discovery left us buzzing—we were eager to tell all our other colleagues as soon as possible. But then the questions and doubts began to creep back in; they made us cautious. News like this would be very embarrassing to retract. Did it make sense? Why were galaxies moving? Was there an overall pattern that would give us a clue? What we had were pieces of paper with plots and numbers: we had no pictures that would allow us to visualize what we had found. Gary and Donald had written the first version of OURV ("our velocity"), an elaborate computer program that would tell us the average peculiar velocity in any part of the sky we specified. It would even make a plot of the motions in that region. But time was running out in Dartmouth, and we only had time to see the one very early plot on the computer screen. We left without any clear overall picture, just the idea that large peculiar velocities had been found.

Also troubling us, in the last days, were two other issues that might be misleading us still. One had to do with systematic errors that would bias the determinations of distances in different regions of the sky, perhaps mimicking the very effect we claimed to have found. In any research project, errors in measured quantities must be weighed carefully to see if the results are statistically significant. Mostly these are *random* errors—for example, we could only measure velocity dispersion to a typical accuracy of plus or minus 5 percent—but the more troublesome effects are the *systematic* errors that random errors can generate. Our sample of isolated ellipticals, and to a lesser extent those in small groups, was susceptible to one such systematic error, referred to as *Malmquist bias*.

Malmquist bias resulted from the imperfect correlation between $D_n$ and velocity dispersion. There was still a considerable scatter from a straight line, "perfect" correlation, so a measured value of velocity dispersion did not uniquely specify brightness (or size), but rather a spread around some average value. This in itself was a random error—one that was just as likely to be too big as too small—but it gave rise to a systematic error. The sample would be biased toward galaxies that were brighter and larger, because they could be seen to greater distances, i.e., over a larger volume of space. Sandy realized that, because of the Malmquist bias, we were, on average, underestimating the distances to our sample galaxies all over the sky, but that the effect might be very small in some regions and much larger elsewhere.

She began to think about how big that bias might be and how a correction might be applied. She was fairly certain that the effect of Malmquist bias was not totally responsible for the large peculiar velocities we had discovered, but just how much it was responsible for she wasn't sure.

One more important idea gelled in the last days in Dartmouth, and this too gave us pause. What was the proper frame of reference with which to measure the motions of the galaxies in our sample? It had been a long time since the Earth had been considered a fixed point of reference. Galileo put it succinctly—"*si muove*" (it moves)—and landed himself in a Catholic jail for pointing out that the Earth turns and orbits the Sun. Astronomers routinely correct the velocities they measure for these motions, which amount to about 30 km/sec. In doing so, they are pretending that the observations were made from the Sun—the corrected measures are called *heliocentric* velocities. But, as was learned at the beginning of this century, the Sun is also moving around the Milky Way galaxy, at a much greater clip of about 200 km/sec. Our group had, in fact, corrected the peculiar velocities we had measured to the center of our galaxy, and eventually added another small correction to simulate observations made from the center of the Local Group of galaxies—the Milky Way, Andromeda, and a few others—all with the idea of getting a wider frame of reference.

But, as we were now learning, galaxies around us were moving too— we had measured large peculiar velocities for many of them. With everything in motion, was there no better frame of reference from which we could judge what was moving where? Remarkably, there was: the cosmic microwave background.

As I discussed earlier, our galaxy and the galaxies surrounding it are like boats on a lake. At first we had measured the velocities of other boats relative to our own, and later relative to a small group of boats around us. However, what we really wanted to know were the motions of all these boats relative to something more substantial—the shoreline, for example. On a street corner in Hanover during one lunch foray, Donald suddenly turned to Sandy and exclaimed: "You know, we really should be plotting all these galaxy motions with respect to the cosmic microwave background, not with respect to our galaxy." Donald was right. There was a remarkable measurement of our galaxy's motion with respect to the "shoreline." All we had to do was look down at the water.

The universe is filled with radio-wave light streaming in every direction, light that arose in the hotter, earlier phase of the big bang—it fills space like water fills the lake. When a boat moves in the water, it creates a wave out ahead and a wake behind. Measurements of the cosmic microwave light had shown just such an effect: the CMB is "bluer" (hotter) in a certain direction and "redder" (cooler) in the opposite direction. Our galaxy is moving against the still water, and so the light is red and blue shifted, an amount that corresponds to a velocity of 600 km/sec. With this measurement of our galaxy's motion relative to a more fundamental frame of reference (basically the entire universe had become our frame of reference), it was a simple matter to transform the speeds of all the other galaxies so that their velocities were measured with respect to the lake itself and not just to other boats. As on the day we "punted" in Cambridge, we had stuck a pole in the mud and we could see how fast we and all the other galaxies were moving.*

Sandy recalled later that the idea struck suddenly, forcefully, that hundreds of elliptical galaxies, and the thousands of other types of galaxy they keep company with, were racing with our galaxy through space. She realized that, for over a hundred galaxies near our own Milky Way, we had measured only small peculiar velocities, the way a boatman might find only small drifts between himself and the other boats nearby. But if the boatman also knew that his boat was moving rapidly in the water, as we did, then it must be that the neighboring boats were moving too, all drifting or powering together. It suddenly became obvious why Allan Sandage had failed to find peculiar velocities for neighboring galaxies, and incorrectly concluded that the Hubble expansion was almost perfectly smooth: galaxies were moving, all right, and rapidly, but over large regions and coherently—together, not randomly. That's why only a handful of galaxies have blueshifts—just a few little drifts toward us as this great flotilla steams on.

And, astonishingly, the EGALSOUTH region, where we had found the largest peculiar velocities away from our galaxy, was moving in the same direction as the Milky Way and its neighbors. This meant that we had to *add* 600 km/sec to find the peculiar velocities of these galaxies with respect

---

*The analogy is incomplete because the universe is expanding, while lakes hold their sizes. The result of the expansion of the universe is that the cosmic microwave background becomes cooler and cooler (it was once as "hot" as visible light but is now as "cool" as radio waves), but it still continues to fill space uniformly and therefore continues to serve as the motionless water in the lake. With this in mind it is easier to understand why the Hubble expansion velocities are not velocities in a strict sense: galaxies are fixed in space like knots on a rubber band and are simply separating as space, like a rubber band, stretches. Peculiar velocities, on the other hand, are true motions *through* space. A galaxy participating in smooth Hubble expansion, with zero peculiar velocity, will see a smooth cosmic microwave background of uniform temperature all around—like still water. There will be no redshift or blueshift of the light. But a galaxy moving through space will measure such a shift because its peculiar velocity is a true motion.

to the cosmic reference frame, which made their velocities even faster and more astounding.

The morning of our last day in Dartmouth we met in a gleeful mood of accomplishment, but Sandy stepped in again to temper our childlike enthusiasm. She reminded us that these results would be considered heretical, and that we had better be absolutely certain that such results could not be due to systematic variations in galaxy properties, or the Malmquist bias that caused us to underestimate the distances in a systematic fashion. She insisted that, before we went forward, we step back and spend the coming months comparing all the properties that could be measured independently of a galaxy's distance—again. We would compare galaxy colors, velocity dispersions, magnesium line-strengths, and ratios of various diameters or brightnesses, to make sure that elliptical galaxies in one part of the sky weren't just different from those in another part.

We had to satisfy ourselves that these effects were well in hand before our next meeting in Pasadena. Once this was done, we needed to recalculate peculiar velocities with respect to the CMB, our cosmic frame of reference. And we would need computer programs like OURV that would quantify the result and make maps and diagrams to replace the maze of plots and pages of numbers that by themselves provided no clear picture of the overall departures from a smooth Hubble expansion—a picture we would need if we were to understand the *cause* of the peculiar velocities, something we had barely thought about. We left Dartmouth with a vow of silence: we would tell no one of our results until these tasks were done.

Dave returned with Sandy to Santa Cruz, where they spent six more weeks combing the data set for errors and making more cross-checks to convince themselves of its purity. Dave made more tests using distance-independent parameters until he was satisfied that there were no discernible differences in the properties of those ellipticals that were moving rapidly and those that had only small peculiar velocities. It had been a hidden blessing that the entire project had concentrated on intrinsic differences; it left us exceptionally well prepared to look for such effects, and had kept us from accepting the evidence for large peculiar velocities until we had exhausted all other possibilities.

On August 28, 1985, Dave wrote a letter to the other members of the group, summarizing his conclusions that "The local velocity field . . . is not uniform on the sky in either [position or distance]. Velocity deviations from uniform [Hubble] flow are large, of order 1,000 kilometers per second . . . for large regions of the sky." Realizing that some of these regions had been covered in the Aaronson et al. studies and by Rubin and Ford, Dave began to investigate their data sets to see if they corroborated our results. This was the beginning of a huge and complicated task, but the first indications were reassuring: where the studies overlapped, the results agreed. In addition to being a great confidence builder, Dave's initiative of bringing all available data to bear would turn out to be vital to making our case, and put us in the unassailable position of being the only workers with "the big picture."

Sandy went to Tempe the last week of September for yet one more "last pass through the data." Dave and Sandy went galaxy by galaxy through the entire data set, assigning quality evaluations for every photometric and spectroscopic parameter we had measured. This was preparation for what was coming, the "final data analysis" we had been expecting for two years. Luckily, Sandy carried a magnetic tape with the information back to Santa Cruz, for Dave's computer disk died a savage death soon after she left and the week's work would have been lost.

While Dave and Sandy were finalizing the data set, Donald was busy analyzing the old data set with OURV, the computer program started in Dartmouth. To use OURV he would input an area of the sky in direction and in depth. The program would retrieve the data for galaxies in that region, apply all the corrections, and then solve for the distances of the galaxies as predicted by velocity dispersion and $D_n$, our best correlation. With the distance for each galaxy in the region, OURV then calculated the velocity expected in a smooth Hubble expansion, then subtracted this expected velocity from the observed velocity to find the peculiar velocity. The output was, then, the peculiar velocity of each galaxy and the average peculiar velocity for *all* the galaxies in the specified region.

Donald had added considerable capability to the program since leaving Dartmouth—scientists like to call such embellishments "bells and whistles." Unfortunately he was a computer novice, so the program had grown into something of a rats' nest. And, like most rats' nests, it was crawling with bugs, the itsy-bitsy program bugs that result in small, important mistakes that were hard to catch. Miraculously, Donald was able to use this clever but faulty program and an old data set to progress in measuring peculiar velocities in different regions of the sky, and, importantly, learning how to calculate the error estimates that told how reliable the results were.

We assembled at the Carnegie Observatories in Pasadena in November 1985. Everyone came but Gary, who had both teaching and observing commitments he could not reschedule. As host, I came to appreciate the efforts Donald, Sandy, Gary, and Dave had put in when we came to visit, and I had my chance to reciprocate by preparing a Thanksgiving turkey with accompaniments for Donald, Roberto, and Roger while Sandy and Dave returned briefly to their families. This was the only break during this meeting. So exhilarated were we by what we were doing that other "extracurricular activities" became unthinkable. Pasadena's charm, the Indian Summer, and the towering San Gabriel Mountains, Mount Wilson—with its Olympus-like connotations for us—in full view, were no more than a pleasant backdrop for our most exciting meeting.

Thanks to intensive effort by Sandy and Dave, our final doubts about the data themselves had been put to rest. We were ready to put in the correction for Malmquist bias and recast the peculiar velocities with respect to the "shoreline" of the cosmic microwave background. It was time to make all-sky maps of the Hubble flow, and to measure quantitatively the "how much and where" of the peculiar velocities. I had been playing around with simpler analysis programs I had written, looking for new ways to display the data in maps, charts, or diagrams that would convey our findings clearly and quickly—again, the picture worth a thousand words. This had always been one of my favorite parts of a science project. Now I wrote a computer program that plotted the sky in the manner of the classic "world map projection," indicating by X's galaxies with peculiar velocities away from us and by O's galaxies coming in our direction. Bigger symbols indicated a larger velocity. Plotting these with respect to the CMB frame, we saw galaxies on one side of the sky coming toward us, those on the other side moving away— this was the pattern we expected if these other galaxies were traveling along with us. It was exciting to see this "coming and going" so easily on a map of the whole sky. On this map we also saw the amazing preponderance of very large X's in the Centaurus region of the sky—these galaxies were really moving. We began to refer to this mass motion of so many galaxies, from those around the Milky Way out to the rapidly moving Centaurus galaxies, as a "large-scale streaming" of galaxies.

The now horrendously feature-laden and complex program OURV was the basis for the quantitative discussion of "how much flow and where." Dave's facility at computer programming became crucial as he charged in to exterminate the bugs in what had become Donald's program. In the process he rewrote much of the code in a frenzy that went on for several days. Donald and Sandy worked out the final mathematical form of the Malmquist correction; for the kind of measurements we had made it boiled down to a rather simple prescription of increasing by 17 percent the predicted distance to each isolated galaxy. This lowered the peculiar velocities of the EGAL-SOUTH galaxies by just about the same amount that the shift to the cosmic reference frame had raised them, leaving the final determination in this region unchanged at about 1,000 km/sec. At last we could reliably determine the peculiar velocities of galaxies in any part of the sky.

Donald used the OURV program to make a different kind of map: while mine indicated each galaxy by its direction in the sky, Donald's sliced space into slablike volumes and marked each galaxy by an arrow whose direction showed whether the galaxy was moving toward or away from us, and whose length showed how rapid was the peculiar motion. Sandy continued to make Hubble diagrams, plots of distance versus recession velocity, for different parts of the sky, in which peculiar velocities showed up as deviations from the straight line that corresponded to smooth Hubble expansion.

Each type of diagram had its advantages and disadvantages. For example, with mine you could see the whole sky at once, but you couldn't gauge the distance to the galaxies. An unfortunate feature of Donald's plot was that all the arrows pointed directly toward or away from our own galaxy—the uninitiated were always distracted by this "radiating" effect. This was only an artifact, of course—an unavoidable consequence of the fact that the Doppler shift only allowed measurement of velocity toward or away from our galaxy. There were undoubtedly motions across our line of sight as well, but we couldn't plot them since we couldn't measure them.

In the end, though, it was Donald's map that was the apotheosis of all our years of effort; we spent an entire afternoon discussing one of the first versions. With it we could see our two remarkable results: our galaxy's large peculiar velocity of 600 km/sec was shared by most other galaxies in our survey—the so-called large-scale streaming—and the fact that galaxies in the Centaurus supercluster were moving even faster than we were. As we had suspected in Dartmouth, adopting the cosmic microwave background as our new frame of reference had left all the galaxies moving in more or less the same direction and at the same speed as the Milky Way, the ones in Centaurus even faster. There was no question that we had succeeded in measuring peculiar velocities, because there were coherent patches of a few hundred kilometers per second here and there, coming and going; for example, we could see the infall pattern of the Local Supercluster. But most of the peculiar velocities we had measured were small *with respect to the Milky Way*, and this meant we were all streaming together.*

---

*It is a fine point, one that was difficult to convey even to our colleagues, but finding small peculiar motions *relative to our galaxy* would have been the least likely outcome if we didn't know what we were doing or our methods were faulty. An experiment that measures a "null result" (a number consistent, within the errors, to nothing at all) has a certain built-in credibility, since a faulty experiment usually finds a large "signal"—that's what errors lead to. It would have been nearly impossible to have recovered the Hubble expansion and found only small perturbations from it if we weren't indeed measuring accurate distances to these galaxies.

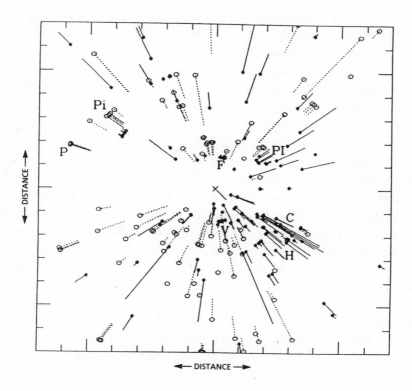

Donald Lynden-Bell's diagram showed the peculiar velocities of galaxies by placing a dot at the position of the galaxy and drawing an arrow in the direction of its motion. The small cross at the center represents the position and peculiar velocity of our Milky Way galaxy; solid lines show galaxies moving away from us and dashed lines show motion toward us. Distances are expressed as expansion velocities, in kilometers per second. Though there are significant uncertainties associated with each point, it is clear that, on average, our galaxy's motion is shared by the elliptical galaxies in the sample. That is, most galaxies are moving from left to right, except for those above and below the cross, for which the left-right motion cannot be detected (the Doppler shift only measures velocity on a line to our galaxy). Especially large are the peculiar velocities at right center, belonging to galaxies in the Hydra-Centaurus (H)(C) and Pavo-Indus (PI) superclusters.

Even more remarkable was that for the one region where we did find large peculiar velocities, the good old EGALSOUTH region containing the Centaurus cluster, these velocities were in the same direction the Milky Way was going. These galaxies were out ahead of us and pulling away, which meant that their speeds were over 1,000 km/sec, an extraordinary amount. This was to have important implications that we couldn't yet appreciate.

For the next few days the group experimented with the best ways to describe this result, and diagrams that would best show what we had found. We decided that the simplest description would have the best chance of being absorbed by our colleagues. The OURV program was asked to calculate the simple average peculiar velocity with respect to the CMB—how much and in what direction—within a large sphere centered on the Milky Way, a sphere about 500 million light-years across that included hundreds of elliptical galaxies. This *average* peculiar velocity we called the "bulk flow." According to the program, the elliptical galaxies within this enormous volume, and by implication the thousands of galaxies of other types mixed in with them, shared a common motion of about 600 km/sec (with a probable error of 150 km/sec) toward the direction in the sky where one sees the Southern Cross. (This, of course, is a constellation of stars *inside* our galaxy; it serves only to mark a direction.)

Now that we had this notion of bulk flow it was time to start thinking about the cause of all these peculiar velocities, the departures from smooth Hubble expansion. We adopted the conventional view that these motions must be the result of gravity, due to the pull of an excessive amount of matter, somewhere. Our result of a huge coherent flow of 600 km/sec was radical because it said that the source of the gravitational pull must be very far away. If the source of gravity were at the center of this sphere, for example, close to our position, then galaxies all over the sky would be approaching us. If the source of gravity were elsewhere, but still within the region we had mapped, we would still expect to see the arrows of peculiar velocity converging toward this region, as we saw in miniature for the infall toward the Virgo cluster. Instead of this, our data were consistent with an even simpler picture: that all of our sample galaxies were moving in the same direction, along with the Milky Way. If so, the source of gravity must be at least as far as the farthest galaxies we had sampled, and possibly much farther. Other astronomers had suggested that the Centaurus cluster marked the center of such an infall, but our results showed that galaxies in this region were moving in the same direction as, and more rapidly than, we are. If Centaurus were "reeling us in," we would have measured it as approaching, it wouldn't have been "pulling away." I had guessed right almost three years earlier, in the Coma-Virgo paper I had written, when I predicted that the ripples in the mass distribution that would give rise to peculiar velocities, like the 600 km/sec peculiar velocity of the Milky Way galaxy, were of enormous scale. I was delighted.

Our findings, it was decided, would be made public in January, first by Sandy Faber at the national meeting of the American Astronomical Society, where Sandy was to give an acceptance speech for the Daniel Heineman Prize, and later by Donald Lynden-Bell in his presidential address to the Royal Astronomical Society. Roger and Dave flew to Houston to hear Sandy's talk; after the excitement of our previous month's work, the three were disappointed but not really surprised when our "announcement" was greeted with near indifference. They knew that few astronomers in the audience were familiar with this complex, relatively new type of measurement, and later we learned that almost all who understood peculiar velocities and their implications were attending a cosmology workshop in Aspen, Colorado. A prepared press release in Houston attracted not a single reporter—they were caught up in the hoopla surrounding official unveiling of the CfA "slice of the universe." The large-scale streaming flow was a fuzzy, hard-to-grasp abstraction compared to the immediacy of this stunning galaxy map. If astronomy stories rated headlines in *Variety*, this one would have read LARGE-SCALE FLOW NO DRAW!

Though the public debut of large-scale streaming of galaxies was a fizzle, our unanticipated firecracker was about to make quite a pop. Dave flew to Kona, Hawaii, to address a conference organized by Brent Tully. Many other pundits of the field flew there, too, including those who had been cogitating and skiing in Aspen. They were looking forward to a three-way heavyweight match—Allan Sandage, Gerard de Vaucouleurs, and Marc Aaronson fighting about the expansion rate of the universe, the elusive Hubble constant that had been a source of debate for so many years. What they didn't expect was a bombshell that the Hubble expansion wasn't even very smooth, with all its implications for large-scale structure of the universe. Worried until the very end that we would discover some horrible mistake in our analysis, we had kept our results completely to ourselves, and our debut in Houston had been such a flop that our work stayed an easily kept secret. So Dave's talk turned out to be the first to an audience that could understand the implications of what we had found, and it caught researchers in the field very much by surprise.

Dave described the reception of his presentation as stunned silence—he could feel attention riveted on him, and a tension he had not experienced in his years of giving such presentations. He reported that the audience froze when he reached the part about the large peculiar velocity of the Centaurus cluster and its supercluster environs, how we expected it to be coming toward us (reeling us in), but, on the contrary, it was moving rapidly away. After he finished, the audience exploded with questions, and, after fifteen minutes, when session chair Gerard de Vaucouleurs attempted to cut off discussion, the audience protested. The question period went on until Dave's talk had run over its allotted time by forty-five minutes, an unheard-of event at such a conference. For the remaining two days our account of widespread, large peculiar velocities was the talk of the conference. Colleagues who attended confirmed that, despite the melodrama of Dave's offputting pregnant pauses, his talk was indeed *the* event of the meeting.

Subsequent presentations by other members of the group went well, too, and soon the community was abuzz with our new and surprising result. We wrote a second press release; this one was picked up by a few dozen newspapers and many of us spent hours with reporters trying to get across the concept of peculiar velocities and large-scale streaming of galaxies. This was quite a challenge since few even understood the Hubble expansion let alone departures from it, and what these might mean. However, even with attention from our colleagues and the media, our work remained at "second fiddle" status compared to the "slice of the universe," the stunning map that showed a bubblelike structure so clearly that it had removed all doubt that the universe had a more complex structure than had been imagined. With two important new results on large-scale structure, one visually stunning and immediate, the other complex and difficult to convey, it wasn't surprising that the slice of the universe remained the center of attention.

But we seven were not overly disappointed. We knew we had brought off one of the largest astronomical collaborations ever attempted, and actually discovered something fundamental—this was satisfaction enough. And we were beginning to sense that our work might have greater long-term impact on the study of large-scale structure, because it could be described in quantitative terms—it was much more than a picture. Already Marc Davis and Michael Strauss at Berkeley and Amos Yahil at SUNY Stony Brook were claiming that the large-scale streaming we had found contradicted what they predicted from an all-sky map they had made of the distribution of galaxies. Other astrophysicists, who were making theoretical models of the growth of large-scale structure, were also raising shrill voices that the large, extensive peculiar velocities just could not be so.

When Dave, Donald, Roger, and I joined Sandy in July 1986 at the

Santa Cruz workshop she had organized—"Nearly Normal Galaxies"—we found ourselves the focal point of a lively debate on how the universe is organized on large scales. We had slashed through what little conventional wisdom had developed on the subject and disrupted the civil discourse—the orderly march to truth. Yahil put it best when he took the podium and chided our group's disrespect for authority. "What are we to do"—he scowled—"with these seven, these, these . . . these Seven Samurai?"

# 8 • HEART OF DARKNESS

Proverbial wisdom tells us as much about human hopes as it does about human experience. "What you see is what you get" is one of those old saws, invoked more often than not by operators of some shell game who would like us to believe that it's true. As all of us have been victims at one time or another, it should have come as no surprise that most of the universe is invisible. But it did.

This ultimate deceit was first exposed in 1936 by Fritz Zwicky, destined throughout his career, it seems, to make unpopular discoveries. Zwicky was measuring redshifts from the spectra of galaxies in the Coma cluster. Because these galaxies reside together—at a common distance—they should have shown the same redshift due to Hubble expansion, but Zwicky found there was a considerable spread in their speeds. He correctly reasoned that the redshifts of these galaxies reflected not only the flight of the Coma cluster in the Hubble expansion, but also additional motions *within* the cluster as galaxies swarmed against the combined pull of gravity from the other galaxies. In a sense, he was making early measurements of peculiar velocities—departures, induced by a cluster's strong gravity, from the Hubble relation of redshift and distance.

Zwicky expected to find such an effect; he had even calculated how big the spread in speeds should be: he knew the distances between these galaxies*

*The distances between galaxies were determined from the angular separation on the sky coupled with an estimate of the cluster's distance in light-years, which came from Hubble's constant relating a certain number of light-years for each kilometer per second of expansion velocity. Although this value of Hubble's constant was far from its modern value, the sense of what Zwicky found was correct.

and had a rough idea of how much mass (in stars) each galaxy contained. From these quantities and Newton's laws of gravitation and motion he expected that the typical speeds of galaxies in the Coma cluster (the velocity dispersion) should be a few hundred kilometers per second.

They aren't. Zwicky found that typical speeds of galaxies in clusters are around 1,000 km/sec. This amounted to an energy of motion ten times larger (kinetic energy scales as the *square* of the velocity) than needed to balance the pull of gravity. Zwicky could have argued, as Armenian astronomer V. A. Ambartsumyan was later to do, that clusters like Coma are flying apart—that he was witnessing the cataclysmic birth and explosive expulsion of galaxies. Instead, he considered the hardly-less-radical idea that the total mass of the Coma cluster was at least ten times greater than what could be seen in the form of luminous stars in galaxies. If so, most of the matter was invisible.

In the decades that followed, other clusters were investigated; high velocities were always found. When studies were made of pairs of galaxies in orbit around each other, an unusual but not rare occurrence, they too were found to be pulling on each other with more force than the mass in stars alone could provide. Observations of these mundane duets eroded what meager support there might have been for the "explosive formation" model, and little by little astronomers began to accept the idea that much of the matter in the universe is *invisible*—dark matter that is either within galaxies or filling the spaces between them.

How could one measure the mass of a galaxy to find out if it contained more matter than that seen in stars, or if, alternatively, invisible matter resides *outside* galaxies? The answer lay in measuring the motions of stars or gas in the galaxies themselves. In 1916 and 1918 Mount Wilson astronomer F. G. Pease, following up on early work done by Vesto Slipher, took spectra of two large "spiral nebulae" (as they were called before it was agreed that they were external galaxies) and found them to be rotating, turning at the incredible rate of once every hundred million years or so. (This is incredibly slow in our perception, incredibly fast if you consider how large a galaxy is.) Pease's early photographic plates were so insensitive that a single measurement required eighty hours of exposure spread over three months. In the late 1950s Geoffrey and Margaret Burbidge realized that observations like this could "weigh" a galaxy—measure its mass, just as Zwicky had weighed a cluster of galaxies. By this time there were better spectrographs, more sensitive photographic emulsions, and larger telescopes—the Burbidges could do better. Following Pease's technique of feeding light into the spectrograph through a long slitlike aperture, they sampled light starting at the center and then out along the face of each galaxy. In order to measure the Doppler shift at different points along the galaxy, the Burbidges centered on the red-orange light coming from glowing hydrogen gas heated by newly formed stars—the *Hα emission line*. Like Pease and Slipher, they too found that, compared to the spectrum of the center of each galaxy, the spectrum from one side showed an additional small redshift, the other side, a small blueshift. The redshift of the center of the galaxy corresponded to the velocity of the galaxy as a whole in the Hubble expansion. The much smaller red and blue shifts on either side showed that the galaxy was rotating: from our point of view, stars on one side were approaching, on the other side receding. The Burbidges mapped the velocity at each point across the face of a galaxy, producing what is called a *rotation curve*.

Again, applying Newton's laws, the mass required to account for the orbital speed at any distance from the galaxy's center could be readily calcu-lated. The Burbidges found that the masses they determined this way were no greater than what the combined mass of all the stars should be—in other words, there was no evidence for "dark matter." But they hadn't been able to reach very far out in each galaxy, even with the better instrumentation. A decade later Vera Rubin was able to go further, using an even more efficient spectrograph at the new four-meter Mayall telescope at the Kitt Peak National Observatory. Rubin, who had worked with the Burbidges before she came to the Carnegie Institution's Department of Terrestrial Magnetism, measured the rotation curves of spiral galaxies right out to their rims. She found that the speed of rotation reached a maximum not far from the galaxy's center and then leveled out all the way to the visible edge.* But her simple calcula-tions showed that the rotation speed should *fall* if all the mass were in ordinary stars. Based on her data, Rubin popularized the idea that there was additional unseen matter that was providing yet more gravity and keeping the rotation speed high far from the galaxy's center. Her model calculations were too simple, it would turn out, but she was on the right track.

The clinching evidence came from radio-telescope observations by Albert Bosma at the University of Leiden, Netherlands, and by University of Vir-

---

*For the sake of historical accuracy, it is interesting to note that Horace Babcock, the Carnegie director who founded the Las Campanas Observatory, measured the rotation curve for the Andromeda galaxy in his Ph.D. thesis research in 1938. He was the first to show that the rotation curve did not turn down, as expected, but continued high out to the limits of the visible disk. But the result was "buried" in the *Lick Observatory Bulletin* and went nearly unnoticed by the astronomical commu-nity, who, like Babcock himself, would have been hard pressed to grasp its signifi-cance at the time.

ginia astronomer Mort Roberts. They were able to use huge radio-frequency antennas to receive and record the radio light that came not from stars but from the cold hydrogen gas within spiral galaxies. Like Rubin, Bosma and Roberts had also measured Doppler shifts of emission-line light of hydrogen atoms, but this time the signal came from neutral (non-ionized) H I gas at a wavelength of 21 cm. Tracking the rotation of cold gas far *beyond* the visible edge of many spiral galaxies (in some cases out to the remarkable distance of 100,000 light-years), each found that the rotation curves remained "flat," as Rubin found over the visible part of the galaxy, but well beyond the visible edge of the starlight.

The radio-telescope observations had provided conclusive evidence of large amounts of matter not in ordinary luminous stars. The reasoning was as follows: beyond the point where the light drops off sharply—what looks in a photograph like the galaxy's edge—there should be very few luminous stars and thus very little additional mass. As a consequence the rotation speed should fall due to the declining force of gravity. (We find this in the solar system, for example, where nearly all the mass is concentrated at the center, in the Sun: the orbital speeds of the planets drop as the square root of distance from the Sun.) But—and this is the remarkable thing—the rotation curves obtained with radio telescopes showed that the velocities do not drop at the edge of spiral galaxies. Gas orbiting the galaxy, even beyond the visible edge, circles at the same speed as material farther in. This led to the unavoidable conclusion that there is a great deal of mass beyond the visible edge of a spiral galaxy, which keeps the force of gravity high and the orbital velocities from falling. Like Zwicky's observation, this was evidence for additional, unseen mass—dark matter.

Now there was evidence not just for the existence of dark matter, but for where it is. At least some of it is associated with galaxies, but little or none of the dark matter is within their luminous boundaries. The picture, then, is one of huge balls of dark matter in which luminous galaxies reside, their stars illuminating only the center of a much larger entity. Interestingly, this model had been proposed by Peebles and Jeremiah P. Ostriker, another Princeton astrophysicist, from completely different reasoning. They pointed out that the thin disks of spiral galaxies would warp into potato-chip-like surfaces and break into pieces unless they were stabilized by the gravity of a larger, enveloping mass. Because, on the contrary, spiral disks appear smooth and stable, Ostriker and Peebles had proposed that halos of dark matter must be present to steady them.

By the 1980s the search for dark matter had expanded to the realm of the supercluster. For the Local Supercluster, Aaronson and his collaborators found a general "infall" of galaxies, including the Milky Way, toward the Virgo cluster,* and Amos Yahil, Allan Sandage, and Gustav Tammann used this information to "weigh" the supercluster. This application of Newton's laws is less than secure because, unlike the cases of rich clusters or individual galaxies, the boundaries of a supercluster are indistinct and the size and shape of the distribution of galaxies is still changing—a stable balance of gravity and motion has not yet been reached. Nevertheless, measurements of peculiar velocities of a few hundred kilometers per second for galaxies in the Local Supercluster left no doubt about the presence of dark matter, minimally ten times as much matter as could be seen in the luminous galaxies themselves. In 1986 the *Seven Samurai*—the name Yahil had given us had stuck—had measured even larger peculiar velocities over a region of space five times as large as the Local Supercluster. Again, the unavoidable conclusion was that most of the matter within that volume is invisible.

All of these studies have found solid evidence for dark matter, but it is important to keep in mind that there is *no evidence* for dark matter within our solar system, and little or none (depending on which studies one believes) within the central, luminous parts of galaxies. Within the confines of the luminous part of a galaxy, the gravitational force can be completely accounted for by the mass of observed stars and gas. On the other hand, *every* measurement that has been made from the edge of a galaxy and beyond into intergalactic space has come to the same conclusion: there is at least ten, perhaps as much as a hundred times as much invisible mass as can be seen within galaxies. There are no exceptions. And much of it, perhaps all, is draped in invisible cloaks around visible galaxies.

The claim that most of the matter in the universe can't be "seen" is an extraordinary one; as such, it must be the object of skepticism and rigorous scrutiny. It is appropriate, then, that a few astrophysicists have been investigating the possibility that evidence for dark matter is a fiction caused by a breakdown of Newton's law of gravitation over enormous distances: the models propose that when gravity is extremely weak, its force falls off more slowly than the "distance squared" of the Newtonian formulation. However, these models have failed to provide a fully consistent description of both the small- and large-scale properties of the universe, and proponents have been unable to construct a counterpart to Einstein's general theory of relativity

---

*Infall is really a misnomer because in most cases the expansion of the universe is carrying galaxies in the Local Supercluster away from each other faster than gravity can pull them back. Except for those galaxies close to the center, which are truly infalling, it is more accurate to say that the expansion of the Local Supercluster has been *slowed* by the pull of gravity in this overdense region.

for their revised universe. Furthermore, no rigorous tests have been suggested that could disprove the hypothesis that Newton's law of gravitation fails on cosmic scales. Identification of a test that could *invalidate* the hypothesis is crucial to the scientific method: without a prediction of new phenomena, a hypothesis is just a description of what was already known, and thus a dead end.

Accordingly, most researchers in the field are prepared to believe that the simple application of the law of gravity has uncovered one of the most remarkable facts yet learned about the universe: what we see is very little of what there is. As such, it is easy to see why many astronomers now consider the identification of *dark matter* as perhaps the central question in cosmology.

What an extraordinary situation that human beings could deduce that their universe is mostly invisible, especially when one considers that there is no direct contact with this larger world. And, even more remarkably, the story doesn't end here. It provides a good example of how, through pluck and ingenuity, we can discover incredible details about a universe seemingly so far beyond our grasp. We can ask questions, make tests, assemble evidence and counterevidence that begins to unmask what seems at the outset to be a totally unknowable secret: what *is* dark matter?

The discussion focuses on two alternatives: (1) dark matter is just ordinary matter, made of the same stuff as our Sun (or the Moon, or the parakeet . . .), but, unlike the Sun, this matter gives off little or no light; or, alternatively, (2) dark matter is wholly different material, for example, an undiscovered relative of one of the exotic particles that emerge when we bash atoms together at extremely high energy.

Dark matter in ordinary form is called "baryonic matter," *baryon* being the collective name for protons and neutrons, the heavy particles that make up the nuclei of all atoms.* Before the invention of telescopes that "see" a broader range of the electromagnetic spectrum—radio waves, infrared and ultraviolet light, X rays—it was easy to entertain the notion of huge quantities of invisible baryonic matter. If they produced little *visible* light, even huge numbers of baryons would go undetected by conventional optical telescopes. For example, a dilute, cold gas around or between galaxies was thought to be a likely form of dark matter, but such gas would give off copious energy in radio waves and would be an effective absorber of the light from distant galaxies. Observations have found no evidence of either effect. Alternatively,

---

*When we speak of baryonic matter we implicitly include electrons, the negatively charged particles that orbit the positively charged baryonic nucleus. Technically, electrons are actually members of another class of particles, called *leptons*, but are so light that they contribute only 1/2000th of the total mass—thus the snub.

it was suggested that the extra matter could be in the form of a dilute, extremely hot gas—*ionized* (separated into protons and electrons) hydrogen. However, if so, this gas would also absorb light coming from the extremely distant quasars and would be expected to give off a detectable, uniform glow of ultraviolet or X-ray light. Again, modern observations have ruled out this possibility.

When confronted with the idea of dark matter, most people immediately think "black hole," that classic example of a lot of mass that gives off practically no light. We know that a black hole is produced when the core of a very massive star collapses, blasting its "outsides" into space. The gravitation field of the collapsed star is so strong that space itself is completely bent around, as was predicted by Einstein in his theory of general relativity. Light trying to leave is forced to follow this contorted space and is unable to get out, thus the hole is "black." This is an effective way to concentrate a large baryonic mass and produce no light that would give its presence away.

Could such black holes be the dark matter? Some astrophysicists speculated that the early universe funneled most baryonic matter into the formation of massive stars, leaving most of the material in black holes when these stars perished as supernova. Interesting, but how could we possibly know if this is true or not? Easy. Calculations have now proven that the sum total of all the heavy chemical elements released by such supernovae eruptions, particularly such atoms as silicon, nickel, and iron, would have far exceeded the abundance of heavy elements found in galaxies today, hence, this possible form of dark matter is also ruled out. However, the idea is still entertained that dark matter is in the form of even more massive black holes, each with the mass of a thousand or even a million Suns. The idea here is that these are primordial black holes—born shortly after the big bang, before ordinary stars were formed. Such massive objects would have collapsed completely—there would be no supernova explosion spewing heavy chemical elements into space—so the previous argument can't rule out this possibility. Even though this model may be unlikely and contrived, it is worth investigating. Experiments are now under way to use the warping of space, the very effect that makes a black hole what it is, to test for the existence of such monsters and, very likely, rule them out.

If not in gaseous form or black holes, could dark matter in baryons be hidden in other compact objects, like stars? After all, that's mostly what we see in the sky. No, for just that reason, dark matter cannot be "ordinary" stars, which are far too luminous to be hidden as "dark matter." However, stars much less massive than our Sun emit only a feeble glow of infrared light and therefore would be *invisible* in the strict sense of the word. Astronomers have been madly searching for these low-mass, barely glowing stars (called "brown dwarfs") by using sensitive infrared detectors placed at telescopes on Earth and in space. Too few have been found to account for the dark matter we see around galaxies, but true believers in this solution to the dark matter "problem" still insist that unseen hoards lie just below the detection limit, and will be discovered with the next generation of Earth-orbiting infrared telescopes. Infrared detectors have already been able to rule out extremely small bodies, like dust grains, as the dark matter; these would produce a detectable glow. So what remains is a very small range in size, from bodies with about ten times the mass of the planet Jupiter—a failed star (or barely successful one, depending on your point of view)—down to boulder-sized rocks. The possibility that these are the dark matter bodies does have its promoters, although most researchers consider it unlikely and ad hoc that nature, with such an enormous range available, would have conspired to put almost the entire mass of the universe in objects within their relatively small mass range.

The resourcefulness displayed in the search to identify dark matter is impressive. A new type of observational program promises to find out how much of the mass in our galaxy is bound up in such compact objects, particularly in objects as small as Jupiter or as "large" as the diminutive brown dwarf stars that give off so little light. In his general theory of relativity, Einstein showed that all compact bodies induce, through their mass, wrinkles in space that deflect passing starlight. The light of more distant stars will appear to flicker slightly as the light is bent by the warped space around matter concentrations, as they pass between us and the background star. This subtle effect will be detectable if most of the dark matter in the universe is in such compact form. The first experiments of this type have actually reported several "events"—stars whose brightnesses are being monitored have been seen to brighten and then dim within several days. This flaring is the signature of a compact body passing in front of the star, and if the brightening happens only once, this rules out the possibility that the star in question is by nature variable in brightness. These early reports provide persuasive evidence that low-mass compact bodies do exist. However, the question remains as to whether there are enough of them to account for the dark matter implied by the motions of stars in galaxies or of galaxies themselves. This important verdict is not in yet, but it should soon be possible to fully raid these remaining "hideouts" and either confirm or deny the possibility that vast amounts of dark matter exist in compact baryonic form.

In a campaign as important as this, it is crucial to have more than one line of attack. While direct observation has been investigating the hypothesis that dark matter is made of ordinary baryons somehow hidden from view, a very different approach has been used to argue that baryonic dark matter is not the answer. The issue revolves around the observation that, as best we can measure, the universe has a near-perfect balance between kinetic energy—the energy of the Hubble expansion—and gravitational energy, the mutual pull of all the matter. Among other tests, the observation that the Hubble relation between redshift and magnitude follows a straight line for very distant galaxies shows that these two energies are approximately in balance. It is extremely unlikely that such parity occurred by accident, given that *imbalance* is incredibly more likely.* This prompts many astrophysicists to conclude that the laws of physics at work in the big bang must have dictated an *exact* balance. As I discussed earlier, for a state of equality between the gravitational and *kinetic* energy the average density of matter in the universe is specified precisely, a value known as the *critical* density. The ratio of the actual density of the universe (measurements of which are still uncertain by a factor of ten) to this critical density is denoted by the Greek letter $\Omega$ (Omega): $\Omega = 1$ specifies the condition when gravity exactly balances the energy of the Hubble expansion.

If we know how rapidly the universe is expanding, we can calculate what the critical density of matter must be for a state of balance. This can then be compared to the amount of matter we actually see or can sense in other ways. Having done this, we find that luminous stars account for about one percent of the critical density, that is, $\Omega_{stars} \approx$ (approximately equals) 0.01. Therefore, the gravity from stars alone falls far short of balancing the Hubble expansion. The dark matter around and between galaxies implicated in studies of the rotation of spiral galaxies, the swarming of galaxies in rich clusters, and the infall of galaxies into the Local Supercluster amounts to ten or twenty times as much mass. Hence, including this "unseen" mass around galaxies, which could possibly be in baryonic form like faint dwarf or failed stars, gives us $\Omega_{galaxies} \approx 0.1$ to 0.2. But this is still substantially less than the critical density, $\Omega = 1$, and furthermore, it is the *absolute limit* (and probably more) of what can be present in baryonic form.

The surprising confidence of this claim that no more than 10 to 20 percent of the critical density could be in baryonic matter comes from comparing predictions of the big bang model with measurements of the relative abundances of the lightest elements: hydrogen and deuterium, helium, and lithium. Along with the stunning prediction of the cosmic background radiation, the big bang model also predicts with remarkable precision the *primordial*

---

*Recall the earlier discussion regarding the extreme difficulty in pitching a baseball from Earth with just the right energy so that it comes to rest at the orbit of Jupiter.

abundances of these elements, that is, the relative numbers of these simple atoms before stars appeared on the scene. Later, stars would alter the chemical mix through their alchemist magic of nuclear fusion, producing large amounts of helium and heavier elements. But long before this, nuclear fusion had first occurred *at large,* when the entire universe resembled the inside of a star. For the first three minutes after the big bang the rapidly expanding universe was hot enough and dense enough to fuse protons into deuterium (hydrogen with an extra neutron), helium-4 (common helium), helium-3 (uncommon helium with one less neutron), and lithium-7. According to very detailed calculations, about 24 percent of the protons were converted into helium-4 nuclei (two protons bound with two neutrons). The predicted fractions of primordial deuterium, helium-3, and lithium-7 produced are tiny in comparison, but significant and measurable. After the first three minutes, the universe had cooled and thinned sufficiently that nuclear fusion stopped, "freezing in" these proportions of the light elements.

The helium-4 fraction has been measured in very small galaxies where the pristine gas clouds have remained nearly "unpolluted" by supernovae explosions. Lithium fractions have been measured in first-generation stars in our own galaxy, and the percentages of deuterium and helium-3 have been best determined from the *solar wind*—the breeze of particles streaming from the Sun. For all these measurements there is stunning agreement with big bang predictions, but only if the total amount of baryonic matter in the universe is well *below* the critical density, i.e., $\Omega_{baryon} < 0.1$. If the density of baryonic matter were any higher than this, collisions between protons in the first three minutes after the big bang would have been so frequent as to produce more than the observed amount of primordial helium and other light elements. Both observations and theory are difficult here, but most pundits agree that the constraint $\Omega_{baryon} < 0.1$ is extremely hard to overcome.

We now put one and one together: if we are convinced that the universe has critical density $\Omega = 1$ (for the universe must be in *perfect* balance if our measurements show it is in approximate balance), and we further accept the evidence from the abundance of light elements that $\Omega_{baryon} < 0.1$, we must conclude that most of the matter in the universe is not in baryonic form.

To review, two types of evidence support the view that dark matter is not in the form of baryons. One is that large numbers of baryons in "invisible" form have not been "seen" at other wavelengths of the electromagnetic spectrum—doubtful contrivances seem to be necessary if such huge quantities of baryonic matter are to be hidden, though some contend it is still possible. The second rationale issues from the measured abundance of light elements compared to what could have been produced in the big bang (nucleosynthesis), a comparison that strongly requires $\Omega_{baryon} < 0.1$. If we believe that the universe has the critical matter density $\Omega = 1$, a point of view that many theoretical astrophysicists (but not as many observational astronomers) find compelling, then non-baryonic matter must dominate the universe.

It is because these two arguments are not trivially dismissed that astronomers have been forced to consider seriously the astounding proposition that most matter in the universe is something other than what we and our world are made of—that galaxies, with their stars and planets, are highly condensed piles of baryons floating in a sea of other "stuff." This fantastic notion has captured the imaginations of explorers of the largest and smallest realms—cosmologists and particle physicists—who hope that properties of the universe-at-large can confirm the existence of very different forms of matter and provide clues to its origin. To some, this has been the unlocking of a treasure chest, to others, the opening of Pandora's box.

Atoms are made of members of one species in the "zoo" of *particle physics*, the study of nature's building blocks. Protons, neutrons, electrons, and photons (light rays) are the most common of these particles. In atomic matter, positively charged protons are bound to negatively charged electrons by electromagnetic force, which is conveyed by the massless photon. Protons and neutrons are tightly lashed together in the atomic nucleus by what is called the strong (nuclear) force.

In addition to these three, however, a plethora of other particles have been discovered. Most are brief visitors to our world—their lifetimes are measured in infinitesimal fractions of a second. Others are long-lived—physicists say *stable*—like the proton, but interact neither through the strong nor electromagnetic force, but only through the very tenuous "weak" force. One such stable, weakly interacting particle is the *neutrino*, a unit of little or no mass capable of carrying great amounts of energy at the speed of light, or close to it. Much of the energy released through fusion reactions in the centers of stars is carried off by neutrinos instead of photons.

Neutrinos and photons behave very differently. Photons interact so strongly with ordinary baryonic matter that they undergo multitudinous absorptions and re-emissions before leaving the star, while neutrinos interact so weakly that those born at a star's center travel unimpeded to the surface and out, at the speed of light. Every second about a hundred trillion neutrinos from the center of the Sun pass through your body; that means that there are almost a hundred thousand, about the number of spectators at a Super Bowl, inside your body at this and every instant. While you sleep neutrinos come up through the bed, entering on the sunny side of the Earth and boring effortlessly through the Earth. They're everywhere.

Neutrinos were "discovered" in 1931 by the brilliant Austrian physicist Wolfgang Pauli, with his *mind*. Pauli was studying the way certain atoms suddenly transform into closely related but different atoms, a process called radioactive decay. He noted that an important component was missing in the process of "beta-decay," the specific process where one of the neutrons in an atomic nucleus spontaneously turns into a proton. Simultaneously an electron emerges from the nucleus, thus, both a positive and negative charge have been lost from the system—physicists say that the charge has been *conserved*. But Pauli realized that another particle must also be expelled if *momentum*—mass times velocity—is also to be conserved. Since charge was already in balance, Pauli reasoned that the mystery particle had no charge—it had to be *neutral*. Enrico Fermi later gave the mysterious particle its name: *neutrino*—little neutral one.

Though neutrinos sound fanciful and exotic, they are real. Neutrinos are routinely detected: their weak interaction with baryonic matter makes snagging one extremely difficult, but not impossible. (We have also learned that there are three types of neutrinos: in addition to the one Fermi described, a partner to the electron, there are two more types that are partners to the tau and muon, the other members of the lepton family.) When, in 1987, light arrived from a supernova in the Large Magellanic Cloud (some 160,000 light-years away, and yet the closest such event in four centuries), neutrinos from the explosion arrived simultaneously. The detection of these ghost particles in huge underground water tanks (built for another purpose—to trap a minute fraction of the neutrinos coming from the center of the Sun) provided a remarkable confirmation of theoretical predictions that most of a supernova's energy is carried off by neutrinos.

That we and our world are made of protons, neutrons and electrons has led to a remarkably chauvinistic attitude about the composition of the universe at large. How could the universe be made of anything else? In fact, we know that in at least one respect it is. By sheer numbers the components of normal atoms—baryons and electrons—are overwhelmed by other particles. For example, for every baryon in the universe there are about one billion photons. The cosmic microwave background that we spoke of earlier accounts for most of these photons, but even the photons released from stars far outnumber the baryons. It is a certainty that neutrinos, too, outnumber baryons, and, if they do make up the invisible matter in the universe (something we don't know) they are even more abundant, about as plentiful as the photons in the CMB. Therefore, though the idea that the universe is mainly composed of "other stuff" seems a strange one, it is undeniably true, at least in terms of *numbers* of baryons compared to other things like photons and neutrinos.

Coming to grips with this reality makes it easier to give a fair hearing to the possibility that one of these "other particles" might dominate *by mass* as well as by number.

The neutrino is a prime candidate for the dark matter that dominates the universe. However, in order to actually *be* the dark matter, each neutrino must have a small mass so that, bit by bit, they sum to the additional mass whose gravity is revealed by the motions of stars and galaxies. Neutrinos have been thought to be absolutely massless, like the photon—just packets of energy traveling at the speed of light. If so, they will not do. But if neutrinos have just a tiny mass, they could be the carriers of most of the mass in the universe, according to theories that describe the production of elementary particles in the big bang. Direct laboratory measurements of the mass of the neutrino (in "particle accelerators," where protons or electrons are forced to collide at enormous energies) have shown conclusively that the neutrino does not have a large mass like a proton or electron. However, to be the dark matter, only a tiny mass—about 30,000 times lighter than an electron—is required for the neutrino. This has not been ruled out by experiment. Of course, if each neutrino has such a small mass, it is easy to see that they must vastly outnumber baryons if they are to dominate the total mass of the universe.

Neutrinos are not the only dark matter candidates to arise from the study of elementary particles. For example, a new and still highly speculative theory of high-energy particle physics called *supersymmetry* predicts that, just as the electron has a "partner," the neutrino, so might the photon have an as-yet-undiscovered partner, the photino. There are many such possibilities for new particles in the supersymmetry model, but the model is far from achieving general acceptance, and none of these particles is actually known to exist (though experiments are currently being designed to try to detect some of these kinds of particles). Nevertheless, the possibility is entertained because these particles *could* exist and, unlike the neutrino, one or more types might be *very massive*, like the proton. This is important because, as we will see, models that start with this kind of weakly interacting, exotic dark matter particles predict different-looking universes, depending on whether the particles are light, like neutrinos, or massive, like protons.

Speculating about a universe filled with an invisible "gas" of weakly interacting particles is all well and good, but how can this idea be tested? The best way would be, of course, the direct way, setting up laboratory experiments on Earth that could actually catch a minute fraction of these particles as they cross through our world. The first such experiments are just now under way, but they are unlikely to succeed anytime soon, not only because these particles are, by nature, extremely difficult to trap but also because the net must be cast very wide to encompass the wide variety of candidates with their different properties. Still, we might get lucky. Another possibility is that one of the huge *accelerators* that create exotic particles from the concentration of enormous energy might discover one of the candidate particles for dark matter, or show that the neutrino does have the infinitesimal mass required. Unfortunately, even with the new, more powerful machines that are being built, this too is unlikely to be just around the corner.

Because direct evidence has been hard to come by, it is well worth looking for other clues. Could the universe itself, as the ultimate experiment in high-energy physics—the big bang—tell us what kind of particle it is constructed from? Are the voids and superclusters themselves evidence of how and from what the universe was built? Consider: if the universe started with an almost-smooth sea of such particles that, over time, became more and more clumped through the concentrating influence of gravity, then its appearance might be sensitive to the amount of such matter and, in a rudimentary way, to its properties. To test this idea one builds a "model universe" in a computer, assembling it from baryonic and/or exotic particle matter and simulating the action of gravity in developing structure. This kind of research became a growth industry in the 1980s, as ideas and technology, i.e., computers, rose to the challenge of modeling a universe.

To model a universe that developed from tiny variations in density, one needs an input pattern of representative fluctuations and a computer to follow gravity's work. We assume that the universe grew from a very smooth, but not perfectly smooth, distribution of matter—the uniformity of the cosmic microwave background is excellent evidence that this is true. For these models we will consider most of the mass is carried not in baryons but in weakly interacting particles: they feel the gravitational pull from each other but cannot bind together to form anything like a star, nor will they ever. A smaller fraction of the total mass is in baryons, too hot at this early time to coalesce into stars but destined to do so when the universe cools sufficiently. Since all matter responds to gravity, any deviations from a perfectly uniform density, for example, ripples left over after the big bang, will grow in amplitude. Regions with a slightly higher-than-average density of matter will resist the expansion of the universe and stay dense while regions of lower-than-average density will become more sparse. Gravity's work will inexorably increase the contrast.

What would be the most general form for the input pattern of fluctuations?

For one thing, it should contain a fair amount of randomness; even though galaxies are highly clumped, they are not arranged in neat patterns like rows or circles or checkerboards. We also know that the initial variations in density must occur over a wide variety of scales because we see structures ranging from smaller than the size of a galaxy (including smaller star clusters and stars themselves) to larger than the size of a supercluster of galaxies. A fair amount of randomness and a wide variation in size is, of course, nature's favorite pattern. For example, every mountain range includes a mishmash of big and small, high and low peaks. The bigger mountains themselves are not smooth but covered with smaller and smaller knolls and valleys.

Any given mountain range can be described mathematically as the relative number of peaks and valleys as a function of how big they are. One such mathematical expression that appears consistently in nature has the characteristic that the total amount of "material" in a fluctuation is directly proportional to its size. For example, if you were to disassemble a mountain range (don't try this at home) you would find that twice as much mass is locked up in mountains 10,000 feet high as in mountains 5,000 feet high. If asked to build a realistic model of a mountain range, from rocks, say, you would discover that a good kit would contain 4 tons of 4-foot rocks, 3 tons of 3-foot rocks, 2 tons of 2-foot rocks, etc.* This popular pattern for nature's construction work is called *1/f noise*, a name it acquired because it was first recognized that the power in a random electrical signal, like the "hiss" on an AM radio, has this form. "Noise" is a term commonly used to describe a jumble of signals, and *f* stands for frequency (so *1/f* is wavelength, which is equivalent to what we've been calling "scale," or simply, size). Traffic sounds and ocean waves are other examples, though less pure, of 1/f noise. Another pattern that pops up a lot is "white noise," for which power is the same at all frequencies, for example, a kit would have four tons of rocks for each size. Compared to 1/f noise, white noise has more of the little rocks. If you built your mountains from a white noise pattern you would find that your model mountain range would not look realistic, but rather otherworldly.

It is perhaps not surprising, then, that large-scale structure in the universe also seems to have grown from initial fluctuations that were close in form to 1/f noise. And we now know that, as a bonus, this pattern of fluctuations arises naturally in detailed models of the big bang. In contrast, computer simulations based on the white noise pattern, which was Peebles's starting

*Note that the *weight* of *each* rock *scales* (increases or decreases) as the *cube* of the size because volume is a three-dimensional property. Compared to a 2-foot rock, each 4-foot rock would weigh $(4/2)^3 = 2^3 = 8$ times as much, so you would have 4 times as many 2-foot rocks to work with, though half as much total weight.

point when he first discussed structure growth through gravity, produce a poorer match to the distribution of galaxies and clusters—too many little peaks for the number of big ones.

Of course, fluctuations in the early universe were not solid constructs like rocks, but slight undulations in the initial density of matter. Waves are a better analogy. What we call *sound* is a traveling wave in air density and pressure. Air is hard to "see," of course, and sound waves travel very fast, but if we could watch one travel across a room, we would see that sound is carried not by air that *travels*, but by a wave that compresses the gas of air molecules as it passes. The cacophony that fills the room as midnight approaches at a New Year's Eve party is nothing more than a mélange of subtle variations in air density. No wonder our ears can barely sort it all out.

This sound wave model is an excellent analogy to initial density fluctuations in the early universe; unfortunately, it is too far from common experience to be easily visualized. A picture that is less faithful to the early universe (because the fluctuations are not in density) but can be taken further (because it *is* very visual) is the pattern of waves on an ocean. From direct experience, most of us know that the greatest "power" is in the huge swells: they may not be very high, but they carry a lot of water. Shorter waves may rise to greater heights, and there are many more of them, but more power is in large, rolling swells.

A snapshot of the ocean, far away from the shoreline with its breakers, will help us visualize the pattern of initial variations in matter density in the early universe, the starting point for computer models of the growth of structure. We choose the pattern to be random—there will be no enormous wake left by a recently passed supertanker. Frozen in this moment are sharp peaks thrust up among the large swells and troughs. Whereas, in the ocean, the pattern continually changes as waves dissipate and re-form, in our early universe model, gravity ensures that the basic pattern is maintained, growing in contrast as time goes on. The amplitude of the density waves will rise, with swells and peaks growing and troughs becoming deeper. The sharpest peaks (the ones covered by whitecaps on the ocean) are places where gravity is strongest; these will grow most rapidly and become more and more dense as gravity pulls the material together. These sharp, small fluctuations are destined to become galaxies, or giant star-clusters that will be the building blocks of galaxies. They might sit on top of a larger wave of more modest height that would itself grow to become a cluster of galaxies, and the whole cluster, with many other *protogalaxies* and *protoclusters*, might reside on a vast, low-amplitude swell destined to be a supercluster. The space between galaxies will empty as matter drains toward the peaks; vast voids will form where troughs were free of galaxy-sized peaks.

This description is merely qualitative. To make a quantitative model requires the input of a specific three-dimensional pattern of density fluctuations, for example, one with the $1/f$ noise pattern, into a powerful computer. The computer runs a program that follows the evolution of such a density distribution under the influence of Newtonian gravity in an expanding system like the universe. This in itself does not sound easy, but what really makes it challenging is that it requires overcoming a formidable computational difficulty called the *N-body* problem. The calculation of the gravitational force between two bodies is trivial—there is just a single interaction, and the trajectories that such bodies will follow under the influence of gravity are also straightforward to calculate. The mutual attraction of three bodies is also a cinch—there are only three interactions—but the orbits have become more difficult to calculate. For four bodies there are six gravitational interactions, for five there are ten, and for 1,000 bodies there are 499,500 forces to calculate. The orbits have become extremely complex, and the evolution of any specific one is now impossible to predict because it requires knowledge of the initial state of the system to an unrealistic degree of accuracy (this is one example of what are called *chaotic* systems). In summary, as the number of massive bodies increases, the problem soon gets out of hand—even powerful computers have great difficulty following the evolution of the gravitational interactions and motions of $N$ bodies when $N$ is a large number.

In the 1970s Cambridge astrophysicist Sverre Aarseth tackled one such N-body problem, the collapse of a "cloud" of matter to form a galaxy. Aarseth made great strides by utilizing the rapid growth in computer power at the time. He developed clever computational strategies and shortcuts that allowed him to follow the time evolution of a system of several thousand "mass-points," the bare minimum necessary to model a galaxy. A decade later, Aarseth's efforts had fathered more sophisticated computer programs that, combined with yet more powerful computers, allowed astrophysicists to represent the mass distribution of an entire section of the universe with millions of mass points and follow its evolution.

The most successful of these early efforts was the collaboration of Marc Davis and three Cambridge-educated astrophysicists—Carlos Frenk, George Efstathiou, and Simon White. The four seized upon the opportunity to test models of a universe dominated by weakly interacting, exotic particles, starting with an input spectrum of the $1/f$ form and a new state-of-the-art N-body computer program. In particular, their models furnished detailed predictions from ideas set forth in an influential paper by George Blumenthal,

Joel Primack, and Sandy Faber at Santa Cruz, and Martin Rees at Cambridge. In this paper Blumenthal and collaborators explored the question of how galaxies and galaxy clusters would form and grow in a universe whose mass was dominated by weakly interacting particles, the ones that are not subject to electromagnetic force and so are oblivious to ordinary, baryonic matter.

One of the issues Blumenthal et al. highlighted was the difference expected for the distribution of initial fluctuations depending on whether the weakly interacting particles were very massive or not. The variation in behavior arises because these particles would move at different speeds: a particle created in the hot big bang that has only a small mass, for example, the putative neutrino we were considering earlier, would travel nearly at the speed of light. On the other hand, the creation of a massive particle, perhaps one of the supersymmetry candidates, leaves it traveling much more slowly than the speed of light. The "low-mass particle" model has been named *hot dark matter*, the "high-mass particle" model *cold dark matter*, the terms *hot* and *cold* borrowed from the description of an ordinary gas, where temperature is a measure of the speeds of particles. Hot dark matter refers to a sea of dark matter particles moving at near the speed of light, cold dark matter describes particles moving much more slowly.

This simplistic division into "cold" and "hot" dark matter is fundamental because, as Blumenthal et al. pointed out, the gross features of large-scale structure depend almost entirely on this simple distinction based on particle speed, rather than on any detailed properties of the particles themselves. In a sea of hot dark matter, small fluctuations would dissipate: the blazing speeds of the particles would simply allow them to vacate the premises before gravity could make any progress in gathering them together. (As in the somewhat inappropriate analogy with ocean waves, these smaller peaks simply vanish.) However, the possibility of escape diminishes for larger fluctuations: the largest would be so big that, even at near-light speed, particles could not "stream away" in time to avoid the crunch—gravity would win.

On the other hand, gravity *always* wins over cold dark matter (CDM) particles. They move so slowly that they cannot escape from even the smallest ripples in matter density. Therefore, the smallest fluctuations—the seeds of galaxies—would survive and grow. Furthermore, because of their higher amplitudes compared to the much broader cluster- and supercluster-size fluctuations (recall the ocean wave analogy), these smaller lumps would have a head start and grow the fastest. As a result, galaxies (or pieces for making galaxies) would appear as the first recognizable structures in a CDM universe. This is what Peebles had in mind in his "bottom-up" universe: small systems forming first and later agglomerating into larger units. The "hot-dark-matter"

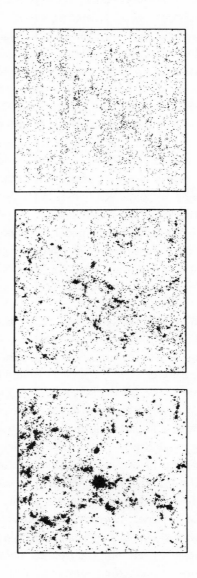

An N-body computer simulation showing the evolution of a smooth model universe into a lumpy one through the influence of gravity. As time passes (top to bottom), the contrast of the pattern grows as gravity works to slow the expansion of the universe in regions of high density. The expansion itself has been removed from the diagram by making each subsequent square larger by the amount the model universe has grown. *(Reference: Davis et al., 1985* Astrophysical Journal, *vol. 292, p. 371.)*

universe should evolve in the opposite way: as small-sized fluctuations disperse, the remaining large fluctuations—future clusters and superclusters—would be the first structures to appear. Only later would these subdivide to form galaxies, proceeding from the "top down." This was the idea behind the model of structure formation popularized in the Soviet Union.

Davis, Frenk, Efstathiou, and White added this information to their computer models of the growth of structure. The most striking result of their early simulations was that when they used the pattern of fluctuations appropriate for hot dark matter, a $1/f$ distribution but with the small ripples wiped out, the universe evolved by their N-body computer program didn't look much like the real universe. The largest structures in this model universe, superclusters and clusters, were too striking—their contrast with the low-density voids appeared too dramatic compared to the real distribution of galaxies as shown by the CfA redshift survey. The model based on cold dark matter, on the other hand, fared much better. With the full $1/f$ distribution of fluctuations that CDM produces, the computer generated model universes that bore a familial resemblance to the real McCoy.

Actually, there was a catch. The N-body simulation showed how the dark matter itself would be distributed, but to compare it to the real universe one needed to know how the *galaxies* were distributed. After all, that was all Davis and collaborators had for comparison: a map of the galaxy distribution. The group had to invent an algorithm that would make reasonable (but somewhat arbitrary) guesses as to where to locate the baryons that would form luminous galaxies. Of course, the simplest thing was to place the baryons in the same pattern as the dark matter and assume that galaxies formed in the densest places. This hadn't worked. Davis and Co. had tried this, evolving the model universe on the computer until it looked as clumpy as the real universe, but at this point the "galaxies" had been accelerated to speeds of about 500 km/sec and were moving in a kind of frantic disarray. This didn't agree with observations. Recall that the Seven Samurai had found only small peculiar velocities for galaxies relative to each other—neighboring boats moving slowly. Peculiar velocities as large as 500 km/sec arose only on very large scales. If galaxies were distributed in the same way as the dark matter, it seemed, the CDM model wouldn't work.

They found a way out, led by an idea from British astrophysicist Nick Kaiser, then at Berkeley. Davis and Co. developed a scheme that made the galaxy distribution more clumpy than the distribution of dark matter. They "biased" the galaxy distribution—put even more galaxies in the dense regions and even fewer in the underdense regions of dark matter. A "biasing factor" of 2.5 was the best fit: for example, if the dark matter was denser-than-average by 10 percent in some region of space, the number of *"galax-ies"* placed there would be boosted by 25 percent above the average. In this strongly "biased" model the galaxies were clustered more strongly than the dark matter, so a match to the appearance of our universe came "sooner"— in a shorter (simulated) cosmic time. This meant less time for galaxies to gather speed, and lower speeds meant much better agreement with the real universe. With "lights installed" in this way, the match of the model CDM universe to the CfA survey—a map of the actual galaxy distribution—was impressive: the clustering of galaxies, and their modest peculiar velocities relative to each other, were reproduced well.

Davis, Frenk, White, and Efstathiou became enthusiastic proponents (one is tempted to say zealots) of this strongly biased CDM model. They published papers showing that the model not only mocked-up the large-scale distribution of galaxies and clusters, but also formed just the right number of little pockets where baryons could collect to form galaxies with the proper range of size and mass. From the "flat rotation curves" to the "morphology-density" relation (galaxy type varying with environment), biased CDM could "do it all," they reported. Even better, few assumptions had to be made— after adopting the 1/f form of the fluctuations, Davis and Co. had only to adjust the bias parameter—yet the results seemed a reasonable facsimile of the real universe over a wide range of scales. Could it be that the universe was this simple: massive, weakly interacting particles, mixed with a small fraction of baryons, together summing to the critical matter density $\Omega = 1$? Was this fundamental piece of the cosmological puzzle solved?

Would that it were so, but in 1986 the fruit of this success began to sour. The first serious challenge to the biased CDM model had been the discovery of the Great Void in Boötes, by Kirshner, Oemler, Schechter, and Shectman. No completely empty regions of its size appeared in the N-body simulations by Davis and collaborators. But further observations by Kirshner's team showed that the Great Void is not completely empty—just a region of very low density. This reduced the challenge to the model, even more so because the extent of the disagreement depended rather critically on the *shape* of the Great Void, which was turning out to be very irregular and hard to describe, not to mention poorly known. An empty, spherical void of the size first advertised would have dealt a critical blow to the biased CDM model, but an emptier-than-normal void of irregular boundaries could be explained away, particularly if it was a rare fluctuation that had been "stumbled upon."

A similar kind of disagreement between model and reality seemed to be indicated by the CfA slice of the universe—the map de Lapparent, Geller, and Huchra had made from their galaxy redshift survey, work sired by Davis himself. The thin arcs of galaxies that could be seen curving around large empty spaces of the CfA slice didn't bear much resemblance to the biased CDM simulations (or any simulations, for that matter). None of the voids in the slice was as large as the Great Void, but there were many of them. The voids seemed a basic organizational feature, and while it wasn't clear if the biased CDM model could reproduce them or not, it certainly hadn't *predicted* this almost "sudsy" appearance of the galaxy distribution. Davis and his collaborators would eventually publish a paper showing "slices" through a computer-simulated CDM universe. There was at least some similarity with those found of the real data, but observers and theorists continued to argue whether they were similar enough—many contended that the simulations looked more choppy, less coherent. Surprisingly, even the addition of new "slices" by the observers, and efforts to slice the model universes in the same fashion, didn't resolve the issue, principally because the comparisons were *qualitative* rather than *quantitative*.

For both the Great Void and the CfA slice, no verdict was rendered on the success or failure of the biased CDM model, chiefly because of the lack of quantitative mathematical tools for comparing irregular or *amorphous* distributions. The old "square peg in a round hole" reminds us that differentiating among regular shapes is pretty easy. And, while everyone can agree that the Grand Teton mountains, with their triangular faces, look something like pyramids, it is another matter to come up with a clear description of something really irregular, even a relatively simple piece of geography like the Great Lakes. Lakes Michigan, Erie, and Ontario are kind of oblong— rather similar and reasonably regular—but how does one describe Huron and Superior? Is the shape of Nebraska more like Kansas or Oklahoma? Which river is more twisty, the Colorado or the Mississippi, and which does the Rio Grande more resemble? Are the lakes in Minnesota randomly distributed or is there a pattern? The activity quickly becomes an experiment in human perception and personality—the basis of the famous Rorschach test—rather than an objective mode of classification.*

---

*A fascinating example of the kind of analysis that can be done goes under the name *topology*, in which the shape of a smooth, continuous surface, like the folds of a blanket or a knot in a rope, is described by its degree of convolution. A topological analysis could be used to examine three paper wads and decide which two are most alike, for example, by comparing the total crumpled area (the size of the paper) to the volume into which it has been squeezed. Or it could be applied to tackle one of the questions about similarity in shape of different lakes or U.S. states: for example, one could calculate a single number for each state or lake, by dividing the total area by its border length, as a measure of its "regularity."

Needed are mathematical tools that can extract *quantitative* information from seemingly irregular forms. Peebles supplied one that is surprisingly robust for its simplicity: the *correlation function* measures the degree to which points (here, galaxies) cluster together, compared to the clustering of a completely random spray of points. The stronger the correlation, the more clumped are the points, that is, the greater the likelihood of finding another point nearby. The correlation function for galaxies played an important role in "testing" CDM simulations; by computing this diagnostic on both N-body models and the real distribution of galaxies, the degree of similarity could be quantified. However, others pointed out that the biased CDM model might pass this test and still not match the structure of the real universe—what a scientist would call a necessary, but not sufficient, condition. They showed that one could rearrange the galaxies in the CfA slice—wiping out the large voids and thin walls—while keeping the exact same correlation function. The correlation function didn't really describe the complexity of the structure very well.*

So the correlation function of galaxies wasn't by itself able to make a strong enough case to validate the biased CDM models. To make matters worse, others were using the correlation function to attack the model. Measurements of the correlation of the rich clusters of galaxies Abell had cataloged showed that they themselves clustered much more strongly than the CDM simulations could reproduce. Here the test was direct, quantitative,

---

*Others were working on topological descriptions that could describe this complex structure. The galaxy distribution is not a smooth surface, but a series of discrete points, therefore some "smoothing" is required. A computer can artificially spread out the galaxies so that their matter runs together like melting wax, then divide space into regions of higher-than-average and lower-than-average density. Simple topological descriptions may then apply. The case where all the low-density regions are isolated from each other, i.e., you *must* travel across a high-density region to get from one "hole" to the next, is called a "bubblelike" or "Swiss cheese" topology. The topology is "meatball" if the high-density regions are isolated.

De Lapparent, Geller, and Huchra had suggested that the galaxy distribution might be bubblelike, but Princeton theorist Richard Gott argued that neither bubble nor meatball topologies are expected for a universe that grew from the kind of fluctuations one expects in a CDM model. Expanding on a science project he had developed for his high-school science fair, Gott pointed out that such a universe would have the topology of a common sponge—both the high-density "body" (the sponge cells) and the low-density "holes" (through which water passes) are contiguous and connected. Gott pointed out that a thin slice of a sponge could appear bubblelike. The results of analyses like Gott's are not yet decisive, but the spongelike topology is regarded by many astrophysicists as the description that best fits the galaxy distribution as revealed so far.

and objective, so the counterattack by the CDM supporters rallied around the position that the catalog itself was tainted by Abell's mistakes. (Abell himself had warned thirty years earlier that his catalog was a dubious source for this kind of statistical analysis.)

In 1986 the jury was still out as to whether biased cold-dark-matter models of the growth of structure had survived these challenges. Thus the advocates of CDM were on the defensive when we assembled in Santa Cruz to discuss new observations of large-scale structure, including the large peculiar velocities—the large-scale streaming—found by the Seven Samurai. In two important ways, our result was the toughest assault yet on the biased CDM model. In the first place, ours was the only study to map the distribution of the dark matter directly, and mass distributions were what N-body models simulated. The other challenges had come from maps of the positions of galaxies; comparison with the CDM model required further assumptions about how these galaxies were placed with respect to the dark matter, as Davis and collaborators had done by introducing the bias parameter. In contrast, measurements of the *peculiar velocities* of galaxies were directly sampling the distribution of *mass*, including both luminous and dark matter. There was little ambiguity in the comparison between models and measurements.

The second important feature of the Seven Samurai result was that it could be described quantitatively because we knew the speed (the peculiar velocity) in one direction for every galaxy in our sample. Rather than facing the daunting task of describing an irregular distribution of galaxies in three-dimensional space, we could do something as simple as boxing off a region and calculating the average peculiar velocity for all the galaxies inside. In fact, this is exactly what we had done—calculated an average speed and direction in a large volume and adopted this as the most straightforward, assumption-free description of our data. Around the Milky Way galaxy we had drawn a huge imaginary sphere which included about 400 of the elliptical galaxies we had studied, and calculated that these galaxies are moving, on average, at about 600 km/sec in a particular direction. The same kind of test could be done for the model universes generated by computer: choose the same size volume and calculate the average peculiar velocity that lumpiness in the mass distribution would induce. Davis and Co. had done just that, made thousands of such model universes, from which they determined a sound statistical estimate of the average velocity, and its probable range. They had found that, for the volume of space surveyed by the Seven Samurai, the expected peculiar velocity for a biased CDM model was only 150 km/sec, much lower than the 600 km/sec we had measured. Worse, the expected deviations for this number were small; if biased CDM were the correct model, the chance that we would live in a region with such rapid large-scale streaming was only one in a million.

After the Santa Cruz workshop, Nick Kaiser would point out that the Seven Samurai had overstated the case. We had thought that our result of 600 km/sec of "average peculiar velocity" applied to a volume reaching out to Hubble expansion velocities of 6,000 km/sec, roughly 500 million light-years in diameter. But, in fact, few of the elliptical galaxies in our sample were so far away, and we should have given the ones that were even less weight because their peculiar velocities were known less accurately. If we had calculated the average peculiar velocity correctly, Kaiser would show, we would have realized that the value of 600 km/sec applied only to a much smaller volume of about 200 million light-years diameter. Still, the biased CDM model was able to reproduce a 600-km/sec "average peculiar motion" over a sphere whose diameter was only about half this size. Kaiser's point was that, while the biased CDM model still might be in trouble, the contradiction between measurements and predictions was less severe than had been claimed. Now it appeared that, if biased CDM was the correct model, the chances of our living in a volume with such rapid large-scale streaming were reduced to around 1 in 500—bad odds at the gaming table, but with such a big universe, who knows?

It seemed strange that the work of the Seven Samurai was regarded by many as a campaign against cold dark matter. Up to this point we had said nothing about the implications of our work for the CDM model; indeed, Sandy Faber could claim to be a parent of the CDM model through her collaboration with Blumenthal et al., and she remained a believer. Nevertheless, the work of the Seven Samurai engendered a fairly hostile response by those members of the audience at the Santa Cruz workshop who were enthusiastic supporters of CDM. Almost all of them were *theoretical* astrophysicists who were not about to give up an elegant, promising concept just because of *observational* data to the contrary. (Members of this group were fond of quipping, half-seriously, that "no observation should be believed until it is confirmed by theory," a famous retort from legendary physicist Sir Arthur Eddington.) For the most part, their reaction was to question our measurements and analysis rather than direct greater scrutiny at the theoretical model. Soon, it was the Seven Samurai who were on the defensive, confronted with the Sisyphean task of *proving* we had done nothing wrong. Fortunately, our own long-standing skepticism about our findings had steeled us for most of their probing. In the end, none of our colleagues could point to an embarrassing oversight or mistake that could invalidate our result of a large-scale streaming of elliptical galaxies. As was appropriate and healthy, we would receive many such challenges in the months to come. The tests they prompted, and the failure of anyone outside to think of a problem we hadn't considered, only served to boost our confidence.

We held our ground. Over the next few years the understanding of both models and data would improve to the point where it became clear that, at the very least, something subtle but important was missing in the simulations of CDM universes, at worst, the whole idea was fundamentally wrong. The biased CDM model as it was concocted in 1986 suffered from a kind of anemia: with too little ''power'' in large, supercluster-sized fluctuations, it could not generate enough large voids, enough long chains or sheets of galaxies, or a strong enough clustering of the galaxy clusters. Most of all, it couldn't create a Great Attractor.

# 9 • THE GREAT ATTRACTOR

"Look, Dave, you can't deny that the peculiar velocities over here in Centaurus and Pavo are much bigger than anywhere else in the sky. My God, some of them are 2,000 km/sec! This isn't just bulk flow, there's acceleration—something is pulling on all these galaxies. We'd better point this out before somebody else does."

Dave Burstein and I were head-to-head, and voices were rising. Often we had found opportunities to disagree, sometimes strenuously, as far back as fifteen years ago when we first met in graduate school here in Santa Cruz. However, this was perhaps the best issue we ever had to argue about. Years of practice prepared us to make the most of it.

An hour earlier we had been christened the Seven Samurai and received a second major hearing of our radical result at the "Nearly Normal Galaxies" meeting. On behalf of the Samurai, Sandy had given a talk on the intrinsic properties of elliptical galaxies and I had presented our results on their surprisingly large peculiar velocities, and how they were streaming together through space.

I gave the party line—the party of seven, that is—and therein lay the problem. It seemed that I alone among the Seven Samurai was unhappy about the description of our result as a *bulk* flow of the more than 400 elliptical galaxies in our survey. Back in Pasadena, the previous November, the group had realized that our results were likely to be very controversial, and that how these results were presented might have a big effect on how they were received. We decided to adopt Joe Friday's approach—"just the facts, ma'am." Let the data speak for itself and keep interpretation brief and on the conservative side. It was this thinking that led to our presentation of the "minimum hypothesis": galaxies in the survey were traveling together at the same speed, about 600 km/sec, a "bulk" flow in which our own galaxy was a participant.

This was, in fact, not a bad description: with such large errors in measuring distance, about 20 percent, we needed to average together the peculiar velocities over rather large regions of the sky, or in clusters, to get an accurate measure of average speed and direction. For example, Dave had used our computer programs to divide the sky into octants (imagine slicing an apple into eight equal pieces by cutting through its middle three times at right angles) and found that for each of these big chunks of the sky the "average peculiar velocity" was about the same. "Peeling" the volume into many nested spherical shells, like the way you take the skin off an apple, gave much the same result. On the other hand, staring us in the face were very large peculiar velocities for galaxies in the Centaurus region and, to a lesser but still significant extent, in the direction of the constellations of Pavo and Indus, the site of another supercluster. When averaged into very large zones like octants or spherical shells, these particular regions contributed to the conclusion that the streaming of galaxies was roughly constant everywhere, but, on their own, their larger peculiar velocities looked like flagrant exceptions.

Now, after our presentations at the meeting, it was becoming apparent that what we considered the most conservative description could also be taken as the most radical interpretation possible—that everything in the huge survey volume was moving. By embracing "bulk flow" we were implying that the *source* of gravity responsible for these large peculiar velocities must be very far away—outside the huge volume of our survey—else we would have seen a convergence of the flow toward the high-density region that was doing the pulling. If far away, a fluctuation in the distribution of mass would have to be very large in order to pull galaxies in this huge region in its direction, so large, in fact, that its existence might be at odds with *all* proposed models of structure formation. This was a worry expressed in two talks at the Santa Cruz meeting, by Canadian physicist Dick Bond, and by Niccola Vittorio from Italy and Roman Juskiewicz from Poland.

Worse, our interpretation of bulk flow could be viewed as evidence for an even greater heresy, that peculiar velocities were not even the result of good-old, familiar gravity, but rather due to some mysterious property of the universe that made it *appear* as if all the galaxies were moving. After all, bulk flow meant only that everything was moving together. This in itself didn't tell us the speed. Like cars on a freeway, they could all be traveling at 60 miles per hour, or 20 or even 0—it would all look the same. The speed attached to this convoy rested squarely on one *absolute* measurement, the velocity of 600 km/sec with which our Milky Way galaxy was moving with respect to the cosmic microwave background (CMB). Few people doubted this interpretation of why the CMB is hotter on one side, cooler on the other—other explanations seemed farfetched. But, if a source of gravity couldn't be identified as the cause of these large peculiar velocities, weren't we driven to abandon this cornerstone of cosmology—the CMB as a cosmic reference frame?

So it was with some enthusiasm that, in the privacy of our group, I had pointed to the larger peculiar velocities in the Centaurus and Pavo-Indus regions, claiming that galaxies were *accelerating* in this direction—it wasn't just bulk flow. My interpretation was that a large mass just at the edge of our survey could be the cause of this acceleration, as galaxies closer to the source of gravity felt an even greater pull. To me, this "speeding up" on the Centaurus side was as good a trail marker as you were likely to find on a trek through intergalactic space. I alone among the Samurai seemed to believe in its reliability, and urged that this interpretation be included in our talks. The others doubted the statistical significance of the "speeding up" and feared that a presentation of this idea would damage our credibility.

For more than a month I had been drafting a "Letter" (a short, rapidly published paper) to the *Astrophysical Journal*; this would be the first formal report of our results to the community. A much more expansive paper was planned, one that would contain elaborate detail of how our analyses were done, but completion of this enormous job was clearly a year off. Driven by my usual desire for speedy publication, I convinced my six collaborators that we should write a summary paper giving the main result, buttressed by the three papers we had already written describing the technique of estimating distances and tabulating the actual data. Not surprisingly, the others suggested that if I were so keen on the idea, I should write the paper myself. Of course, all of us would have to sign off on what the paper said. Now that this "Letter" was almost finished, we discovered that there was one crucial issue on which we didn't all agree—the characterization of galaxy streaming as a simple bulk flow versus a speeding up of the flow in Centaurus. Ironically, it was the first author—me—who alone wanted to stray from what had been considered the conservative tack and include the more complicated interpretation as well.

In my Santa Cruz talk I had dutifully chanted the Samurai mantra—bulk flow, bulk flow, bulk flow—but questions from the audience both during and after made me realize that bulk flow was not regarded as a conservative description after all. Jim Gunn, a valued colleague and certainly a brilliant one, made the biggest impression. After the talk he came up to the lectern chortling: "I didn't believe your large-scale streaming result when I heard about it, especially when you called it bulk flow. But now I see there is this obvious acceleration pattern. The mass you need to move all these galaxies is right here." With that pronouncement he brought his finger to the view-graph and tagged the region near the edge of our survey, just beyond Centaurus. "This is really astonishing," he concluded, and he was *smiling*.

That was enough. After the session, I dragged Dave, Sandy, and Donald into an adjacent classroom. Emboldened, I again pleaded with my colleagues to include the idea of an attracting mass in our analysis. I got nowhere. Donald and Sandy mostly stared into space as Dave and I got into it. Finally, Dave waved pages of computer output from Donald's OURV program in my face, claiming that it showed conclusively that the apparent speeding up of galaxies in the Centaurus and Pavo-Indus regions was not statistically significant. I was disarmed. By this point, I was too far out of touch with OURV to understand all its complexities; challenging Dave's declaration seemed out of the question. No matter what it *looked* like, I didn't have the numbers to back up my case for a large, attracting mass, so it died right there. The paper would stand as is. Bulk flow.

◗◖

A good mystery story is what science is all about. Astronomy, in particular, is the kind of activity Hercule Poirot would have appreciated. You examine a set of "facts" in search of a unifying cause, trying to reconstruct an event beyond your reach in space and time. The clues are usually scarce, frequently indistinct, often contradictory. If you are lucky, a number of them point in one direction; this encourages you to develop a hypothesis that might explain most of what you have observed. But any good detective knows that the process does not stop there: explaining any given set of observations *after the fact* is altogether too easy—the wrong culprit might well be fingered. (Consider the tragic cases of Hamilton Burger and Lieutenant Tragg, the legal adversaries of Perry Mason who, for years, deduced plausible, logical, and utterly wrong scenarios for crimes.) The key step is to use the theory to guess the nature of as-yet-unobtained information, a process that is unlikely to succeed unless there is some truth to the hypothesis. Then one is back on the case, looking for additional evidence that will support one explanation to the exclusion of all others.

In the fall of 1986, after the Santa Cruz workshop, I was back on the case, looking for new evidence to convince my colleagues that the motions of our sample of elliptical galaxies were due to the gravitational pull of something just beyond the Centaurus clusters. I began by looking for a huge supercluster of galaxies in the general direction of the streaming, figuring that galaxies, like city lights, would betray the presence of a vast metropolis of dark matter. If no excess of galaxies were to be found, well, this might be even bigger news—a vast, invisible mass unmasked.

I drew my plans to conduct a survey of galaxies in the region: first, a redshift survey to add the third dimension into the map of the galaxy distribution, later, distance measurements for some of these in order to determine peculiar velocities. Pointing the way were the little arrows on the Seven Samurai map, to a region of the sky in the Southern Hemisphere encompassing the now familiar EGALSOUTH region. This was both one of the best studied and least studied parts of the sky. The direction the elliptical galaxies were moving in lay only 10 degrees up from the *plane* of the Milky Way, the band of stars, gas, and dust that defines our galaxy. This was a great place to study the contents of the Milky Way itself: only 50 degrees in longitude from our galaxy's center, the region was a Galápagos of star formation, replete with glowing gas and dark, cold clouds, and the young and old star clusters they had given birth to.

This playground for astronomers who study the Milky Way galaxy is, alas, a quagmire for those of us with extragalactic persuasions. Within 20 degrees of the galactic plane, stars are so plentiful that remote galaxies hide like tigers in jungle thicket; within 10 degrees of the plane, they vanish completely behind the cover of Milky Way dust. Not surprisingly, then, there had been no extensive redshift surveys near the galactic plane: the CfA survey and its counterpart in the Southern Hemisphere had been aimed away from the plane, toward the comparatively star-free and dust-free "poles," to avoid these very problems. Fortunately, astronomers at the European Southern Observatory had persevered in at least "counting the tigers in this picture," and their catalog provided a list of potential targets. It was extremely unlikely, I decided, that a huge supercluster could be completely hidden behind the thin band of dust, so I cast my net wide and marked out a squarish region that covered almost 10 percent of the sky. Based on the results of the Southern Sky Redshift Survey conducted by Brazilian astronomer Luiz da Costa, Marc Davis, and their colleagues, I knew that if I followed their procedure in selecting all galaxies in the ESO catalog larger than seventy-two arcseconds, the "typical galaxy" in the survey would have a Hubble expansion velocity of about 5,000 km/sec. This corresponded to the distance where the attracting mass might lie, near the limits of the Seven Samurai survey. There were 1,403 galaxies in the search area, and only for a few hundred of the closer ones had redshifts been measured. There was much to do.

I arrived at Las Campanas in January 1987 with twenty pounds of finding charts; Basil Katem, a veteran employee of Carnegie Observatories, had generously relieved me of the tedium of making 1,403 Polaroid enlargements of the sky atlas photographs. Meanwhile, the other six Samurai were assembling in southern England, at the Herstmonceux Castle headquarters of the Royal Greenwich Observatory. I was literally half a world away as the last of the Samurai group meetings got under way and, in truth, I had made only a halfhearted attempt to exchange the observing time assigned to me so that I could join them. I would have arrived late anyway, since I was the chief organizer of the national meeting of the American Astronomical Society, that month, in Pasadena. Just "out of the frying pan," facing a grueling week of work and the possibility of more bickering about the "attracting mass" hypothesis, I opted instead for the peace and majesty of Las Campanas. What was needed, I thought, was more data, not more talk, and more data was what I could best provide.

On January 22, the first night of the observing run, redshifts were rattling off at about one every five minutes—spectra good enough for a redshift determination were a lot easier to come by than those the Samurai had collected for measuring velocity dispersions. Better still, the "answer" could be gotten almost immediately: I could estimate the redshift merely by laying a ruler along the computer screen after the spectrum had been plotted, and then recording the positions of the absorption lines—the dark bands at known colors caused by certain chemical elements in each galaxy's stars. Fernando Peralta and I, the telescope, and the computer, were chugging along like some infernal contraption; charts flying, motors whirring, computer beeping, and, at the heart, two humans piloting their way through deep space at a madcap pace. I can think of no better image than Toad in his motorcar in Disney's animated version of Kenneth Grahame's *The Wind in the Willows*—eyes lit up like headlights, bouncing, exulting, merrily on his way to "nowhere in particular."

Each time the massive steel frame of the telescope sailed to its next destination, I had just enough time to record the new redshift. The "map" I was making was a crude, one-dimensional type: a straight line parsed by tick marks from 0 to 10,000 kilometers per second. For each galaxy, as its speed was determined—for the first time ever—I added a tiny box at the appropriate place along the axis. Slowly this *histogram* rose like an emerging city skyline, and with it my excitement built as well—there was something new here. True, as predicted, a plurality of the galaxies had the velocity (the Doppler shift, or redshift) of the Centaurus clusters, about 3,000 km/sec—these clusters were thought to dominate this volume of the universe. But, in addition, a new peak was rising farther out; box after box landed in the interval from 4,000 to 5,000 km/sec and built a second skyscraper. The second peak looked somewhat less important than the Centaurus peak, but, in fact, galaxies so much farther away were discriminated against because they were fainter than the closer Centaurus galaxies. When allowances were made for this biased selection, it was clear that the farther concentration was the dominant one, the urban center of a vast metropolitan complex of galaxies.

The pattern continued to grow. By the end of the third night I was sure that my hunch about the greater mass beyond the Centaurus clusters was correct. At first I was puzzled why the Samurai survey hadn't reached far enough to see this vast concentration; many of the galaxies in the Centaurus clusters had velocities of around 4,500 km/sec, as had been found by British astronomers John Lucey, Malcolm Currie, and R. J. Dickens. Suddenly, I remembered that, of course, the Samurai had found very large peculiar velocities for the Centaurus galaxies. This meant that they were not as far away as their velocities indicated, because there was a large peculiar velocity added to the Hubble expansion. After subtracting the peculiar velocity in order to leave only the expansion velocity, their distance according to the Hubble relation would be much less.

Ah, so, I reasoned, the Centaurus galaxies are much closer than one might think on the face of it. But suppose the galaxies in my second peak, the ones with measured velocities of 4,500 km/sec, have little or no peculiar velocity—they shouldn't, in the "attracting mass" model, because they would be at the *center* of a converging flow, pulling in surrounding galaxies. Then it all made sense. Galaxies with measured velocities of 3,000 to 4,500 km/sec could be at a wide range of distance, depending on how much of that velocity was a peculiar velocity and how much was due to the Hubble expansion. The more distant peak, probably a vast supercluster reeling other galaxies in with its gravity, was *twice* as distant as the clusters in Centaurus, and largely beyond the distance the Seven Samurai had sampled. Seen in this perspective, the rich Centaurus clusters and their surroundings were just populous suburbs of a much larger, more distant metropolis, the possible source of the extra gravity that gave rise to the large peculiar velocities.

After five nights of observing, more than 300 new redshifts had confirmed my first impressions. I decided to defer the big push in this project to an upcoming observing run only three months away, confident that I already had the evidence I needed to persuade the other Samurai that the gravitational pull of a giant supercluster beyond the Centaurus clusters was the major cause of the large-scale streaming of galaxies. With feverish enthusiasm I calculated the effect of gravity expected from a giant wall of galaxies—this was the

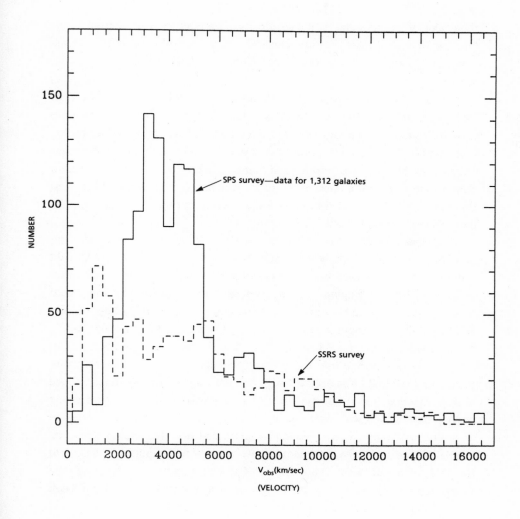

The distribution of galaxies in the direction of the large-scale streaming found by the Seven Samurai. I called my sample the Supergalactic Plane Survey (SPS). Compared to the Southern Sky Redshift Survey (SSRS) by da Costa and collaborators, which serves as a calibration for a smooth distribution of galaxies over space, this region shows a great "overdensity"—a greater number of galaxies per unit volume than the average. *(Reference: 1991* Astrophysical Journal Supplement Series, *vol. 75, p. 241.)*

picture I had in my mind for the distant supercluster. Using this model, I estimated how many galaxies (and their accompanying cloaks of dark matter, in the ratio of 10:1 dark/luminous matter) would be required to generate the peculiar velocities we had measured. It would take, I figured, some tens of thousands of galaxies and their dark cloaks to account for a gravitational pull that had to extend from the supercluster's center, roughly 200 million light-years away, all the way to our galaxy—this was architecture on a cosmic scale.

In early February I returned to Pasadena, data in hand, but, more than this, paper in hand. The last few days at Las Campanas I had raced to write a short "Letter" to the journal *Nature*, anxious that we stake a claim to this idea that I had promoted for more than a year. Of course, including the other Samurai in this effort was both good manners and good science, so I called Sandy to tell her that a copy of the paper was in the mail. I was shocked to find that, while I had toiled in seclusion at Las Campanas, the other Samurai had reached the same conclusion, finally.

Their path had been a crooked one. Sandy had carried with her to England new misgivings about the large-scale streaming flow altogether, and Donald Lynden-Bell soon joined her in a duet of worry and doubt. In Santa Cruz, before she left for England, Sandy had been dividing the sky into small sections again. She had not been able to assemble a coherent picture, despite what the OURV computer program had been saying about smooth "bulk" flow. In October a "bug" had been found in this program; this was one of the reasons OURV had told us that any gradient in the flow (the departure from bulk flow that I had been touting) was not statistically significant. Since it was this that had made me back off the previous summer from going beyond bulk flow, I was more than a bit annoyed to learn that it was a couple of faulty instructions in a computer program that had shut me up. Now the bug was fixed and Dave had additionally made some important changes to the data for about twenty galaxies in the crucial direction of the flow—he found that the obscuration of their light by dust in the Milky Way had been overestimated the first time around.

But these were minor mistakes compared to what was coming. Sandy's new and confusing findings propelled the six Samurai to reopen all the old issues. At Herstmonceux they discovered a more fundamental conceptual error in the operation of OURV, one that none of us had caught earlier. When we had set the boundaries over which the program examined peculiar velocities, we had settled on dividing the sky into two volumes defined by the distances of galaxies from our own Milky Way: one was the sphere including all galaxies with measured velocities of less than 3,200 km/sec, the other a thick shell surrounding that sphere in which velocities ranged from 3,200 to 6,000 km/sec. With this division we had found that the average peculiar velocity was about the same for these two volumes; this was one of the clearest demonstrations of the "bulk flow" picture.

Wrong! By using measured velocity directly as an indicator of distance, we had "forgotten" our major finding that these velocities included large peculiar motions as well as the Hubble expansion. Our spheres were not spheres at all, but wobbly balls whose dimensions were affected by peculiar velocities. We needed to use true distances, not just redshifts, to define our volumes. Since we had made measurements of the true distances to these galaxies—this is what the correlation between velocity dispersion and the donald diameter $D_n$ was designed to do—we were able to use these to put our spheres round again. When the six Samurai did this, most of the galaxies in the Centaurus region, whose large velocities had placed them in the outer shell, moved to the inner sphere—because of their large peculiar velocities, OURV had (at our instruction) incorrectly assigned them to the outer shell.* After this correction, the computer analysis showed that the average peculiar velocity of galaxies in the inner sphere was unchanged at around 600 km/sec, but that the average for galaxies in the outer shell had dropped to a lower value. The two were no longer the same; the bulk flow model had taken a serious hit.

The Samurai now used the computer program to slice the sky into cone-like volumes—this, too, had been done incorrectly before. They discovered that the average peculiar velocity varied substantially from region to region, from 1,000 km/sec in Centaurus to virtually nothing in some parts of the sky. This was a different result than before and another blow to the picture of bulk flow. The six Samurai were aghast that such revisions would surface so late in the game, and there was much disagreement about what the conclusion of this phase was, and what to do next.

To round out the Herstmonceux meeting Roberto Terlevich had organized a workshop where the Samurai would discuss and compare their results with other invited researchers in the field. Neither he nor the other Samurai could have guessed that, the day before the first session, the Samurai would be

---

*It was embarrassing to have made such an elementary mistake. It demonstrated that, in science, as well as all other endeavors, "what you learned at your mother's knee" (Allan Sandage's way of putting it) can be pretty hard to shake. So well-entrenched was the notion that expansion velocity (redshift) gave distance—the smooth Hubble expansion—that our minds had continued to hang on to it even after our brains had disproved it. Now it was gone, forever. From now on redshift would be only an *approximate* indicator of distance.

reduced to a state of demoralization and irritable bickering. Donald was ready to abandon the bulk flow model and describe the flow as chaotic. Dave clung steadfastly to the idea that bulk flow, with sizable departures from region to region, was an adequate description. Sandy thought maybe they should present no interpretation at all.

They argued until they hammered out a compromise: at the workshop, each would give a talk describing the data itself, holding interpretation to a minimum. In the end, Sandy gave the summary talk for the workshop. She showed that the bulk flow model worked up to a point, but that there were significant motions left over, and that these seemed to be systematic, not random, around the sky. Without knowing exactly why or how, she guessed that these deviations might be explained by another simple model.

After the workshop, a rare, two-day snowstorm trapped the participants and disrupted plans for further work. Dave and Roberto had intended to spend several days working on a Samurai paper on the properties of elliptical galaxies—the original goal of the project back in 1981—but Roberto's great white van capsized in the snow, injuring his wife, Elena, and leaving them without transportation. The study of the intrinsic properties of the elliptical galaxies slipped into limbo, where it has remained ever since. The collaboration of the Seven Samurai had reached its low ebb.

While Dave returned, discouraged, to Phoenix (an omen?), the universe of the Samurai was about to rise from *its* ashes. The Samurai were about to discover a better model—one that made more sense than bulk flow—that of a large, attracting mass. Sandy and Roger plowed their way to Cambridge when the roads were cleared. Donald and Sandy had set aside the next two weeks to work on the major paper of the collaboration that would give the detailed results. But the shakeup in the bulk flow model had opened minds again about trying new things and, as it turned out, there was only one thing left to try that could be done easily and quickly.

It came from OURV. As a scout on our journey, Donald's computer program had led us into a minor ambush on the battlefield of bulk flow, but now, in the final hours, this novella of about 2,000 computer sentences redeemed itself handsomely. Donald had endowed the program with the ability to model the peculiar velocities that would arise from another attracting mass, one in addition to the Local Supercluster. This option had never been exercised, but after the other Samurai had gone home, Sandy had the good sense to throw the switch. She supplied the computer with educated guesses as to the amount of mass, its direction and distance, that might generate the flow pattern, then she adjusted parameters by hand in trial after trial, looking for a better model. It took a few iterations to find that the inclusion of just one new mass center, in addition to the concentration of mass in the Local Supercluster, produced an *astonishing* agreement with the pattern of peculiar velocities we had measured, including the exceptionally large flows in the Centaurus and Pavo-Indus superclusters and a number of smaller flows around the sky. The vast pile of matter that accomplished this feat was spread over at least 200 million light-years; the computer program located its center at a distance that would correspond to a Hubble expansion velocity of about 4,500 km/sec, exactly where the second peak in my galaxy map had appeared. Furthermore, it was a mass comparable to that of the largest known superclusters, a number equivalent to tens of thousands of galaxies with their accompanying dark matter, a number close to the rough estimate I had made at Las Campanas.

Donald and Sandy could hardly believe they had overlooked such a simple solution to the deviations from the bulk flow model. But they were troubled. Such a powerful concentration of mass would surely include a vast collection of galaxies that could not be missed—why wasn't this supercluster well-known? They desperately needed reassurance and confirmation that there was something visible in the sky that matched up with the huge attracting mass that OURV was calling for.

As luck would have it, just the right person was in Cambridge at that crucial moment. One of Donald's students, Ofer Lahav, was working with galaxy catalogs and making maps, so Donald and Sandy asked Lahav to make a galaxy map of the sky *centered* on the direction of the streaming flow. By merging three galaxy catalogs, Lahav produced the largest coherent picture yet seen. His map showed half of the sky, as a black circle, with each of the thousands of galaxies represented by a single white dot on a black "sky"—again, that remarkable abstraction of replacing a *galaxy* with a dot. What made the picture even more amazing was that it showed clearly, for the first time, a huge swath of "light" stretching across nearly the entire hemisphere, a glow made up of the "dots" of many hundreds of galaxies. The clusters of galaxies in Hydra and Centaurus, previously thought to dominate the region, looked like mere suburbs against the nighttime glow of this megalopolis. Just as I had found in a different way—from redshift measurements—the major concentration had previously been missed, probably because no one had made a picture big enough to fully apprehend this broad swell of galaxies that traversed fully one quarter of the sky. The clusters, with their dense concentrations, had been easier to pick out than this diffuse background, but it was, plain to see, far richer.

Sandy and Donald had become converts. Dave resisted at first, but when he ran the program for himself, he too was convinced that a vast attracting

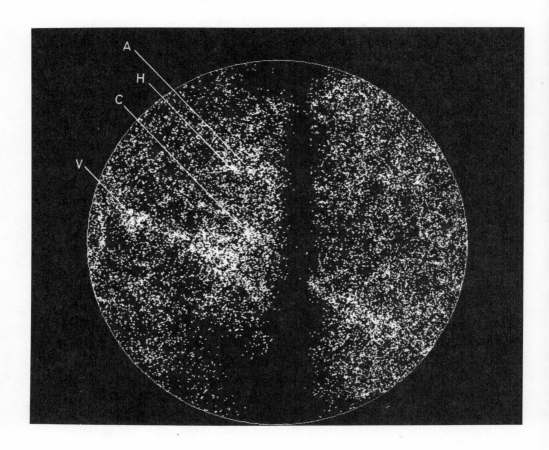

Ofer Lahav's map showing the concentration of galaxies in the direction of the large-scale streaming found by the Seven Samurai. The map shows half the sky, with each galaxy indicated by a dot. (The black vertical band is the absence of galaxies caused by absorption of their light by dust in the plane of our Milky Way galaxy.) The Virgo (V), Hydra (H), Centaurus (C), and Antlia (A) clusters are small concentrations of galaxies compared to the swath of light cutting across from Virgo down and to the right.

mass was a much better model for the large-scale streaming velocities of elliptical galaxies. Meanwhile, my commitment to the same conclusion had been strengthened by my redshift survey of galaxies in the direction of the streaming flow. From different trails we had reached the same point in our journey, an overlook from where, we believed, we could see the source of the cosmic flow. It was the climactic moment of our journey.

I received the news of the conversion of the Samurai with mixed feelings. For a week I had basked in smug satisfaction that I had been right all along about the attracting mass and had found the first evidence of it. Suddenly, I learned that the point I had reached through my own initiative had been attained by the others as well, and I wondered how Scott and Amundsen would have felt to have arrived at the South Pole simultaneously? (A grandiose analogy, to be sure, but such is vanity.) On the other hand, these were *my* colleagues, working on *our* project, so, although I now regretted missing out on the excitement of the final Samurai meeting, I could still feel a part of what they had achieved.

Now I didn't know what to do with the "Letter" to *Nature* that I had written. Sandy had made suggestions that had greatly improved the paper, and it was now ready to send to the other Samurai. To me it was important that we get the news of a revision to the bulk flow picture out quickly—I still regretted that the original "Letter" to the *Astrophysical Journal*, with me as first author, had not laid claim to this "attracting mass" idea. Here, I thought, was an opportunity to repair that mistake, and, of course, I didn't mind a bit that I would be first author on a paper that would help disassociate me from the dreaded "bulk flow" model. There was no doubt that Sandy should be included as a coauthor, and, even if the other Samurai hadn't contributed directly, it was equally clear that the new survey was a direct result of our joint efforts. But now I also knew that the major paper of the collaboration, taking shape under the leadership of Donald Lynden-Bell, would also include a detailed analysis of this important idea. I had hoped that the two papers would harmonize, but discord lay ahead.

It seemed to me that my "Letter" would be valuable only if it came out rapidly, well before the Lynden-Bell et al. extravaganza. Therefore, on February 25, I sent a copy to the other Samurai and included an offer to put all seven names on it, on condition that we wrap it up quickly. Wary of how our other papers had dragged on, I warned that I would require "pretty strong arguments" before I would be talked into "wholesale changes or expansions." I soon learned that several of my colleagues were offended by the tone of my letter, which to them read like an ultimatum. Lynden-Bell was more than offended—he was furious. He thought I was selfishly trying to preempt the main paper, and, although Sandy had tried to explain the circumstances, Donald was unsympathetic to my attempt to distance myself from the bulk flow model I had never liked. He sent back a curt note more than a month later, asking whether I was "prepared to be generous" and let Sandy be first author on the *Nature* "Letter." My visceral response to this (to me, completely inappropriate) suggestion was an angry, defensive letter that lashed out at Donald, suggesting through sarcasm that his own motives were more self-serving than mine. My first mistake was not putting my letter in a drawer for a week, until I cooled down, the second, directing my barbs not just at Donald, but rather parading my immaturity in front of all the Samurai. Unfortunately, this was how I released the frustration I had felt in trying to adjust my independent style of research into the discipline that a large, collaborative effort required. At the end of the letter I thanked the other Samurai for a rewarding experience, but then concluded: "I guess I am not a big fan of 'socialized astronomy' . . . it's just not my style. Best to you all, but sayonara."

Of course, tantrums rarely have satisfying effects; this one put off my colleagues even more than my previous feisty letter. Ironically, shortly thereafter Donald wrote me an effusive apology—one of the kindest and most generous letters I have ever received, and I responded in kind to him, and eventually to the other Samurai, with the sincerest apology I could construct. But, even if Humpty-Dumpty was put back together again, the cracks showed; the others may have forgiven, but they did not forget. When all the rumblings had settled down, it was too late to submit the "Letter" to *Nature*, so I ate the paper and swallowed my pride as best I could. The Lynden-Bell et al. paper was a masterpiece; I wrote to Donald telling him how proud and grateful I was to be included on what was bound to be regarded as seminal work.

The Seven Samurai collaboration had been a success. The project had blended the different talents of seven scientists—this was an experiment in itself, one that ranged from the smoothness of a rowing crew to the awkwardness of a three-legged race. There had been difficulties with logistics and personality conflicts—for example, the frustration of working with "so many cooks" left me looking for a graceful exit on more than one occasion. By her own admission, Sandy had been stubborn, unwilling to accept Dave's persistent claim that "things are moving!" and my insistence that "there's something pulling over there!" until she slowly rediscovered these things for herself. These difficulties aside, though, the payoff of the collaboration was nowhere more evident than in the Lynden-Bell et al. paper, the main product of seven years' work. The contribution by each of the seven to the project was clear—it was the most balanced effort that any of us had participated in or even knew of. No one of us, or two or three, could have accomplished what we had done together. Sandy later wrote to me: "This paper represented collaboration at its best. What this group did was provide the supportive yet challenging atmosphere in which we all felt free to express any idea and ask any question, expecting that our colleagues would take the time and effort to weigh our idea and give incisive criticism. These group discussions, and the way each of us had his own favorite way to prove things, provided the forum for the learning process. Thus our final conclusions were proved not once, but many times. In this we were all together."

The Lynden-Bell et al. paper was a group accomplishment, to be sure, but it especially showcased Sandy's keen judgment and extraordinary scientific tenacity, and Donald's physical intuition and exceptional proficiency in mathematical analysis. Together, they brought our journey to its intellectual summit.

In May 1987, at a conference of the American Physical Society in Washington, D.C., I gave an invited lecture on the large-scale streaming of galaxies and its implication for cosmology. The proceedings included a press conference, where several researchers involved in this kind of work were to bring science reporters up-to-date on what had become the hottest field in astronomy. While trying to explain the enormity of a supercluster mass capable of pulling in galaxies on a cosmic scale, the name "Great Attractor" slipped out, as I waved my hands groping for words grand enough to describe the universe. These words were not grand enough, of course, but the press corps delighted in having "a hook"; the next day Walter Sullivan's story in *The New York Times* made certain that the name would stick. The other Samurai were surprised when they heard it, and not all of them were pleased. Although they were glad to see our work get public attention, they were justifiably worried about the consequences of attaching a flashy label to a serious piece of scientific work, one that required more than a few catchy phrases to be explained to a lay audience. In fact, the "Great Attractor" label gave us both wings and an albatross, and my fellow Samurai were happy to pass on both the credit and the blame.

Setting out to explore the properties of elliptical galaxies, hoping to learn more about galaxy formation, we had instead found the Hubble expansion was not smooth, but distorted. The redshift surveys had found that the galaxies were distributed in "lumpy" fashion; we had found that the expanding universe was a "bumpy" ride. We attributed the largest of departures from smooth Hubble expansion to an enormous swell in the density of matter—the Great Attractor. We imagined this huge overdensity to be made mostly of the same dark matter that had been found each time the universe was examined on large scale. As best we could tell, this vast wave on the dark ocean rose to a peak density several times "normal" at its center, and trailed off slowly, returning to near average somewhere around our own distance from its center, some 200 million light-years.

We could not say much about the form of dark matter in the Great Attractor—many of the candidates discussed earlier could provide the gravity attributed to this "extra mass"—but there might be clues in our work that would eventually help find the answer to this fundamental question as well. It was clear that many "extra galaxies" were also found in the vicinity of the extra mass, but we had insufficient information to decide if this swell in the matter distribution contained more dark matter than usually accompanied galaxies, or whether it was just a conglomeration of "normal" galaxy super-clusters whose dark fabric weaved together to form what amounted to one enormous mass. Although we had found only one Great Attractor, we assumed that such large structures had to be a common feature of our universe, so common that we had found one right on our doorstep.

We had come to the end of the beginning of our journey. We had returned from our trek into the cosmic outback, telling wild stories of mountains with a breadth and height not seen before. Many would doubt the credibility of this tall tale. Soon plans would be laid for new expeditions that could bring back confirming, compelling, and irrefutable evidence, or discredit the whole notion. But, for a while, there was time to pause and investigate the implications of this discovery for theories of how the universe, starting with the smooth sea of hot matter left by the big bang, developed the complex structure we see today.

As 1986 began, when the Seven Samurai were first formulating the bulk flow picture for the streaming of galaxies, the biased cold-dark-matter model was by far the most successful model in explaining how structure had formed in the universe. Its basis was a universe whose mass was dominated by weakly interacting particles, in which structure grew solely through the amplifying effect of gravity, and where galaxies were "biased" to preferentially inhabit the denser places. This model had apparently solved the fundamental problem: how the smooth, hot big bang evolved to the clumpy, cold universe of today.

Previous models, which were built around ordinary baryonic matter (protons, neutrons, and electrons) only, hadn't succeeded. The difficulty arose because ordinary baryonic matter interacts strongly with light, and, for the first few hundred thousand years, the energy of the universe was dominated by a blinding light that uniformly filled all space. If left alone, the baryonic matter would have started to clump and form structures, by the influence of gravity, but the intense light pushed baryons around and prevented fluctuations in the matter density from growing. Astronomers describe this by saying that matter and radiation were *coupled*. This state of affairs persisted for the first few hundred thousand years, but then the universe cooled sufficiently and the light and matter *decoupled*—when protons and electrons combined to form neutral atoms, their interaction with the light became much weaker. From this point on, matter was free to clump as it wished—gravity could do its work. But such a delay in "growing" the initial fluctuations would be fatal: calculations showed that if fluctuations were allowed to increase in strength only after the epoch of decoupling, they could not have grown big enough to form galaxies and clusters by the present day.

Enter dark matter. Non-baryonic dark matter, the kind that interacts weakly with ordinary matter and photons, would not be bound to this "no-growth ordinance." Because, by definition, they do not respond to the electromagnetic force, weakly interacting particles would have been oblivious to the intense light energy. Thus, fluctuations in the density of non-baryonic dark matter could have grown from Day One by the coalescing effect of gravity. The dark matter could do the hard work, growing bigger and bigger lumps for the first few hundred thousand years, ready and waiting with deep pockets for the baryonic matter to slide into, as soon as baryons were set free from the photons. The computer simulations showed that this head start was the feature needed to ensure the formation of today's galaxies and clusters in the available time of about 15 billion years.

Both cold-dark-matter (CDM) and hot-dark-matter (HDM) models shared this attractive feature, but the CDM model produced a better match to the observable universe. CDM's strongest proponents, the team of Davis, Frenk, White, and Efstathiou, reported success after success in accounting for all sorts of clustering properties, and even some aspects of galaxy formation. They had to make only one addition—biasing—to the simple picture that Blumenthal, Primak, Faber, and Rees had put forward when they suggested that the slow-moving, weakly interacting particles would provide excellent seeds for the growth of structure. Biasing was a way to install galaxies, like lights, on the dark superstructure below. By preferentially placing them where the density of dark matter was highest, the CDM model succeeded in matching both the gross features of galaxy maps that had been made, and the small (relative) peculiar velocities that teams like the Seven Samurai had found for galaxies with respect to their surrounding neighbors. The unbiased model, where galaxies were placed in fair representation of the dark super-structure, seemed unable to pass both of these tests.

Before the work of the Seven Samurai, the challenges to the biased CDM model had come from observations of the distribution of galaxies, which offered sheets and chains of galaxies, large areas devoid of galaxies, and strong clustering of galaxy clusters. Some thought that various of these features were not well enough reproduced in the N-body simulations—that the model didn't produce enough large-scale "power" or the right topology. But without good tests for comparing model and observations, the biased CDM model withstood the challenges, or at least it hadn't succumbed.

The Seven Samurai finding of large peculiar velocities, described first as a "bulk flow" streaming in our region of the universe, had sent rumblings through the CDM world. Here was a result that was describable in quantitative terms, and it referred to the clustering of dark matter, the very thing the N-body models were simulating. Dick Bond had insisted that either the biased CDM model, or the interpretation of large-scale streaming, was wrong—they couldn't peacefully coexist because there was no way for the model to develop such large, coherent departures from smooth Hubble flow. But Nick Kaiser had countered that the conflict was not as serious as the Seven Samurai had suggested, because it was still fairly uncertain that the large-scale stream-ing covered as large a volume as had been claimed.

The challenge had now been framed in a new way—the Great Attractor—but the contest was like instant replay. According to the biased CDM model, the largest coherent structures, a giant supercluster or void, could be no bigger than about 150 million light-years across. The Great Attractor was supposedly almost three times as large. This was the same old criticism that had been aimed at CDM—not enough "power" on large scales to make really big structures. Ed Bertschinger, a physicist at MIT who was rapidly becoming a "player," and Juszkiewicz had compared the Great Attractor model directly to biased-CDM N-body simulations they had made, and de-cided that such a large overdensity as the Great Attractor would be less than a one-in-a-million shot. If the biased CDM simulations were a good model, one shouldn't expect to find even one structure as large as the Great Attractor in the entire observable universe, let alone on one's doorstep. But again Kaiser would inject that the disagreement between model and data was being overstated.

There had been some continuing doubts about data quality, methods, and techniques, but the Samurai had been very careful, and we were much more knowledgeable than our critics, of course, about what we had done. Berkeley theorist Joe Silk suggested that perhaps the properties of elliptical galaxies varied from one volume of space to another—what we ourselves had labored so long to prove, and eventually rejected. But this was an arbitrary, a posteriori attempt to explain large-scale streaming as an illusion, and it found little sympathy among other theorists. This challenge, along with sporadic efforts to show that we had done something really wrong in our analysis, was ineffectual.

The most vigorous assault on the Samurai results came from Marc Davis and his graduate student Michael Strauss. They took a different tack altogether and tried to show that the large-scale streaming *couldn't be right*, never mind what mistake we or nature had made that no one could quite put a finger on. The basis for their disclaimer was an important new galaxy catalog, one made with an orbiting satellite-telescope, *IRAS*, that viewed the sky in far-infrared light. Two features made the *IRAS* galaxy survey unique: it was the first catalog of objects made uniformly over the entire sky with one telescope (it was not limited to one hemisphere like an earthbound telescope). Furthermore, far-infrared is much less absorbed by Milky Way dust, so the galaxies could be detected down to "latitudes" very close to the plane of stars, gas, and dust that defines our galaxy. Both these aspects made it the most uniform, most completely sky-covering galaxy survey yet made. Of course, there was a catch. Not all galaxies are strong emitters of far-infrared light, so for every three seen in visible light only one made it into this new catalog. But it was hoped that these "infrared-bright" galaxies were sufficiently representative that this would be the best map to date of the distribution of *all* galaxies in the local universe.

This new map started out as a flat map on the sky, as all galaxy maps had. Some of the galaxies had measured redshifts, which gave an estimate of their distances, but many were unknown. Strauss and Davis went back to the earthbound telescopes and spent two years collecting about half of the 2,500 redshifts that would turn their galaxy catalog into a three-dimensional map.

With this three-dimensional map of the whole sky (out to a limited distance, of course), Strauss and Davis were able to do something Marc couldn't do with his earlier CfA and Southern Sky redshift surveys, which covered only the polar regions of both galactic hemispheres and were probably not perfectly calibrated to each other. With this new map they could *predict peculiar velocities*, as follows: Strauss and Davis started with the assumption that the *IRAS* galaxies were fair tracers of all the matter, luminous and dark. In other words, they assumed that a higher density of *IRAS* galaxies marked the location of a crest in the density of dark matter, and a dearth of *IRAS* galaxies signaled a trough. They then calculated the net gravitational pull on each *IRAS* galaxy from all the galaxies in the survey (which by assumption was to represent the pull of all the mass, including dark matter) and in this way predicted which way each galaxy would move and how fast— its peculiar velocity.

Strauss and Davis wrote a computer program that generated such a map of *predicted* peculiar velocities, for comparison with the actual measurements by Aaronson et al. and the Seven Samurai. The agreement between prediction and observation was good for the spiral sample studied by the Aaronson collaboration, which was generally confined to galaxies within 80 million light-years of the Milky Way. However, for the elliptical sample of the Seven Samurai, which reached three times as far (thereby covering a volume some thirty times bigger), the agreement was poor. For example, the Strauss-Davis map predicted only a small flow of at most 500 km/sec in the Centaurus region, a peak less than half that found by the Samurai. According to them, the Great Attractor was not so great, and its influence in accelerating our own galaxy was offset by the pull in the opposite direction by another supercluster, in the direction of the constellations Perseus and Pisces. Our galaxy's peculiar velocity of 600 km/sec, they argued, arose not from the pull of distant concentrations of matter like the Great Attractor, but from matter closer in. With the influence of gravity thus reined in, the Strauss-Davis *predictions* of peculiar velocities were in no great conflict with the biased CDM model that Marc was so fond of.

It was hard to know how to respond to this criticism, since it included no suggestion of why our result might be wrong, only that it *had* to be wrong. Peculiar motions were a relatively new element in the large-scale structure game, and few of our colleagues even appreciated the difference between actual measurements of peculiar motions and *predictions* of them, as had been provided by Strauss and Davis. (Many times over the next year, I was asked about the reliability of measurements of peculiar velocities in light of the fact that two of the sources, Strauss-Davis and Samurai, didn't agree!) Mostly, we simply repeated that we had made measurements of what the peculiar motions are, not what they *should be*. If the two didn't agree, most likely it was because one of the assumptions made by Strauss and Davis was incorrect, for example, that *IRAS* galaxies were fair tracers of the location of dark matter. The possibility that *IRAS* galaxies might not closely follow the distribution mass would be, of course, an important discovery in its own right.*

In the end, only one volley lobbed at the Great Attractor model really hit its mark, the criticism by Kaiser that we had not "seen" the whole structure. Our expedition, like the Donner party in a blinding snowstorm (an analogy that had better not be taken too far), had pulled wagons up a steepening hill. We knew for certain only that the terrain was rising ahead of us, and that to the left and right the ground was level. With less certainty, we had made an educated guess where the center of the mountain lay and how high it was, but we had not actually climbed it on our first expedition. Because of the choices we made when we drew up our first plans, we were destined to turn back before we had taken its full measure. So it was a fair warning when Nick Kaiser, who had previously emphasized that the "bulk flow" might only describe a smaller volume than the Samurai had originally proposed, made a similar point about the Great Attractor. He agreed that a vast, spherical overdensity might be the *simplest* description, but it was not the *only* one—the actual data did not *compel* us to the vast, spherical Great Attractor. We would need to venture out again, to climb the summit of the Great Attractor and descend its far slope, to be sure.

The Great Attractor lay south, its heart about 50 degrees below the celestial equator—out of reach for Northern Hemisphere telescopes. Because of my access to Las Campanas, it was clear that among the Samurai, I was the one best positioned to mount a second expedition to cross the summit of this majestic peak. One foray had already begun—my survey of 1,403 redshifts would provide the first good map of the galaxy distribution in this direction. Blessed by good weather, my observing run of ten nights in April 1987 pushed this ambitious program past its midpoint faster than anybody— even I—expected. The new data confirmed what I had found in January—

---

*Later, this supposed disagreement between the Strauss-Davis predictions of peculiar velocities and the Seven Samurai measurements largely evaporated when Amos Yahil, in collaboration with Strauss and Davis, found the Achilles' heel of their method. The pitfall was that gravity is an effective force over long range; one needed to know the undulations in galaxy distributions out to very great distances in order to predict accurately which way and by how much each *IRAS* galaxy would be set in motion. The *IRAS* sample of Strauss and Davis simply ran out of steam before it reached far enough—it was too sparsely sampled at great distances. As a result, only "local measures" of the peculiar velocity, for example, galaxy motions relative to our own Milky Way, could be well determined; *absolute* peculiar velocities—those referenced to the cosmic background radiation, as the Samurai had done—would have to await a deeper, more complete sample.

the great concentration in galaxies lay *beyond* the Centaurus clusters. Bob Kirshner, recently relocated to Harvard, was utilizing the forty-inch Swope reflector to study the Great Void, and I was using the du Pont telescope to sort the galaxies in the Great Attractor. We were driving ourselves to ecstatic exhaustion as we delighted in a long spell of clear skies.

Sometimes, euphoria is beheaded by careless fate. Our exultation was demolished when Carnegie astronomer Ian Thompson told us, over the short-wave radio, that Marc Aaronson had been killed in a gruesome, freak accident at the Kitt Peak National Observatory. He was an admired colleague and a pioneer in the very work in which we were so passionately engaged. But, more than this, he was a close friend to each of us—a person we knew would always be around. His departure would leave a void as great as any we would find in the skies.

In summer 1987 I processed the data for the redshift survey, replacing the estimates I made at the telescope with accurate measures of each galaxy's velocity. The paper I wrote included some 900 of the 1,400 targeted redshifts, enough to sketch a fair picture of the swell in galaxy numbers that accompanied the Great Attractor. The midstream of the river of light that flowed across Lahav's map of the southern galaxy sky lay at a distance whose Hubble expansion velocity averaged about 4,500 km/sec, the exact distance predicted by the Great Attractor model. It was the first confirmation that our basic idea for the source of the streaming flow of galaxies was correct.

This map would be a prerequisite in planning a more ambitious journey—crucial new measurements of peculiar velocities that would confirm the large peculiar velocities in Centaurus and the convergence of the flow toward the Great Attractor center. This is the evidence that the theorists demanded. I invited Sandy to collaborate with me in the more challenging task of obtaining peculiar velocities for many more galaxies in the Great Attractor region. The original Samurai sample included only a few dozen ellipticals in this region, but in short order we had a candidate list of nearly 100 more. It seemed surprising that so many targets in the region had been missed by the Samurai, but we had since learned that, despite the team's concerted effort to choose a uniformly deep sample over the entire sky, our survey had been considerably more shallow in the south than in the north. This defect was traced to a problem with the ESO catalog: the fainter elliptical galaxies had been systematically misclassified as S0 galaxies; because of this, more distant ellipticals had been kept out of the Samurai sample. If we had known about this four years earlier, our first expedition would have probed far enough to reach the Great Attractor center.

Sandy and I had paid the usual dues of inspecting each galaxy on the photographs of the Southern Sky Survey. This time we also included S0 galaxies, cousins of elliptical galaxies that also have disks. I had just written a paper including data for thirty-six S0 galaxies in the Coma cluster, which reported that reliable distances to them could also be found using the same relation between $D_n$ and velocity dispersion. When we headed south in March 1988, our goals were to obtain both the spectra of these galaxies, using the du Pont telescope, and the photometry, with a CCD (charge-coupled device—a high-precision TV camera) at the Swope telescope. Again, we enjoyed good weather; in ten nights we collected data for more than one hundred galaxies—about one-quarter of the entire work of the Seven Samurai.

It was Sandy's first trip to Las Campanas and a treat for me, her student, to show her my turf, this extraordinary site that had meant so much to my career. Sandy, now a Trustee of the Carnegie Institution, could appreciate firsthand the value of this remote outpost, whose establishment had engendered personal sacrifice and struggle, and financial hardship for the Carnegie Institution, not to mention the divisiveness that led to the dissolution of the long-standing scientific marriage between Carnegie and Caltech (but that's another story). It seemed wonderful and, in a way, unbelievable that we could be here doing what we were doing, so single-minded in our purpose to reach the center of the Great Attractor and beyond, while the world carried on—somewhere else—in its mad ways. We could look into the night sky, the grand Milky Way a roof over our heads that said "home," and know that we had received the supreme prize—the freedom to look up into the sky and wonder.

It was late fall of 1988 before I had cleared the time to immerse myself in the processing of these new data. A solid month of computer work was needed to prepare the spectra and extract measurements of velocity dispersions, and to turn CCD pictures into reliable maps of the brightness of each elliptical galaxy. Dave Burstein was very helpful in evaluating the quality of the photometry, and for each elliptical in the sample he provided the important estimate of obscuration by dust in our galaxy. The work with the Samurai had been good preparation, and Dave's counsel was valuable, but I was still surprised how much work there was to do before galaxy distances and peculiar velocities could be derived with confidence.

Finally came the fateful day, between Christmas and New Year, when, in the quiet of the nearly empty Santa Barbara Street offices of the Observatories, I finished entering the completed measurements of velocity dispersions, velocity, and $D_n$ diameters into a computer file. The program to combine these data and calculate the distances and peculiar velocities was written and tested, but I had been careful not to look at any preliminary "results" that might tempt me to alter my data-processing prescriptions. As much as possible, I wanted this to be a "blind" test, because there was so much at stake. Now it was time to set in motion a sweep through data for more than 136 elliptical and S0 galaxies that would tell if the Samurai had been right or wrong. I took a deep breath—for the first time apprehensive that this wonderful adventure was about to take a bad turn—shrugged, hit "return" on the keyboard, and tensed in my chair while the plot began to materialize, point by point, on the screen.

Twenty seconds later, I clenched my fist and drove it skyward. "Yes!" There was the Hubble diagram, the simple plot of distance versus velocity, but with a twist that Hubble himself had never seen—points that ran well above the line representing smooth, uniform Hubble flow, then crossed over and dived below. On the left-hand side, representing the closer galaxies, dozens of points lay above the line—that is, they had too high a velocity for their distance. These were galaxies streaming at the same high velocity the Samurai had found for the Centaurus clusters, but there were many more of them now and they were broadly distributed over this region of sky. The large-scale streaming flow was confirmed. Even better, we had pushed beyond the frontier opened by the Seven Samurai, to more distant galaxies that drifted back to the Hubble line: at a distance corresponding to a Hubble expansion velocity of about 4,500 km/sec, near the center of the diagram, elliptical galaxies had an average velocity of 4,500 km/sec—no peculiar velocity. This is just what the Great Attractor model predicted, that the center of the giant mass would be at rest with respect to the cosmic microwave background. As the Great Attractor slowed the Hubble expansion, it would generate peculiar velocities coming toward its center from all sides, but the center itself would be stationary, on average, as galaxies converged and crossed from all directions. Finally, on the right-hand side I saw the first traces of infall from the far side, beyond the center of the Great Attractor where galaxies were being pulled back and, thus, coming our way. This showed as a few points *below* the Hubble line for the most distant galaxies.

The new evidence for large peculiar velocities on the near side of the Great Attractor was now compelling. A substantial number of points showed the decline in peculiar velocity at the putative center of the Great Attractor, even if the few points farther out were only suggestive of "backside infall."

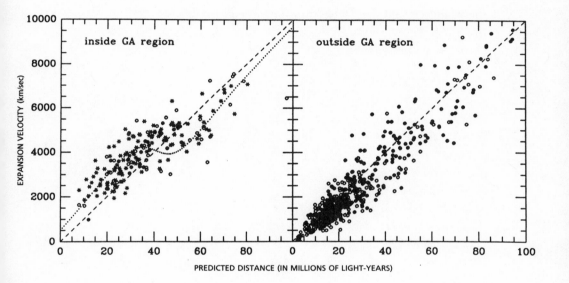

Hubble diagrams from the data by the author and Sandy Faber, showing the predicted "S-Wave" curve for the Great Attractor region. The diagram at the right shows the "average" relation between distance and velocity, but in the Great Attractor region the large peculiar velocities cause the data for both spiral and elliptical galaxies to lie above, then dive below, the average Hubble relation. *(Reference: 1990* Astrophysical Journal Letters, *vol. 354, L45.)*

But the triumph was that the Seven Samurai had *predicted* just this pattern, and the new data confirmed these predictions as well as they could have. Opportunities to make a prediction and gather new data to test it are unfortunately rare in astronomy compared to other sciences—the impossibility of performing classic experiments on our subjects has that sad consequence. So, in our role as detectives, we had to solve the mystery and look for other clues left by a careless suspect. This, then, was a rare moment, one to be savored, but, more so, to be shared. I rushed to the laser printer, made a "hard copy" of the plot, and immediately sent fax copies to Sandy and Dave, and called them on the telephone. This was news: We had been to the mountain; we and the Great Attractor had survived.

# $IO$ • ROAR FROM THE PAST

So many of our fathers' stories begin, "When I was a boy," that we groan almost before all the words are out. But, just this once, the lead is too good to resist. "When I was a boy," it goes, "there were two universes."

One was an eternal universe without boundaries. Though expanding, this universe grew no emptier. Like a thousand clowns tumbling out of a tiny circus car, the "steady state" universe accomplished this cosmic miracle by invoking the continuous creation of atoms that built new galaxies to fill the spaces left by those outward bound. So the universe had been, and so it would be, forever. Agreeably, there was no beginning and there would be no end.

The other universe was a big production number—the true show of shows. This universe began with a bang, a *big* bang, as Fred Hoyle, the author of the steady state theory, derisively named it. Before the bang, there was . . . who knows what? Was there even *time* before the big bang, was there even a *before*? At any rate, the universe had been incredibly dense and small when young, very different than it is today, and in its remote future it would be different again—this was a universe of change, of evolution.

The smart money was on the "Bang," but the heavy sentimental favorite was the "Steady." Even though some of the world's brightest scientists knew that the evidence was tilting in favor of the big bang, they candidly admitted that they found the model naive and unappealing, perhaps even frightening—who liked the thought of descending from violence and chaos, or from nothing at all? They preferred the steady state—it had style and sophistication, and skirted those embarrassing questions about the beginning and end of time—

you know, *creation* and all that. Journalists had an insatiable appetite for this rare example of scientific controversy. The public delighted in the double pleasure of being so close to the secret of the cosmos and having a colorful contest to choose the winner.

Things are a lot duller today. The big bang has reigned as champion for more than two decades now. The Super Bowl of cosmologies didn't live up to its hype and, curiously, the public has not caught on that the vast majority of scientists declared "game over" a long time ago. A great deal of detail has been added to the big bang model—thirty years ago the big bang was little more than a conceptual framework—and there is much left to understand, but most scientists have come to believe that the big bang is a remarkably successful model of the story of creation, and far more instructive than the steady state. What made them choose?

The key observation, without doubt, was the discovery in 1965 of light waves at radio and microwave frequencies, streaming in from all directions. It's a wonderful story. The discovery of this *cosmic microwave background* was serendipitous and completely independent of predictions made by theoretical physicists that a glow of radio light would be an enduring remnant of a universe that was once extremely hot and dense—the *primordial fireball*, as it's sometimes called. This was clearly not a case of adjusting the model to fit the data, rather the model had predicted that the greatest energy in the universe was yet to be discovered. The power of prediction is what gives a scientific idea weight and credibility.

The story of the discovery of the cosmic microwave background begins in the 1940s. George Gamow, a particularly imaginative physicist, had seized upon the idea of Belgian mathematician/priest Georges Lemaître that an expanding universe was a "cooling universe"—as such, it must have been much denser and hotter in the remote past. Gamow was well equipped to pursue the implications of this notion. In particular, he wondered if the early universe had been hot enough and dense enough, like the center of a star, to fuse hydrogen into helium and the heavier elements that astronomers collectively call "the metals." Could creation itself have delivered all the "parts" from which our universe is built? Alas, Gamow concluded, it could not. In its outward rush, the universe would have remained extremely dense and hot for only a few minutes. This was enough time to produce a fractional helium abundance of 24 percent, by the process of two protons fusing into deuterons (a proton bound to a neutron) and two deuterons fusing into a helium nucleus. This is, in fact, just the fraction of helium that has been found in the most pristine gas clouds—"primordial" gas—so this prediction was a great success. But, because particle collisions became less energetic and less frequent as the universe cooled and thinned, fusion was unable to build the other common elements such as carbon, nitrogen, oxygen, and iron.*

Now comes the fantastic part. In the course of these investigations, Gamow's colleagues, Ralph Alpher and Robert Herman, discovered a subsequent event with observable consequences, one that would have happened eons after the first few minutes of *the nucleosynthesis* Gamow had been thinking about. If, as they believed, the universe passed through an extremely hot, dense phase, there would be a faint glow—no less than the twilight of the big bang itself—still with us today. Finding the fading glow of the big bang would be a strong demonstration that this extraordinary event actually happened.

What led to this remarkable prediction? Alpher and Herman were investigating the interaction between light and matter in the early universe, an interaction whose character changed markedly a few hundred thousand years after the big bang. Before, free electrons and protons roamed the universe. These are the components of ordinary hydrogen gas, but with a temperature of over $10,000°K^{†}$ the electrons were too agitated to settle into stable orbits around protons. Formation of cold hydrogen gas clouds and their collapse into stars and galaxies lay far in the future.

Mixed with this screaming sea of protons and electrons—an *"ionized"* hydrogen gas—was an intense glow of light, the main repository of the energy of the big bang itself. These photons dominated the energy content of the universe for the first few hundred thousand years. Matter was a sideshow, but an important one, because photons—packets of electromagnetic energy—interact strongly with electrons. In a process physicists call Compton scattering, photons were continually deflected from their intended straight paths to slightly altered courses and energies. The situation was not unlike the struggle of light to fight its way out of the Sun, where light is absorbed and reemitted at longer and longer wavelengths as successively cooler layers are crossed. As in the Sun, then, the light energy of the early universe was tightly coupled to the temperature of the ionized gas of electrons and protons: when the temperature was billions of degrees, the light was in

---

*In the 1950s, work by Geoffrey Burbidge, Margaret Burbidge, Willy Fowler, and Fred Hoyle solved the puzzle of the origin of the heavy elements. In one of the classic papers of the century, $B^2FH$ (as the paper is reverently called) showed in great detail how the proper amounts of the chemical elements heavier than hydrogen and helium are produced in the centers of stars which, in contrast to the big bang, maintain high densities and temperatures for extraordinarily long times.

†Recall that Kelvin degrees are like Centigrade degrees except that the zero point of the °C scale equals 273°K. For very high temperatures, like 10,000° or more, this difference is so small that the Centigrade and Kelvin scales can be thought of as equivalent.

the form of extremely energetic gamma rays; when the universe had cooled to millions of degrees, most of the photons had been degraded to less energetic, but still powerful, X rays.

The result of this strong coupling of photons and electrons was that no photon traveled very far before its direction and energy were altered. In other words, the universe was opaque—had there been anyone around to look, the universe would have appeared like a fog. No information could be propagated for any significant distance or time. (In a "real" fog water droplets scatter the light, so the images of light sources, or objects they illuminate, are diffused or dispersed altogether.) However, the universe was continually cooling and thinning, so the uninterrupted paths of photons grew longer as time passed.

Then came an "event." As the universe cooled below a temperature of 10,000°K, electrons began to drop into their orbits around protons, which vastly reduced their ability to scatter the omnipresent photons. Suddenly (in astronomical terms, actually over the next hundred thousand years), the universe became "transparent," and photons that had been scattered countless times were never harassed again. They would sail forever across the universe—unimpeded, unless they ran into a rock, a star, a rockstar, or a radio telescope. They are still traveling today, and, by their color (energy), they record the temperature of the universe when they were set free.

From well-understood physics, Alpher and Herman were able to estimate that the temperature when the universe went opaque to transparent was about 5,000°K. Matter at this temperature radiates most of its energy in visible light, like the Sun, for example, whose "surface" is about this hot. However, this *decoupling* of matter and light took place when the universe was 1,000 times smaller than it is today, and the rapid expansion of the universe *redshifted* this light a thousandfold. (One way to think about this is that each of these cosmic photons that finds its way to us today had to fight its way back through a rapidly expanding space, its energy degraded as its wavelength was stretched by a factor of 1,000.) Alpher and Herman predicted, then, that the "roar" would be reduced to a whisper: the photons crossing today's cold space would appear to have come from a gas at a cool 5°K instead of 5,000°K, their energies redshifted from visible light all the way down to far less energetic radio waves.

That was back in the 1940s. Few people took the big bang idea that seriously, and anyway the radio waves Alpher and Herman were predicting were far too weak to be detected with the technology of the time. Whatever the reason, the work, and the paper Alpher and Herman published, were soon forgotten.

The story moves ahead twenty years, to 1965, when Arno Penzias and

Robert Wilson, physicists working at the AT&T Bell Laboratories, accidently discovered the *cosmic microwave background* (CMB) that Alpher and Herman had described, a prediction they knew nothing about. Reports of an unidentified source of "noise"—a hiss degrading the quality of telecommunications using microwaves—had sifted through the Bell Labs for years. Given a radio antenna shaped like a giant squared-off earhorn, Penzias and Wilson were asked to investigate this and other problems associated with relaying radio signals using the new Telstar satellite, and were promised that they could do some radio astronomy as an encore. They soon found the annoying interference and spent months testing to see if it came from the equipment itself (up to, and including, contamination by pigeon droppings). They found nothing wrong, but still they were reluctant to identify the "noise" as *extraterrestrial*. If this energy was coming from space, why did it arrive from all directions with equal intensity—*isotropically*, rather than from the direction of a certain astronomical source, like the Sun, or the center of our galaxy?

It took a visit from next-door neighbors—a group of Princeton physicists led by Bob Dicke—to convince Penzias and Wilson that they were really onto something. The Princeton team had independently rediscovered and refined Alpher and Herman's prediction of a cosmic microwave background, and were busy planning their own search when they learned of Penzias's and Wilson's result.* Jim Peebles, then a young postdoctoral fellow whom Dicke had steered into investigating big bang cosmologies, was astonished to learn that these audacious theoretical speculations about the birth of the universe might have more than a toehold in reality.

This was a defining moment in the human search for understanding the universe. What had begun as a fanciful notion, and a wild one at that, had been mixed with good solid laboratory physics that had taken a hundred years to piece together. The big bang model, added to now well established rules of how matter behaves on an atomic level, had produced the astounding prediction that the universe would be filled with a glow of radio light. When humans devised the means to look for this light, they had looked, and, lo and behold, there it was.

A new window on the universe had been found, a window that looked back into the previously unimaginable past. And, again, far from an ending, this discovery would lead to more subtle, sophisticated observations. These

---

*Even though Bell Labs is within thirty miles of Princeton, Dicke's group knew nothing of the work by Penzias and Wilson, until a young radio astronomer, Bernie Burke, visited both groups in 1965 and made the connection.

would provide the first measurement of a peculiar velocity, that of our own galaxy, and the notion of a cosmic frame of reference. And, eventually, the very seeds that galaxies and superclusters had grown from would be revealed in the glow of this ancient light.

Alpher and Herman did more than just predict the *existence* of the cosmic microwave background, they provided a description of its *spectrum*—the intensity of light over a wide range of radio frequencies. They were able to do this because of their understanding of the electromagnetic interaction between light and matter, and their knowledge of the temperature of the universe when the CMB was emitted. When matter and light are in a state of equilibrium—matter emitting the same amount of light energy as it is absorbing—intensity of light varies with "color" according to what physicists call a "blackbody spectrum." This is a lopsided bell-shaped curve that exactly describes at what wavelength (or equivalently, energy, or color) most of the light emerges, and then how the intensity falls off at longer and shorter wavelengths.* The temperature that Alpher and Herman predicted, 5°K, corresponded to a blackbody spectrum with peak intensity (the most photons) at millimeter wavelengths, with declining intensity—in a rigorously defined way—at longer and shorter wavelengths.

In the years that followed Penzias's and Wilson's detection of the cosmic microwave background, physicists shifted their attention to measuring the CMB spectrum to see if it followed the blackbody shape. Dicke's group had refined the calculations and revised downward the predicted temperature to around 2.7°K. This meant that the present-day universe was even colder—closer to absolute zero, so the blackbody spectrum should be shifted to slightly longer wavelengths than Alpher and Herman had thought. Subsequent, difficult measurements by this Princeton group and others of the intensity of CMB at other radio wavelengths little by little confirmed that the spectrum had the blackbody shape at the predicted temperature of 2.7°K. The *pièce de resistance* of this measurement came in 1990, when NASA launched the Cosmic Background Explorer satellite, *COBE*. The *COBE* satel-

---

*German physicist Max Planck found this form of spectrum for macroscopic bodies that are "black," that is, absorbing all the energy incident upon them. A clay pot being fired in a kiln is an example: after the pot reaches oven temperature, it glows with the same color as the kiln walls. Even a body whose temperature is different from its surroundings, like an Eskimo sitting in an igloo, or the Sun radiating into cold space, has a spectrum very similar to that of a blackbody. According to the big bang model, the early universe expanded slowly enough to maintain near-equilibrium conditions, thus the cosmic background radiation is predicted to be that of a (red-shifted) blackbody spectrum at the temperature of the universe when matter and light decoupled.

lite was able to make measurements with unprecedented accuracy and for radio wavelengths that were blocked by the atmosphere of Earth. Within days of operation it showed that the spectrum is a blackbody to exquisite precision—it follows the predicted blackbody spectrum to one part in a thousand—another stunning confirmation of the big bang model.

It is hard to overemphasize the accomplishment of the big bang model in making these predictions about the cosmic microwave background. The CMB photons dominate the energy density of the universe—every cubic centimeter of space contains about 400 photons left over from the big bang.* The steady-state model made no such prediction. Even though its few remaining proponents have suggested other ways to explain the cosmic microwave background, whether plausible or not, these explanations are after the fact—a crucial difference.

As I discussed earlier, the big bang model has had another remarkable success, in the area of nucleosynthesis—the production of the chemical elements. Twenty years after Gamow's work showing that the primal abundance of helium could be produced in the big bang (but that most of the "metals" could not), more detailed calculations showed that other light elements, among them deuterium and lithium, would also be assembled in the first few minutes. Though a billion times less abundant than hydrogen or helium, these rare elements have subsequently been measured in spectra of "first-generation" stars, confirming the theoretical predictions of the big bang model with phenomenal accuracy.

Finally, another type of evidence attests to the most obvious implication of the big bang, that the universe is evolving. The big bang model requires the universe to have been very different in its youth, and demands that the elements, the stars, and the galaxies all date back to the time of the beginning of the expansion. The "nuclear timescale" comes from the fact that some of the heaviest chemical elements are not stable—they break down into lighter atoms at exceedingly slow rates, lending themselves for use as "atomic clocks." Studies of these rare elements in our solar system indicate that they were manufactured about 10 billion years ago. The "stellar timescale" comes from computer models of the evolution of stars and their comparison with observations of temperature and luminosity. This analysis gives ages for globular clusters—the oldest star clusters in the Milky Way—of about 15 billion years. The expansion age of the universe is determined by "running

---

*One used to be able to make the CMB less abstract by pointing out that one percent of the blips of "snow" on a blank TV channel are big bang photons, but it is hard to find a blank channel nowadays.

the universe backward'': the rate of expansion—the *Hubble constant*—is still a subject of controversy, but nearly all workers in the field agree that the expansion age of the universe is between 10 and 20 billion years.

The agreement of these three ages, determined in very different ways, is more than encouraging. Of course, the astronomers who support the view that the expansion age is only 10 billion years are legitimately worried about the implication that the oldest stars predate the expansion—these two ages must be reconciled if the model is to make any sense. However, it is crucial not to let this concern cause us to miss the point: the approximate agreement of the "atomic clock age," the ages of the oldest stars, and the expansion age of the universe is astonishing. These three numbers could range over orders of magnitude: the fact that they agree within a factor of two is compelling evidence that the basic tenet of the big bang—a creation event—is correct. Age discrepancies of a factor of two, if they persist, will drive us to dig more deeply into theory and observations, but they certainly do not force us to abandon the big bang in favor of a model like the steady state, which offers no insight at all as to why these "ages" would be so similar.

The big bang seems to be the correct framework within which to model the universe. It tells us that, if we look far out into space, we will see back to a time closer to the birth of our universe. The younger universe should have been measurably different; again, this prediction is in marked contrast to the steady-state model. Of course, the CMB itself is the best example, a look back to a time when the universe was filled uniformly with hot gas. However, we can also survey the more recent epoch of galaxies to look for signs of evolution. The light from distant galaxies arrives with the image of the remote past (the simple consequence of the finite speed of light), so younger galaxies can be seen directly—like living fossils. Analyses of the spectra of such galaxies, at "lookback times" of 5 to 10 billion years, show that the average galaxy was forming stars more vigorously at this earlier time, and was composed of a noticeably younger population of stars. Recent observations of distant, rich clusters with the Hubble Space Telescope, by a team led by the author, show clearly that these star-forming galaxies are spiral galaxies, a type that is absent from clusters today. There is no doubt about it: the galaxy population is evolving.

However, the most striking example of evolution has come from studies of quasars, bright beacons that inhabit the centers of rare galaxies. Because they are extremely luminous, quasars can be seen at enormous distances, at lookback times up to 80 percent of the age of the universe. Quasars are predominantly found at high redshift, which is evidence that they were much more common when the universe was only a few billion years old.

Apparently, these beacons have been fading, or fewer of them have been born, as time has progressed.

The cosmic microwave background, the agreement between the expansion age and the age of the oldest stars and chemical elements, the cosmic abundances of helium and other light elements, and evidence that objects in the universe are evolving—all these testify to the big bang's astonishing success as a model for the birth of the universe. No other model even comes close.

∞

The mere existence of the cosmic microwave background confirms that the universe was hotter and denser in the past—the basic principle of the big bang model—but, more than this, precise measurements of its spectrum and intensity hold a treasure of information about the distant past and the present. However, capturing that treasure has required technical capabilities that have emerged only in the last decade, so only recently have we begun to realize the phenomenal payoff from this signal from early times.

The key word is *precision*. Measurements are made, in science and in everyday life, with a certain degree of precision. It usually isn't necessary to gauge one's position better than a couple of inches when going through a doorway, but when wielding a carving knife, such an error can be disastrous. It takes little precision to cut the lumber for a tree house, but quite a bit more to build a parson's chair. The ruler one uses for making the chair is completely inadequate for measuring the steel pieces of an automobile, and the devices used for that task aren't nearly good enough to check the size of the turbine blades in an aircraft engine. Each represents an improvement in measurement accuracy that is crucial for achieving the desired result.

Many of us are not "builders," but all of us rely on the measurement of time to go about our daily routines. Relying on a faulty watch is, for many things, inconsequential. You won't be expected to show up at *exactly* eight o'clock for a party (unless it's a surprise party), but the half-an-hour margin for error you have there could cause you to miss the opening curtain for a Broadway show, and the five- or ten-minute accuracy you need to make that event won't be good enough to catch the train into the city or the bus to the airport (one hopes). People who run the evening news need to know the time to the second, and the person who designed your television set had to be able to measure time to a *millionth* of a second, since the electron beam in a television tube is changing its intensity that rapidly as it paints the picture. Even though accurate measurement is not something most of us think about in everyday life, it is there.

I have labored a bit with these examples because the ability to measure things accurately is crucial to how we learn about the world through science. In our story of the exploration of intergalactic space and elliptical galaxies there have been many examples—for example, measurements of spectra. The modest dispersion of the Sun's light into colors by water droplets—a rainbow—is sufficient to determine that the peak intensity is in yellow-green light and that the intensity declines to the red and blue. These measurements provide all the accuracy necessary to decide that the Sun has the classic "blackbody spectrum" and that the temperature of its surface is about 5,000°K. But the light must be spread into much finer divisions of color if one wants to see the dark absorption lines—the missing light at certain colors—due to hydrogen, iron, carbon, magnesium, and other elements in the Sun. Measuring the redshift of a galaxy amounts to detecting such absorption lines in the blended light of millions of stars in the galaxy, and finding out how much they are shifted in color due to the velocity of the galaxy away from us. But, in order to measure the velocity dispersion—the typical speed of the stars within the galaxy—the intensity of the light at every color must be known more accurately still, in order to measure how wide (in color) the dark lines are. What was good enough for *detecting* the lines may not be good enough for measuring their widths, the way it is hard to read in very poor light even when you recognize that you have print in front of you.

The case of the cosmic microwave background is much the same. Penzias and Wilson first *detected* the light, at a certain radio wavelength (color), but they and the others who followed had difficulty in measuring the intensity of that light very accurately, particularly at other radio wavelengths for which the Earth's atmosphere partially or fully blocks these radio waves from reaching the ground. (In fact, no measurements could be made at all at the wavelength of peak intensity.) So for many years physicists knew only that the spectrum was approximately that of a blackbody at about 2.7°K—this was in good agreement with the theoretical prediction of the temperature from the big bang model. But it was important to know whether the spectrum followed the blackbody shape *exactly*, because the simplest models for how this light arose in the big bang predicted a precise match with the blackbody spectrum. Any significant departures would force revision of the model and a reconsideration of how things really happened. The measurements were not precise enough for this crucial test: it was clear that the intensity fell off to both longer and shorter wavelengths, but it was impossible to tell, since each measurement of intensity was uncertain by several percent or more, whether the spectrum followed the predicted shape.

In order to overcome one of the limitations, the absorption of radio waves by the Earth's atmosphere, instruments were taken aloft in balloons and into space by rockets. One such measurement, a collaboration between Japanese and U.S. physicists in the late 1980s, actually reported a second intensity peak in the spectrum of the cosmic microwave background. This is not expected for a blackbody spectrum. The discovery precipitated a great deal of excitement and speculation that another source of light from the early universe—perhaps the first stars to form—had been found that *added* to the glow of the big bang.

It wasn't real. The challenge of accurate measurement had gotten the best of this difficult experiment. When, in 1990, the *COBE* satellite was finally able to make precise measurements of the intensity of radio light over a very wide range of wavelengths, the spectrum of the cosmic microwave background was found to be *exactly* that of a blackbody. To within one percent, the intensity at each color was that of a blackbody at a temperature of $2.735°K$—the second "bump" on the spectrum was nowhere to be found by *COBE*. Calibration problems in the previous experiment were suspected, but, regardless of the cause, there was no question about which experiment had achieved the higher accuracy.

The measurement of the spectrum of the cosmic microwave background has told us a great deal. By confirming that the spectrum has the exact form of a blackbody spectrum (within what are now very small measurement errors), our confidence that we understand what conditions were like a few hundred thousand years after the big bang has soared. Furthermore, these much more accurate measurements reveal no strong signal from the formation of galaxies or from other processes that might have released vast amounts of energy early in the universe's history. This negative result, therefore, is very valuable in itself; it will allow us to further refine the big bang model and obliges us to look elsewhere for the first signs of star formation.

But the most precise measurement of the cosmic microwave background, the most challenging, and the one with the biggest reward, was yet to come.

The cosmic microwave background is both boon and bust for learning about the early universe. On the plus side, this ubiquitous glow is a direct signal from the universe at the time of its infancy. On the minus side, because the CMB represents a transition from an opaque to transparent universe at an age of a few hundred thousand years, the "fog" will forever block a direct view of even earlier times.

We can learn much from a "picture" of the universe when it was a few hundred thousand years old. For years astrophysicists have dreamed of seeing

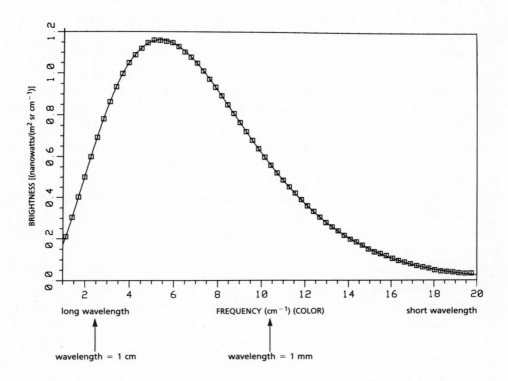

The spectrum of the cosmic microwave background (CMB) light, as measured by the *COBE* satellite. The intensity of the glow (brightness) is plotted against frequency, which is analogous to color. The measured data, shown as open squares, fall with near perfection on the predicted "blackbody spectrum," shown as the curved line. *(Reference: J. Mather et al., 1990* Astrophysical Journal Letters, *vol. 354, L37.)*

the seeds of the great structures of today: embryonic galaxies, clusters, and superclusters. The big bang model tells us that the universe was extremely hot in its early moments, and extraordinarily smooth. Our thesis has been that the structures we see today must have been present even then, as extremely gentle ripples in the density of matter and energy—what we have been calling *fluctuations*. This is hard to visualize, but it means that in this smooth sea of particles and photons, there were undulations—like waves—in the amount of matter and energy in a given volume. We are not yet certain where these fluctuations came from, but it is certain that gravity began to work early on to amplify them, to make the crests higher and the troughs deeper. The process continues still.

Today we see galaxies, and these galaxies are surrounded by space that is virtually empty—the contrast between the crests and the troughs has become enormous. Even the contrast in the matter density between a supercluster of galaxies and a neighboring void is impressive. Our computer simulations show us that, in order to reach this high level of contrast, the concentration of matter into these structures must have been well advanced when the cosmic microwave background light began its journey to us.

The cosmic microwave background gives us the means to look back in time so that we can "see" these embryonic structures, much in the same way we look back in time for distant galaxies. Recall that for the first few hundred thousand years the tremendous energy in light—photons—was scattered repeatedly by free electrons, a condition much like a fog. But, when the universe had cooled sufficiently, electrons combined with protons to make hydrogen atoms, releasing these photons to sail freely through space forevermore—the universe had became transparent. These newly released photons, the same ones that arrive here today as the cosmic microwave background, carried away with them a record of the pattern of fluctuations in density of the region where they were last "scattered." This is because photons departing denser-than-average regions lost a slight amount of energy as they "climbed out" against the pull of gravity, while photons leaving regions less dense than average actually got a slight kick to higher energy. These photons have crossed the universe untouched, so these differences in energy have remained to the present day. Astrophysicists realized years ago that, in principle, precise measurements of the energies of these cosmic microwave photons should allow us to make a picture of the pattern of fluctuations in matter and energy as they were 15 billion years ago.

What kind of camera takes such a remarkable picture, and where do we point it? The answers are: a radio (microwave) telescope; anywhere in the sky. Remember how earlier we took photographs of regions of the sky and looked deeper and deeper into space? Our most shallow photograph showed stars in our own galaxy. These are fluctuations, in a sense, because they are places where the matter density is extremely high today, surrounded by almost complete emptiness. When we looked to fainter objects we saw galaxies—larger, less contrasting fluctuations in matter density. The more deeply we looked, the more galaxies we saw. With each step we were looking at a picture of space at an earlier time, at 3 billion years ago, at 5 billion, at 10 billion. In the "background" of all these superposed pictures of earlier times is a final picture of "the day the fog lifted," 15 billion years ago, but we must take our photograph in radio, not visible light, because the light from this time has been redshifted all the way from visible light to radio light. If we look in any direction, in a picture made with radio light, we will see the places where matter was collecting to form galaxies, clusters, and super-clusters, the gentle undulations in the density of matter that were recorded by the cosmic microwave photons when they left on their journey to us.*

But, as photographs go, this last one is about as "faded" as you can imagine. The differences in energy CMB photons got as they "climbed out" or were "kicked out" of these early fluctuations in matter density were minuscule—a change of only a few thousandths of one percent. Imagine a picture in which the intensity levels differ by less than 1 part in 100,000.† How remarkably sensitive a camera with its detector would have to be to record such a pale picture. When the cosmic microwave background was discovered, Penzias and Wilson were impressed by its uniformity—wherever

---

*Again, it may seem puzzling that we can look *out* into space, back into time and see, all around us, a universe that was much *smaller* in the distant past. Recall the answer to this riddle is that our universe has more dimensions than three. If we go back to the analogy in which the universe is like an expanding balloon in three dimensions, the sky appears to us "flatlanders" as a ring (instead of a sphere). Likewise, the place where cosmic microwave photons appear to come from, at any given moment, is a ring—for simplicity, if we are at the north pole of this sphere, the light arriving today came from, say, the 48th parallel of latitude (later from the 47th parallel, etc.). This ring is a horizon—the farthest that we can see. Light left this ring when it (and the balloon) were much smaller, but it appears to be more distant than anything else, because, in order to reach us, these cosmic photons had to traverse all the intervening space (the skin of the balloon), passing first young galaxies, and then older ones, as they sped toward us and toward the present time.
†For example, the white of the paper on this page varies by about a percent in the intensity of light that bounces off it. Imagine reducing the blackness of the *print* to even less contrast than this, to where it stood out from the white page only 1 part in 100 as much as the variation in intensity you see *between* the lines. If the secrets of the universe were printed that lightly on this page, how hard would you work to read the message?

they pointed in the sky the intensity varied by less than 1 part in 100, the best accuracy they could achieve. At that time, then, there was no hope of seeing the slight variations in intensity that would correspond to the density fluctuations in the early universe recorded by the CMB photons.

Ten years later, in 1976, experimental physics groups at Berkeley and Princeton developed a new generation of balloon-borne detectors that could detect variations of 1 part in 1,000 in the intensity of the CMB light, greater precision by far. With this more sensitive "microwave camera" they did indeed find that the intensity was not exactly the same in all directions. But this variation was not patchy, as was expected for the slight undulations in intensity corresponding to embryonic superclusters. Instead, these researchers found a gentle variation from one-half of the sky to the other: the CMB was more intense in one-half of the sky, reaching a peak of about 1 part in 1,000 greater intensity in a particular direction, and less intense in the other half, reaching its low point of 1 part in 1,000 in exactly the opposite direction. The intensity varied smoothly from peak to valley over the entire sky, according to a common mathematical function called a *cosine*.*

There is only one simple explanation for this variation in intensity of the CMB over the sky: it is the Doppler shift caused by the motion of our own galaxy through the sea of CMB photons. As our galaxy swims in this sea, photons met head-on have their energies boosted, by 1 part in 1,000, while those in the wake lose the same amount of energy. For any other direction the boost or loss is proportional to the cosine of the angle between that direction and the direction the galaxy is moving, exactly the form of variation observed in these careful measurements of the CMB intensity. The speed implied is 1 part in 1,000 of the speed of light, which is 300 km/sec. This is, of course, the speed recorded from our position on Earth, orbiting the Sun. When a correction is made to what this speed would be if measured from the center of the Milky Way, the number rises to almost 600 km/sec. Again, the ability to make more precise measurements had uncovered something totally new, that our galaxy has a large peculiar velocity through space. This was the discovery that allowed the Seven Samurai to measure the peculiar velocities of all the galaxies in our sample to this cosmic reference frame, the CMB.

Except for this slow variation around the sky, the CMB still looked smooth—like a perfectly white wall—no sign of the budding clusters or

---

*The cosine is a mathematical function that describes, for a triangle, the ratio of the lengths of the adjacent side and the hypotenuse. As the angle between these two sides changes from 0° to 180° to 360°, the value of the cosine varies smoothly from $+1$ to $-1$ and back to $+1$, passing twice through 0.

superclusters. Not enough precision! With extraordinary effort, detectors were built that could record variations in intensity to 1 part in 10,000 in a patch of the sky, and still, no variations were found—the CMB picture still appeared perfectly smooth. This was a particularly important result because it ruled out several models of structure formation. One model tried to make today's structure from ordinary baryonic matter only—no exotic, weakly interacting particles to start things off—and in the other class of models the total density of matter was much less than the "critical" density, the value at which the universe is "pitched" at the perfect speed to eventually stop expanding. In these models, fluctuations in density destined to grow into clusters and superclusters would have been well developed when the CMB photons were "released," and their energies would record changes in intensities of more than 1 part in 10,000 from place to place in the sky. Such large variations weren't seen.

There was a great deal of nervousness in the astronomical community in the 1980s, the period during which more and more sensitive "picture taking" failed to reveal the embryonic structures that most believed had to be there. Some worried openly that this whole idea of the growth of structure might have a serious flaw, and that somehow this might reflect on the big bang model itself. Indeed, a few scientists who regularly campaign against the big bang gloated over this failure to find the expected fluctuations. More seriously, some papers were written about some very exotic, not very appealing alternatives that could be considered if no variations in intensity to better than 1 part in 1,000,000 were found. Those of us with confidence in the big bang model just waited.

In 1991 Berkeley astrophysicist George Smoot reported that the *COBE* satellite—the labor of love of hundreds of scientists and technicians—had taken a year-long picture of the CMB sky, a picture so accurate that it could show variations in intensity at the phenomenal level of 1 part in 100,000 from place to place. The gentle ripples finally surfaced, but only just barely, from the smooth sea of microwave light. It had taken this final step in precision measurement, an amazing technical accomplishment, to show that the intensity of CMB light varies about 2 parts in 100,000 over patches of about 10°. These "super-supercluster" sized regions are not well-developed structures in our era (they will become more apparent when the universe is several times as old as it is today), but we know from theoretical models that this variation of 2 parts in 100,000 is just about what we expect for models of structure formation based on weakly interacting dark matter and a universe at the "critical" density. A subsequent observation by MIT physicists has confirmed the landmark *COBE* result.

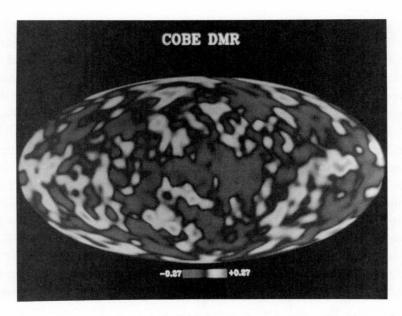

Two pictures of the sky in microwave (radio) light taken by the *COBE* satellite. The top picture shows the smooth variation in intensity around the sky produced by our own galaxy's peculiar velocity of 600 km/sec. With this gross variation removed, the more sensitive measurements of the bottom picture show a faint, blotchy pattern believed to be giant embryonic matter concentrations in the early universe. *(Courtesy of NASA-Goddard Space Flight Center and the* COBE *Science Working Group.)*

These are early days for this kind of work, and really to learn about the pattern of fluctuations in the early universe that grew into today's galaxies, clusters, superclusters, and voids will require new "cameras" to take radio pictures of the sky to a precision of 1 part in 1,000,000. But I, for one, am filled with admiration for the generation of scientists who accomplished the marvelous feat of taking baby pictures of the universe.

Like the universe, the consciousness of each and every one of us grows from a primal seed and expands, structuring our brains with networks of neurons that encode, in a complicated, nonlinear way, each experience gathered by our senses. Night and day our minds sift through these patterns, sorting them, arranging them, above all, *comparing* them. From infancy we search out associations, similarities, recurrences; as children, we are busy concocting "theories," testing what we have "learned" against new experiences. Each child performs countless experiments of discovery on the objects of his world: as a consequence, for example, we (generally) try not to walk through walls, do not grab glowing coals, and selectively fluff kittens but not porcupines. These choices may seem obvious, but unconsciously, our minds are obliged to sort through hundreds of analogies because rarely are the situations we encounter perfect copies of those we have met before. It may be your good fortune never even to have *tried* to walk through a wall or pick up a porcupine, but still you can anticipate the results by analogy with other less consequential adventures. To anyone who has sipped orange juice when expecting milk, the power of this process—what is going on "behind the scenes"—is startling.

This is how we learn and, of course, how we teach. Thus, the introduction of this idea is itself couched in an attempt to elicit your memories, to draw parallels with your childhood, when you were learning what in your world was "safe" and what was not. However, this ability—really the tool of "thinking"—goes far beyond navigating through a potentially hazardous world. Abstraction, no matter how complex, is the true achievement of "connection-making"—the analogies, comparisons, parallels, and the building of new neural patterns, *assembling puzzles*, that link these together. The mechanism of higher thought is a soaring hierarchy of such models, built from the most basic experiences. Of all nature's "inventions," none we have discovered is more miraculous than this. There is no area more fertile for human exploration than "the way we think," ironic when you consider that what could be the universe's most remarkable accomplishment is capable of being perceived only by itself.

Despite stereotypes to the contrary, particularly those portrayed in B-grade sci-fi movies, scientists "think" in exactly the same way: scientific theories are also hierarchies of analogies. Furthermore, and also contrary to colloquial myth, no scientist, no matter how brilliant, ever dreamed up something out of thin air. Rather, by forging new analogies, sometimes astounding, usually subtle, always complex connections with past experience, new "understanding"—a new *pattern*—emerges. A deep reservoir of experiences and the ability to discern intricate, elusive analogies are the chief ingredients of a creative scientist.

I grew up knowing just such a person—Robert Dressler, my father's kid brother, who had been born with that near-obsession to "see the patterns," to find out "how things worked." He himself grew up a tinkerer and experimenter, in my grandparents' Bronx Park East apartment during the Great Depression, in a loving family that, nonetheless, had absolutely no idea of what he was about. Frightened by the small explosions, fuse-blowing, and lesser fires from his "playing," my grandmother could only hope against hope that he might do something useful, like run a radio repair shop—this was the extent of her understanding of his passion. Robert's public school education was good enough to prepare him for a master's degree in physics at Columbia, which he pursued during World War II, supported by the U.S. Navy as he worked on the development of radar. After the war, he moved on to Paramount Corporation, where he worked on the burgeoning technology of television, and later to Raytheon, Inc., where he was swallowed by the macabre world of national defense. His opportunities, and, of course, his own choices, took him further from science and into management—he has an extraordinary ease with people—and as the years passed I saw a deterioration of his spirit as his contact with his true love, science, became more and more vicarious.

About once a year Uncle Robby came to town. He was the circus and Merlin and "Mr. Wizard" rolled into one. My dad would fetch him from the airport and I would patrol the living room picture-window, waiting to rush out to see him tumble from my dad's car onto the lawn—his usual routine was to feign impossible clumsiness and then pick himself up and readjust his rumples. All the time, a boyish grin, a perfect complement to soldiers' rows of strawberry-blond hair but seemingly at odds with the sobriety of the traditional Dressler cleft chin, decorated his ruddy face.

His frumpy briefcase was his treasure trove. Uncle Robby would pluck strange chemicals, electrical gadgets, crystals, tops, radiometers, and other treasures with great suspense and panache—nature's own miracles were the props of his magic show. With his toys he unveiled the mystery of electricity, chemistry, and mechanics. Each year his nomadic travels—journeys of escape from his problems at home (the kind he didn't know how to solve)—added more creases and strain to a face that only wanted to be Santa Claus. But the eyes always glowed when the magic show began. In 1959 he returned from Japan with a Sony transistor radio—no one I knew had even heard about them—and I peered for hours at the guts of this miniature wonder. It replaced a twenty-pound, foot-and-a-half-long, battery-powered "portable" Philco tube radio in the bathroom, placed there because my father wanted music but thought it safer than plug-in radios that carried the risk of electrocution (nonetheless, its 67.5 volts of *direct* current from two bulky batteries gave me the worst shock of my life). Another time, Uncle Robby arrived empty-handed (I was devastated), but took me to a hobby shop where we bought an electric motor kit that consisted of only a handful of simple metal parts, a couple of tiny magnets, and some spools of wire. He led me patiently through its assembly, showing how the electricity would flow and how the magnetic force would be generated and "flipped" as the contacts spun around. All day he shuttled his attention effortlessly between me and my parents. By nightfall, it was time to connect the battery, and the motor spun to life immediately, smoothly, and perfectly. I knew this was how the world was meant to be.

I was a teenager when Uncle Rob showed up with a four-inch Erlenmeyer flask filled to the half-inch level with an ivory-colored solid, sodium acetate in water solution, he told me, a substance with the amazing property of being either solid or liquid at room temperature. We commandeered my mother's stove, where we gently melted the solid sodium acetate; several times Uncle Rob raised the flask and swirled it around to make sure the thick, clear liquid contained no unmelted chunks. Then he declared that we must place this flask where it would not be disturbed, not even jostled, and then leave it alone for a couple of hours. I hadn't the vaguest idea what he was up to.

When we returned, the flask and its contents had cooled, but it still contained a clear, viscous liquid—the sodium acetate had not returned to the solid form Uncle Rob had brought in the flask, even though it had cooled back to room temperature. Puzzling, I thought; I had expected it to "re-freeze." Then, with great ceremony, Uncle Rob removed from the pocket of his sport coat a small vial and tweezers. Plucking a tiny flake of sodium acetate from the vial as cautiously as if it were nitroglycerin, he handed the tweezers to me and told me to carefully drop the flake into the flask with its persistent liquid, and watch carefully. I obeyed. The instant the flake hit the liquid it radiated triangular sheets—in a few seconds the entire mass had crystallized! "Wow!" I exclaimed. After waiting about thirty seconds, Uncle Rob ordered: "Pick it up." I reached for the flask and grabbed it close to its base, then nearly dropped it: to my utter surprise, it was hot. (I was startled and nearly dropped the flask because my inner "modeling" had failed—I wasn't expecting any heat.)*

Uncle Rob steadied the flask, and, chuckling, began to explain that we had just witnessed the phenomenon of supercooling, and the release of the heat of crystallization. Supercooling, he explained, had to do with the temperature transition from the liquid to the solid form, or "phase." Many substances could be cooled below their freezing points—even common water could be "supercooled" slightly before it turned into ice, though the effect was more dramatically demonstrated with sodium acetate. If a liquid could be carefully and quietly coaxed into this supercooled state, the transition to the solid form could be loosed like an avalanche and finish with astounding speed. This meant that the molecules, previously free to roam around in the liquid phase, were quickly organized into the regular lattice that characterized the solid. Normally this would occur gradually, as each atom was directed into its proper spot in the grid, but supercooling forced the process into a huge game of musical chairs—the molecules were poised to dive into their "seats" as soon as any one of them made a move.

The amazing release of heat also had to do with the difference in energy per molecule in the solid state compared to the liquid state, he explained. In order to go from liquid to solid, it was first necessary to "catch" every molecule—reduce its kinetic (motion) energy to near zero so it could be tacked onto the lattice. This is what was accomplished by "cooling" the liquid. But, once the liquid was at the proper temperature, and ready to freeze, each molecule had to be properly *oriented*—turned to match the pattern. (For example, the water molecule $H_2O$ looks like a Mickey Mouse head—the "ears" are the hydrogen atoms. If you can imagine an Andy Warhol lithograph of Mickey Mouse heads, you've got the picture.) The "disorderly" state, with the freedom to face and move in any direction, is a higher energy state than the "aligned" state of "order," in which choices are fewer. The technical term for this is *degrees of freedom*: a system with more degrees of freedom is a higher-energy state. Freezing a liquid is like the transition from recess to homeroom for a class of first-graders. First you must "cool them"—stop them from running around the classroom—and

*Those wishing to duplicate this experiment will obtain anhydrous sodium acetate in powdered crystal form—perfectly safe unless you eat it—and dissolve it in distilled water at a ratio of, say, 30 grams of powder for 60 milliliters of water. Add a bit more water if the solution tends to crystallize before it cools to room temperature. Keep everything clean—contamination will prevent the supercooling. Crystallization is most dramatic if started with a single, tiny crystal.

then, somehow, drain even more energy so they will stop twisting every which way in their seats (until they face forward and only fidget—vibrating like the atoms in a crystal lattice). The liquid state of running around in any direction is a highly *symmetrical* state—all directions look equally good to a rambunctious first-grader—but in order to confine a child to a chair the symmetry has to be "broken."

When our sodium acetate solution had been chilled to below the freezing point, the kinetic energy had been reduced to the point where the molecules could be "seated." Then, in the cascade of crystallization triggered by the seed flake, the energy stored in the degrees of freedom was released as each molecule quickly assumed its proper orientation. This substantial energy release that accompanied this "broken symmetry" is called the *heat of crystallization* (or heat of *fusion*, but having nothing to do with nuclear fusion).* This was the sudden source of heat that surprised me.

On this trip Uncle Rob had taught me about the difference between solids and liquids, about energy and temperature, and, the most powerful idea of all, that *symmetry*—a seemingly abstract notion—could have very tangible consequences. He was building my intuition about the natural world, with analogies, connections, patterns, and models. But even Uncle Rob could not have dreamed that we were nosing around a model for no less than the birth of the universe.

There is no heavier burden than a great potential. Not only children, but theories, must suffer the stress of having their accomplishments taken for granted and, worse, raising expectations for even greater achievements. Take the big bang. Conceived only to account for the expansion of the universe, it did us proud by successfully predicting the emergence of a dominant energy field a few hundred thousand years after the primal moment, and amazed us in its account of the synthesis of the light chemical elements when the universe was a matter of minutes old. But we were hardly satisfied. Just one more thing, we said: could you please tell us about the ultimate nature of matter, and what the universe was like in the first trillionth of a trillionth of a trillionth of a second? Oh my.

---

*The heat of crystallization for water is 80 calories (not the *kilo*-calories of diets) of heat-energy per gram. We see its presence most dramatically when, as we heat ice water, the temperature stays at 32°F until all the ice is melted. The temperature of the ice-water mixture does not rise, because the added heat-energy is all going into restoring the degrees of freedom of each molecule—giving it the ability to fly off in any direction, as the liquid state requires.

When it came to predicting the CMB, or describing nucleosynthesis, physicists were on familiar ground, applying what they knew about the behavior of matter and energy at densities and temperatures within the scope of earthbound laboratories. The physics was "known," and its consequences could be checked in the context of a big bang to see if this extraordinary event actually took place. The predictions checked out, and confidence grew in the model. But, some wondered, could the process be taken back to an earlier, extremely brief time of even more stupefying temperatures and densities?

By the 1970s, theoretical physicists had stitched together some quiltlike models of what matter and energy might be like under such extraordinary conditions. Unfortunately, validation of these models required experiments that were beyond our most outrageous dreams.* Hold on, some said, why pursue such fantasies when the experiment has already been done? Let's turn the process on its head. Before we used well-known physics to check the big bang. Now that we have confidence in it, let's use the big bang to check new physics—a bold proposition, to say the least.

Today's universe is a stripped down version of what came before. Protons, neutrons, electrons, photons, and neutrinos are the last survivors of a rich ancestry of particles that could survive only in "hotter times"—this more colorful cast of characters disintegrated, annihilated, and decayed. Once, these minions were continuously created and destroyed in the boiling matter-energy sea of the past, according to Einstein's famous prescription, $E = mc^2$—mass and energy are equivalent and related by the speed of light squared. But, as the universe cooled below the energies necessary for creation, the vast majority were annihilated or fell apart, most species became extinct. Only a few diehards have survived "the big chill."

Not only has this impoverished collection provided our entire world, but, happily, these few surviving varieties make it possible to probe the world long gone. By accelerating protons or electrons to near-light speeds and bashing them together, brief instants of the high-energy universe of the past are re-created, and scores of additional particles are revealed. For example, it has been found that the electron is a member of the "lepton family," which

---

*The Superconducting Super Collider (SSC), begun but now canceled, was expected to cost in the neighborhood of $10 billion. Its powerful magnets were to ring an area the size of a small city, yet the tremendous energies it would have explored, by accelerating protons and antiprotons and bashing them together, would have been nowhere near those of the early big bang. A scaled-up version of the SSC powerful enough to approach big bang energies would have a diameter about that of our solar system, a project too big even for Texas.

includes two other negatively charged, but more massive particles, the muon and the tau. Just as an "ordinary" neutrino accompanies the creation of an electron from pure energy, a muon neutrino emerges with a muon, and a tau neutrino arrives with a tau.

Protons and neutrons are built of sterner stuff—quarks. These building blocks come with a variety of "qualities," and it is the various combinations that result in different heavy particles (collectively called *hadrons*). Of these, the proton and neutron are the only "living" examples—each of these is composed of three quarks, and the difference between a proton and a neutron is that one is made of two "up" quarks and a "down" quark, the other, two "downs" and an "up."* But long ago, when the temperature of the universe was hundreds of trillions of degrees rather than the frigid $2.735°K$ it is today (this is the temperature of "space" today, as defined by the photon energies of the cosmic microwave background), many other quark combinations abounded. (Keep in mind, though, that this entire phase lasted for what we would perceive as an incredibly brief moment.)

In this earlier time of enormously high energy density, particles were created out of pure energy. The rules of particle creation were, as they are now, very specific. For example, because charge must be "conserved" (kept in balance), the creation of each charged particle was accompanied by an *antimatter* partner, a particle with the same properties but opposite charge. So the creation of a proton always included an antiproton, and when protons and antiprotons ran into each other they annihilated, returning the pure energy from which they came. Protons and antiprotons were "in equilibrium," which means, for every pair created, another pair was destroyed, until the universe was a few seconds old. After this, the universe cooled below a trillion degrees—too cold (not enough energy) to produce any more protons or antiprotons. Fortunately, at the very end of this phase, nature had an extremely slight preference for producing more protons than antiprotons, to the tune of one part in a billion. As a result, after the annihilations were finished there were $10^{80}$ protons left over, to build the world we know.

This is half of the picture, that the types of particles at each epoch depend on the temperature; *forces* make up the other half. Forces mediate the interactions among particles, for example, the creation of a proton-antiproton pair is arranged by the electromagnetic force, which holds the energy and then transfers it, like a stockbroker, to another form. The photon is the messenger of electromagnetic force and also its energy bank, the currency of

---

*An "up" quark has ⅔ of a unit of electrical charge, a "down" quark has $-⅓$ of a unit. With two of one, and one of the other, you can get either a $+1$ or $0$ charge.

the transaction, so to speak. In addition to the electromagnetic force, we know of three others: the *strong force* that binds protons and neutrons together in the atomic nucleus, the *weak force* that governs (among other things) the creation of neutrinos, and the gravitational force. Particles like the photon that carry these forces are collectively called *bosons*.

Nowadays, the principal task of particle physicists is to try to specify all possible particles and describe their interactions and transformations through the four fundamental forces. To make things more familiar, let's imagine each of the particles as some kind of financial instrument and the forces as brokers: *electromagnetic force* is the business of a stockbroker, *strong nuclear force* is handled by a commodities broker, *weak force* by a bond trader, and *gravitational force* by a real estate broker. As in the world of high-energy physics, in our country these very similar activities are governed by different rules. Each broker is a specialist in that type of transaction, so, for example, there is no system in place to trade IBM stock for wheat futures or swampland.

Starting in the 1930s, physicists undertook the task of describing and *understanding* the "system" of particles and forces. Using the analogy with the investment world, this might be compared to challenging a musician to figure out the world of investments and the maze of trading rules, his only preparation that he speaks the same language and is able to do arithmetic. *Understanding* this world, as opposed to just describing it, means that if successful he would be able to outline a valid transaction or financial instrument that he had not previously seen, just by extrapolating from known rules. Likewise, physicists were looking for simple rules underlying the complex-looking world of particle interactions, simple rules that might be combined to yield a more complicated pattern, the way, for example, a six-pointed star is constructed out of twelve identical equilateral triangles. But, by the 1950s, the number of particles had grown substantially and new ones were still being found, and theories that attempted to explain their interrelationships appeared tangled, their rules arbitrary if not capricious.

The breakout from this mire came when physicists Chen Ning Yang and Robert Mills, working on a small corner of the theory, proposed that the rules of particle physics that we observe at low energies might be examples of *broken symmetries*—as if a beautifully cut diamond had been shattered into pieces that carried little trace of its former elegance and proportion. The essence of this idea is that the four fundamental forces, apparently distinct in today's low-energy world, might be "broken" versions of one or two "superforces" that functioned in a simpler, more symmetric fashion. These superforces would allow interactions and transformations among all the extant kinds of particles, including processes that are now forbidden.

To put this in terms of our analogy with finance, one might imagine that at an earlier time there were "superbrokers" who could deal both in real estate and stocks, or bonds and commodities, or maybe even a super-superbroker who could deal with three or even all four. In this earlier version of the brokerage world, simpler rules might have applied that allowed any type of transaction—trades from any financial instrument to any other. Only the "breaking of symmetry" in later times would prevent these transactions and evolve a system that would mask the underlying principles that once provided simpler rules and more options.

The idea of unifying forces was not a new one in the physics world. In fact, British physicist James Clerk Maxwell sparked a physics revolution in 1864 when he showed that the electric and magnetic forces are manifestations of a single electromagnetic force. Now, spurred by the insight of Yang and Mills, Steven Weinberg and Abdus Salam, and independently, Sheldon Glashow, proposed that the weak and electromagnetic forces are united into a single *electroweak* force, but only at much higher energies than exist in our everyday world. Subsequent experiments with the latest generation of particle accelerators have confirmed the detailed predictions of their theory, verifying that at very high energies these two of the four fundamental forces unite as one. The theory precisely predicted, and experimentalists confirmed, the discovery of a previously undetected particle that carries the electroweak force, the W-boson, which lives in the high-energy world.

Physicists are continuing to wrestle with different expressions of the basic rules of interactions between particles and forces that could unite the strong force to the electroweak force. These grand unified theories would apply to even higher energies. Some have even dreamed of finding a theoretical structure that would envelop the gravitation force as well, thereby uniting all four of nature's forces into just one elemental force that decomposed into separate manifestations as the universe cooled.

The goal of a "theory of everything" remains elusive. But the exploration of various symmetrical descriptions of the "rules," and the way these symmetries might have been broken to form our world, has led to certain recurrent themes that would have profound consequences for cosmology. It is possible that the composition of dark matter will yet emerge from such investigations. That is, in the same way we have been able to understand the production of protons and neutrons in the early universe, knowledge of the complete picture might dictate that the mass of the universe was largely deposited in some weakly interacting particle, for example, a neutrino with small (but non-zero) mass, or a heavy, cold-dark-matter particle. When one considers the myriad possibilities for particles that could have survived, the prejudice that the universe is made only of baryons because that is all we "see" looks rather silly, like a modern reincarnation of Ptolemy's "Earth-centered universe."

But there is a more profound and subtle notion that has surfaced from all this deep thinking. It involves symmetry breaking, and what the universe might have been like when the electromagnetic, the strong, and the weak forces were united, a time of unimaginably high temperature that lasted for only a trillionth of a trillionth of a trillionth of a second (which can be written and thought about but not truly comprehended). Such models describe the earliest moment of the universe as highly symmetrical, that is to say, the rules of the game permitted all manner of transformations and interactions, like trading preferred stock for some municipal bonds, a treasury bill, and a few hog bellies.

It is thought that the early universe might have been controlled by one such superforce named Higgs, who, like a superbroker, could handle any transaction—stocks, bonds, commodities—except real estate (gravity). (The Higgs boson conveyed this superforce, much in the same way the photon carries electromagnetic force today, that allowed conversions of all manner of hadrons, leptons, and other bosons.) Higgs could do such diverse work because of the extraordinarily high energy that was available at that time. However, the universe was cooling rapidly and it was inevitable that the highly symmetrical world of Higgs transactions would end with the Higgs force breaking up into the strong and electoweak forces, the latter destined to further split into the electromagnetic and weak forces. In a colder universe there wasn't sufficient energy to continue the highly flexible system, so the superforce "broke down." This process was very much like the *phase transition* liquid water goes through when it is about to "break symmetry," that is, freeze into solid ice. There simply isn't enough energy to support the ability of water molecules to run around in any and all directions.

The demise of the Higgs force was like our superbroker leaving separate trading systems of stocks, bonds, and commodities that would carry on as best they could in the lower-energy world with its "restrictions" on trading. But, in setting up the new, less-elegant system, it was critically important to get the rules straight for everyone, that is, a smooth pattern of the new rules must apply everywhere. This is like the formation of a perfect solid—a crystal. If the transition didn't go so smoothly, and new instructions were passed on erratically from one region to the next, there would be fractures—zones where things didn't fit together, the way an ice cube is full of fractures rather than clear. In our analogy, unless we laid down an even pattern of new brokers with all their infrastructure, there would be places where two stockbrokers would go head-to-head, and there would be no commodity broker to carry on her specialized function.

In the late 1970s, Alan Guth was working at Cornell on unified theories like the Higgs. He was encountering just such a problem, specifically, that the Higgs force wouldn't break up smoothly and evenly throughout space into weak, strong, and electromagnetic forces. Instead, he found there would be "faults" or "fractures" where the ordering would break from one volume of space to another. The "freezing" of the Higgs force into the less symmetric state of three separate forces would apparently be more analogous to an ice cube than to a pure crystal, like a diamond. This meant that the model was in trouble, because, had this happened, the consequences for the universe would have been catastrophic (at least from our point of view). At each "fracture" a *magnetic monopole*, a single pole of magnetic force (something that we have looked for but not found) with enormous mass would have been formed; added together, the mass of monopoles would have been sufficient to collapse the universe into a big black hole only moments after its birth.

While contemplating the high-energy Higgs world and the magnetic monopole disaster that it apparently included, Guth realized that monopoles could be nearly wiped out if the liquidlike state had been "supercooled" before it made the transition to the solidlike state. As I had witnessed in my Uncle Rob's flask of sodium acetate, when a liquid is supercooled and "freezing" is initiated from just one spot, the crystal spreads evenly and smoothly throughout the volume—fractures are kept to a minimum. Normally, as an ice cube forms, the freezing begins at many different centers; as a result, the crystal pattern runs helter-skelter and collides in disarray at multitudinous locations. But, if the freezing process is prepared for, but *delayed*, as happens in supercooling, then crystal building sweeps rapidly (like a huge swath of dominoes) across the material, leaving a uniform, coherent crystal. Guth found that if the Higgs force survived to a lower energy than its nominal "breakup" temperature, then "broke" from a supercooled state, its transition would be a smoother and more uniform one, and it would leave few or none of the magnetic monopole "fractures" that might have dominated the universe if the symmetry breaking had been more haphazard.

This idea would do more than "save" the universe from instant death; it would also solve troubling problems that had persisted with the big bang model. Guth didn't have much experience with cosmology, and little confidence in its ability to connect with particle physics—considerable prodding from his partner in the work, Henry Tye, was required before he investigated the Higgs field and monopoles in the context of the hot big bang. In particular, Tye urged him to explore the consequences of a supercooling episode in the expansion of the universe. The consequences, Guth found, were monumental. Einstein's equations of general relativity implied that the universe would expand smoothly with steadily increasing volume and, therefore, steadily decreasing energy density. In contrast, a period of supercooling would throttle the universe into an era of *inflationary* growth. This would happen because of the enormous energy buried in the symmetry field—the energy of crystallization that should have been released when the universe passed through the symmetry-breaking temperature, but wasn't. Unable to tap and release this energy, the universe would continue to expand with too high an energy density, a situation that exacerbated as the volume grew.* Because of this "hitch" in the energy density, the universe would grow exponentially— doubling and doubling and doubling—for each fixed interval of time. As long as the Higgs force "held on" in the supercooled state, the universe would "inflate," rapidly reaching a vastly greater volume than its previously tame expansion had achieved. Exponential growth, the period of inflationary expansion, could amount to some 100 successive doublings of the size of the universe.

In a very real sense, this proposed event of inflationary expansion, driven by the supercooled phase transition of the Higgs force, *was* the big bang. It would have so radically altered what came before that "initial conditions" were made irrelevant. At the completion of the symmetry-breaking phase transition, the universe, given this whopping kick, would sail into the expansion epoch we find it in today. When finally released, the enormous energy from the symmetry breaking (like the heat of crystallization released in Uncle Rob's flask) would reheat the universe, providing the energy from which a zoo of particles was "created." An extraordinarily hot, expanding cloud of mass-energy would be the result of "inflation." This was the big bang.

Guth soon discovered that the inflation hypothesis automatically solved the two outstanding riddles of the big bang model, conundrums that he had read about in a famous paper by Dicke and Peebles. One was the amazing uniformity of the universe, best exemplified by the minute variations in the

---

*The equations formally describe this as a situation of *negative pressure*. For positive pressure, the kind we associate with the gas in a balloon, an increase in volume corresponds to a decrease in pressure—we all know the expression "the pressure is off," meaning that a space or time constraint has been relieved. Negative pressure, which some scientists like to think of as a "tension," increases as the volume grows. It has no analogy in "balloon world": it is not equivalent to reducing the pressure in a balloon and having it contract—this is just a lowering of positive pressure.

intensity (or, equivalently, temperature) of the cosmic microwave background over the sky. In order to reach such stunning uniformity, there must have been contact between all regions, that is, sufficient time for exchange of energy—at the speed of light—to even out the temperature. For example, stirring water in a dish leaves the surface disturbed; the water will not return to its level equilibrium until the few seconds it takes for waves to travel across the dish. But this seemed impossible. In the standard big bang model, the universe effectively divided itself into myriad regions that were not in "causal contact," that is, there was not enough time to propagate a signal from one region to another. How could such a system have settled into a uniform, smooth distribution of energy?

Guth's inflationary model provided an answer: before the inflationary epoch began, the universe had been much smaller than we had thought. In this smaller universe there had been time to "get everything in order," then send the smooth universe on its explosive way. More fundamentally, inflation had seen to it that the symmetry of the Higgs force was broken in the same way over the whole observable universe. The same feature that rid the universe of the troubling magnetic monopoles had guaranteed that the *laws of physics*, which are really the parameters that define what we recognize as *the universe*, spread evenly over a vast volume of space-time. The universe visible to us today could, then, be just a small part of this vastly inflated volume, which is now often referred to as a "bubble."

The other puzzle emphasized by Dicke and Peebles concerned the geometry of the universe, a matter I discussed before. The universe sits at an improbable balance between the extremes of "open" (a forever expanding universe where the force of gravity has been overcome) and "closed" (a universe that collapses due to gravity's dominance). (This approximate balance is what Sandage and his colleagues had found when they tried to measure the curvature in the Hubble relation between redshift and velocity: the relation remained linear out to a fair fraction of the size of the universe.) When, as earlier, we associate terms of geometry to these conditions, we say that for the "open" universe space is negatively curved, like a saddle shape, and for the "closed" universe space is positively curved, like a sphere. The improbable condition of balance corresponds to "flat" space, an outcome that Dicke and Peebles said was extraordinarily unlikely given the enormous range of possibilities. Guth's hypothesis of inflationary growth provided an explanation of this too: by "blowing up" a small, highly curved bubble into a vastly bigger space trillions of light-years across, the entire region we identify as "the observable universe," about 20 billion light-years across, would be but a small, almost flat patch on the huge curved space of the "bubble." That is, in the same way we perceive our local surroundings as flat because the sphere of the Earth is very large, the observable universe would be flattened by the enormous expansion factor.

Guth had not been the first to contemplate the idea of an inflationary epoch, but he developed and promoted the concept and its implications far beyond the efforts of his predecessors. He and his colleagues delighted in musing about our universe literally erupting out of nothing. They suggested that we viewed only a small fraction of our "bubble," and speculated that even this unobservable bubble was likely to be only one of a frothy manifold—each one with its own laws of physics, its own "reality" separated forever by unbreachable "walls" of primal energy. The naive model of a solitary universe—the search for its destiny, as if it were all there was and would ever be—disappeared over this new, expanding horizon.

The idea of inflation had offered explanations for the two most perplexing riddles of big bang cosmology—the "flatness" of the universe and its astounding uniformity. It could rid the world of monopoles and level the "rules" over a vast empire, even start the present era of expansion. On further investigation, it was found that inflation could also spawn the fluctuations in matter density that would grow into structure, with exactly the form of the spectrum (the amplitude of fluctuations vs. their size) preferred by those who made N-body computer simulations of the universe.

These are its possibilities, but we don't know if inflation actually occurred. We can't yet specify the characteristics the Higgs world would have needed for the proper supercooling episode, or even say if nature actually provided these conditions. But most physicists agree that a model that can do so much is bound to contain at least the seeds of truth.

The big bang model has proven fertile territory for cultivating ideas about the birth of the universe, from the cosmic microwave background, through nucleosynthesis, to the inflationary paradigm and its attempt to link the shape and sense of our universe with the physics of elemental particles and forces. In each of these, the big bang framework has succeeded in growing our insights, fostering what is perhaps the most ambitious human intellectual endeavor. Though still in its infancy, the big bang model, judged by its ability to lead us to new, more challenging questions, and to open our minds further in the search for their answers, has no peer.

# $II$ • HALFWAY TO CREATION

From the time of our remote ancestors, it has been a human dream to know "where we came from." Many of our written treasures are stories that aspire to this question, and we can be certain that the oral histories of our forebears, reaching back tens of thousands of years, shared our preoccupation. The question fires the hearth of religion, and permeates many of its rituals. Our desire to "have an answer" seems to be bound up in what we call "peace of mind," a hard-to-quantify but strongly felt need for a sense of purpose and a feeling of belonging, and desire for protection from the unknown, especially the ultimate one we all face.

Science is a relatively new lamp in this ancient quest, and it offers a qualitatively different approach: at its core is the belief that we can learn much of the answer on our own, and that "how well we are doing" is self-evident in the method. Often we hear the fluffy phrase "just another theory" when the big bang or Darwinian evolution are compared to stories of religion or folklore that purport to explain the origin of the universe or life on Earth. But the comparison is specious because the "theory" in its scientific context means much more than an explanation of a series of facts or events—fundamentally, it is an explanation that can be *falsified*. A theory cannot be proven to be true; contrary to this popular notion, only a mathematical *theorem* can be subjected to such rigor. On the other hand, a scientific theory *can* be shown to be false with a single stroke, and most eventually are. Because of this, the ones that survive worthy challenges deserve their stature as correct descriptions of the way things are. Gravity may be just a "theory," but no one in his right mind looks up to the ceiling to find his lost keys, even after all other plausible locations have been repeatedly checked.

Now it is time to collect the pieces presented in previous chapters into a "theory" of the birth of the universe, and to consider how robust are its different ingredients. That is, how thoroughly has the story told by cosmologists of "how we got here" been tested—subjected to attempted falsification—and how have the ideas fared? Since much of what has been discussed in this book is work on the scientific frontier, a fair amount will someday turn out to have been wrong or misinterpreted. This is the nature of any push into uncharted territory: there will be many dead ends and misreadings of the terrain. But the ease with which the general picture incorporates new observations and concepts, and sometimes even *predicts* them (by far the strongest endorsement of a theory), shows that the story has some validity and is worthy of vigorous pursuit.

Without a doubt, this is the craziest story humans have ever concocted to answer the question "Where did we come from?" This is exactly how it should be—the true story should amaze us utterly. The shattering of our preconceptions and the stretching of our imaginations is what will assure us that we are on the right path: along the road to creation.

Here is the story, in a borrowed, familiar format:

The First *Day*—Let there be "light." Today's universe traces its roots to a time of incredible density and temperature. "Matter" as we know it didn't even exist, only some ultraheavy particles that were born and perished in an instant. The universe was mostly composed of force *fields*. The term *field* is a convenient way to express the energy stored in forces over a volume of space, for example, an iron nail in the vicinity of a magnet is sensing the magnetic field, and the electrons in a TV antenna bounce up and down as they sense the radio waves (photons) that carry electromagnetic energy from the TV station. In the earliest moments of the universe there were two such fields, one associated with the Higgs force, the unification of today's electromagnetic, weak, and strong forces, and the field of gravitational force. What makes the situation especially interesting is that the energy of the Higgs field could have been exactly balanced by the energy associated with the gravitational field, which is properly thought of as *negative* energy.* Quite possibly, the (negative) gravitational energy exactly balanced the positive

---

*Gravitational energy is negative because it is *potential energy*. Imagine two massive bodies exerting powerful gravitational force on each other. Suppose we work very hard (put in energy) to separate the bodies. The farther apart they get, the smaller the force of gravity, and, therefore, the less energy stored in the gravitational

energy in the Higgs field, meaning that the universe could have started out with *zero* net energy—creation out of nothing. In this picture we imagine the gravitational and Higgs energies (which themselves may have been united into an even more symmetrical, singe field) fluctuating randomly around zero. For a moment, the Higgs field would release some of its energy (in effect, cool) causing the universe to expand slightly; then it would snap back, perhaps overshooting to a much higher Higgs energy (hotter) and much smaller size. Swelling or receding, what became today's entire universe was all contained in the unimaginably small size of $10^{-38}$ cm.

As the universe danced dreamily in this configuration, it waltzed on the edge of a five-hundred-foot cliff: one step too far, and the Higgs field would cool below its symmetry-breaking temperature. A chance flirtation—an expansion of space lowering the Higgs temperature enough to break symmetry—could have been innocent enough. The universe might simply retreat (sort of the way cartoon characters manage to scramble back to the cliff before they fall), by shrinking to reclaim gravitational energy and restore the symmetry of the Higgs field. But, if symmetry-breaking occurred in a supercooled state, there would have been no way back. The exponential growth of the size of the universe would have driven it past the point of no return, because the burgeoning gravitational energy of the inflating universe could never be recycled to restore the symmetry of the Higgs field. According to our current understanding, this event would have erased all knowledge of the prior state of the universe, forever.

The Second *Day*—A big bang: The Higgs field, in its supercooled state, is hoarding energy to maintain symmetry, thereby developing a negative pressure that causes the universe to inflate in size at an exponential rate. This *inflation* continues until the symmetry of the Higgs field is finally broken. The universe, about the size of a softball, is pitched into the steadier expansion phase we find it in today. Fluctuations in density that will become galaxies and clusters appear and begin to grow.

The Third *Day*—The particle zoo: symmetry has been broken. The heat of crystallization of the Higgs field, finally released, is "dumped" into the creation of vast numbers of particles that come in about one hundred varieties. Particle creation is balanced by particle destruction, and enormous energy

field. Eventually, if separated by infinity, the energy in the gravitational field will be zero. Since we were putting in energy, but the energy only *rose* to zero, the system must have started out as a large negative energy. In all other situations, like the electromagnetic field or the Higgs field, the energy of the field is positive, i.e., it can be used to *do* work. In these cases, when all the energy is withdrawn, the field has *declined* to zero energy.

remains stored in the strong and electroweak fields, the latter destined to break into the electromagnetic and weak fields as the universe continues to cool. As the expansion continues, energies continue to fall and, as a result, the heavier particles—most combinations of quarks (the hadrons) and the heavier leptons—become extinct. Protons and electrons survive, though only one out of every billion, as the others annihilate with antiprotons and positrons. Some kind of dark matter also survives, possibly a stable, weakly interacting massive particle (WIMP) that will dominate the mass of the universe.*

The Fourth *Day*—Nucleosynthesis: Rapid cooling has stripped the universe of its great diversity. Only the dark-matter WIMPs, two kinds of stable quark matter (protons and neutrons), four stable leptons (electrons and the three types of neutrinos), and photons are left to roam the expanding space, now governed by four effectively separate forces. Matter densities and temperatures are still high enough to promote fusion of protons and neutrons into substantial numbers of helium nuclei; trace amounts of deuterium and lithium are also produced, but, because of the rapidly dropping temperature and density, the process dies away before heavier elements can be synthesized.

The Fifth *Day*—Decoupling: Since the end of the second *Day*, high-energy photons (and before them, the analogous Higgs bosons) have dominated the evolution of the universe with their prodigious energy density. Gravity has been amplifying the small fluctuations in the distribution of WIMPs—these will be the sites of future galaxies, clusters, and superclusters. However, photons have enslaved the baryonic matter by continually "bumping" the electrons, thereby frustrating gravity's attempt to pull them and the protons mixed among them into the clustering of the WIMPs. Finally, the universe cools enough that electrons bind with protons, making neutral atoms. This gas, which is 76 percent hydrogen and 24 percent helium, is transparent to light, i.e., the strong interaction of light and matter is ended. With further evolution of baryons and photons *uncoupled*, baryons are free to cluster, and photons are freed to wander endlessly. By today, the expansion of the universe has redshifted this cosmic background light down to microwave frequencies (radio waves).

The Sixth *Day*—Galaxies: the fluctuations in density of the WIMPs and baryons have grown large. Gradually, cluster-sized lumps and supercluster-sized walls, filaments, and voids begin to emerge as matter flows to the

---

*A massive particle would automatically fulfill the requirement of moving slowly compared to the speed of light. However, a light particle, like the hypothetical *axion*, can be "born cold," so it would also serve as cold dark matter.

denser regions. Where the density is highest, the contraction of the baryonic gas begins to outpace the swells in the dark matter. This happens because hydrogen atoms begin to bump into each other more frequently, and these collisions result in the emission of photons that carry away some of the "thermal" (heat) energy of the gas. A reduction in thermal energy gives gravity the upper hand, and it compresses the gas to even higher densities, leading to even more frequent collisions of atoms and even more thermal energy lost from the system, and so on. This process of *dissipation* of thermal energy "runs away" in the densest places—these contracting gas clouds become the galaxies, and they "snug in" their dark-matter halos with their increased gravity. Within these galaxies much smaller "droplets" of gas condense out of the whole—these become the stars. When most of the gas in a galaxy has been converted into stars, the newborn galaxy stops contracting. This is because there are no longer collisions of atoms that can lead to cooling and further contraction. The first galaxies to form are likely to be the densest fluctuations, and they probably grow massive black holes at their centers, which gobble vast quantities of residual gas. The release of gravitational energy, as the mass of millions of suns tumbles down the endless stairs of the black hole, powers the bright beacon we call a quasar.

The Seventh *Day*—the universe today: The galaxies become fully formed, while clusters, superclusters, and voids continue to grow in prominence. The Sun forms and, along with it, planet Earth. Life begins—but that's another story.*

∞

If this version of creation was underwritten with a guarantee of authority, like the story of creation in the Judeo-Christian Bible, it would end here, a perfectly consistent tale. But this is science, so for better or for worse, this account can be checked with new observations and for its predictive power and implications. This is where its shortcomings will show like bad joinery, and this is where we listen for the ring of truth, or the hollow thud of contrivance. Let's examine how the picture is holding up.

The first *Day* has a startling, oddly appealing (or appalling) idea at its

---

*In our own way of reckoning time, the lengths of these different *days* vary enormously, from $10^{-35}$ seconds to 10 billion years (42 powers of 10). But, perhaps, in a truer sense the duration of each of these "eras" was comparable, because the processes that ruled in each occurred on a comparably faster pace in the shorter *Days*. The brief life of a Mayfly is compensated somewhat by the feverish pace of its activity; all animals live for about one billion heartbeats. So, in some sense, is each period in the universe's history a full epoch.

core—a universe springing from nothing, pitching back and forth between the gravity field and the Higgs field. Unfortunately, at the present time there is no detailed theory that successfully elaborates this idea. The proposition has been made that a "grand unified theory" (GUT) of matter, energy, and forces can be constructed to describe this epoch in detail, a model that can be "broken down" at lower energies to the particles and fields we see today. However, the particular models that have been tried have failed, in a way that exemplifies the scientific method at work.

Starting from the proposition that the simplest theory is the best theory, physicists investigated the simplest GUT model consistent with what has been learned from particle-accelerator experiments. This model proposed a unification of the electromagnetic, weak, and strong forces within the framework of a certain symmetry transformation.* It made several predictions of previously unobserved phenomena, but, unfortunately, most of these were well beyond the energy range of present (or even planned) particle accelerators. However, one consequence of this simplest GUT model was the prediction that "protons are not forever," that the typical proton would live $10^{31}$ years before decaying into other particles. Happily, it is not necessary to wait for the warranty on the universe to run out in order to check this prediction, because it also means that for every collection of $10^{31}$ protons, on average one will decay *every year*. By placing sensitive detectors in underground mines ($10^{31}$ protons amounts to about 500 tons of rock), where they are shielded from particles from outer space (called cosmic rays) that confuse the issue, physicists have already found that this is not the case: the lifetime of the proton is at least $10^{32}$ years. This is how science works: the simplest GUT model has been eliminated, leaving the theorists to contemplate other models and their observable consequences.

While some physicists search for a successful GUT model to unify

---

*The goal is to find a complete description of the laws of nature at high energy with a minimum number of particles and forces and a maximum degree of "symmetry." A trivial example of the symmetry principle is the die, the little cube whose six faces are marked with different numbers of spots. The die itself presents the same appearance in six different orientations, reflecting the sixfold symmetry of a cube in three dimensions, but the expression of the die varies depending on which face, with its unique number of spots, is presented. In an analogous way, the unified force may express itself in many ways (different particles and interactions) depending on the "way it is viewed." These symmetries are expressed not in real space, but in a mathematical space that can have any number of dimensions, though the aim is to find the lowest dimensional space that "works" to explain the phenomena. Such sets of symmetrical configurations in many-dimensional space are called *gauge transformations*.

the three forces, others are looking for even more grand descriptions—"supersymmetric" theories (SUSY) that could unify all four forces, gravity included. A promising theoretical framework—*superstrings*—could provide the high degree of symmetry needed. The basic idea of superstring theory is that particles are not "points" with charge, mass, spin, and so on, but higher-dimensional entities, for example, infinitesimal loops of string, or surfaces.* The theory predicts the existence of additional particles complementary to the ones we know—phot*inos*, *s*electrons, *s*neutrinos, and the like. Not as frivolous as it sounds, this extension greatly enhances the possibility for symmetry and offers a rich conceptual structure. Many physicists find the superstring theory "beautiful," and there is much sympathy with the idea, based on previous experience, that the ultimate laws of nature will display such an elegance, symmetry, and simplicity. However, as attracting and promising as the superstring theory may be, none of its predictions have been confirmed, so, although the idea has promise, it has not yet made much of a contribution to our understanding of the early universe.

The second *Day* is the inflation era. Inflation is a conceptual triumph in that it provides explanations for some of the universe's basic characteristics: improbable flatness, extraordinary uniformity, suppression of magnetic monopoles, cause of the present expansion, and the source of primordial fluctuations. For most theorists, the inflation model's ease at addressing such a wide range of phenomena qualifies it as a sort of master key. Few theories succeed so well on issues other than those for which they were developed—this gives inflation the ring of truth.

Though the generic idea of an inflationary epoch has been a major breakthrough in the study of cosmology, the development of a *specific* model has been anything but easy. The sticking point has been whether the episode can leave the universe with a uniformly broken Higgs field. In Guth's original model, the exponential expansion of the universe "outran" the progress of the Higgs field in its attempt to flop over to a broken symmetry. Symmetry breaking began spontaneously at myriads of locations, and from each the conversion to the broken symmetry state grew outward as a spherical wave,

---

*This fits naturally into the notion that space-time itself has more dimensions than the four we perceive. The idea here is that during the epoch of inflation the four dimensions of space and time ballooned to what are now gigantic scales, while the others remained ultracompact, at about the size of the universe before inflation—$10^{-38}$ cm. Even though the submicroscopic scale of these other dimensions would hide them from perception in our inflated world, their effects could still be substantial. It has been suggested that a perfectly symmetrical world of ten dimensions "broke down" into the complex world of particles and forces we observe today.

like the sodium acetate crystal in my Uncle Robert's Erlenmeyer flask. These spreading centers would have joined up, leaving the universe in a uniform state of broken symmetry, had it not been for the expansion that carried them apart at a phenomenal rate—as if the sodium acetate crystal never reached the edge of the Erlenmeyer flask because the flask itself grew faster than the crystal. According to Guth's calculations, the universe should have been left in a highly nonuniform state: a Swiss cheese with "bubbles" of broken symmetry separated by zones of unbroken symmetry. This is definitely not the universe we inhabit today.

Enormous efforts have gone into "fixing up" inflation; some subsequent incarnations repaired this original defect but developed new troubles. A recent rendition, called *extended inflation* by its promoter, University of Pennsylvania physicist Paul Steinhardt, solves the "bubble problem" by requiring that the universe expanded a little more slowly at first, allowing time for the "bubbles" of broken symmetry to catch up with each other and merge. As a specific example of why this might happen, Steinhardt suggests that gravity was a stronger force (per unit mass at a given separation) "in the beginning." A stronger gravitational field could have retarded the expansion just long enough to give the bubbles time to collide.*

Extended inflation might have other consequences. One might be the boosting of the largest-scale fluctuations in density—this would help the apparent "anemia" that current models seem to have in growing the largest superclusters and voids. Another possible consequence is that the bubble collisions could have been a strong source of "gravitational radiation," tidal waves that send shivers through the local gravity field as they pass. Searching for gravity "waves" is a challenging new form of astronomy. These tremors are expected from many violent events in the universe, for example, from supernova explosions. The minute disturbances of the local gravitation field from long-ago collisions of Higgs symmetry-breaking bubbles are far beyond present technology, but in principle they could be searched for.

Whether or not the aforementioned problem can be solved, the greatest misgiving about inflation is that it requires "physics" to be "just so." Not all GUT theories have the right properties to set inflation going, and further restrictions are apparently required to avoid the "bubble problem." The odds appear to be against discovering the "correct" GUT theory, the one whose elegant symmetries describe all the complex physics we observe today, and

---

*Physicists have recognized the necessity of a new description of gravity for the extraordinary conditions of the first moments of the big bang. It is possible that this as-yet-undeveloped theory of *quantum gravity* might specify the stronger gravitational force that would extend the epoch of inflation, as Steinhardt envisions.

finding that it also sets the proper conditions for a supercooled symmetry breaking of the Higgs field at $10^{-35}$ seconds. But there might have been other epochs when inflation could have occurred, driven by other "symmetry-breaking events." Many cosmologists continue to believe that there is something right about this idea, even if we may still be far from specifying the specific epoch and its properties.

The third *Day* is the time of particle creation. The details of this epoch—figuring out the which, when, and how of particle genesis—are probably not crucial for understanding the dynamic evolution of the universe, with one glaring exception: the production of dark matter. Recall that inflation leaves the universe at critical density, but, according to a very successful model of nucleosynthesis, no more than 10 percent of the mass of the universe can be in baryonic form, the ordinary matter we are familiar with. For this story to make sense then, at least 90 percent of the mass in the universe must have been left in non-baryonic form.

At this time, no GUT or superstring theory is sufficiently detailed and well understood to predict how and which dark matter particle(s) were produced. But even more discouraging is evidence that, by itself, either of the two generic types may be insufficient to generate the observed large-scale structure of the universe. Models based on hot dark matter, for example, neutrinos with a small mass, can account for plenty of large superclusters and voids, but they produce far too little clustering of galaxies with their near neighbors. This is because, for hot dark matter, initial galaxy-sized fluctuations die away as neutrinos (or their like) stream away at near-light speeds. An additional unfortunate consequence of this is the delaying of galaxy formation to a very late time, a prediction seemingly at odds with the existence of quasars.

Cold dark matter got off to a much better start. In the early 1980s there was widespread optimism that it could reproduce all the basic structures of galaxies and galaxy clustering. A key step in this success had been the introduction, by Marc Davis and his collaborators, of Nick Kaiser's notion of *biasing*, the idea that galaxies were clustered more strongly than the underlying dark matter. Although some questioned the ability of the biased-CDM model to account for the Great Void, or the tendency of rich clusters of galaxies to themselves cluster, proponents brushed these challenges aside, citing selection effects in the data. However, with the coming of the "slice of the universe" and the Seven Samurai's "large-scale streaming flow" in 1986, this weakness—not enough power on large scales—grew more apparent. The distribution of galaxies in the "slice" was hard to quantify, but it left the distinct impression that the galaxy distribution was cellular, with many large voids and surprisingly thin cell walls. The large-scale streaming flow, which led to the Great Attractor model, indicated a huge swell in the distribution of dark matter. At best, these were exaggerations of anything that showed up in the CDM simulations; at worst, they were virtual impossibilities.

The apparent contradiction has spurred new observations of large-scale structure. Many additional "slices" have been added by the Harvard-Smithsonian group; they confirm the prominence of large voids defined by thin sheets and shells of galaxies. When Geller and Huchra stacked three contiguous wedge-shaped slices, forming a bellowslike volume, they found a continuous "wall" of galaxies stretching across the entire survey. Israeli astrophysicist Avishai Dekel nicknamed the structure the Great Wall, after its Chinese counterpart. Although it is not certain whether the Great Wall is one continuous structure several hundred million light-years long, or a chance butting together of smaller structures (which would be easier to account for), the challenge to CDM models was clearly raised a notch by its discovery.

One of CDM's leading proponents, George Efstathiou, led an ambitious effort to make a uniform, deep map of the galaxy distribution over a very large patch of sky. He and students Will Sutherland and Steve Maddox used photographic plates from the Southern Sky Survey and a leviathan scanning machine at Cambridge to transform the photographic images into computer data. With the most carefully constructed two-dimensional map ever attempted, these three British astronomers were able to extend the study of galaxy clustering to enormous scales. They, too, found that galaxy clustering over great distances exceeded the amount predicted by the biased-CDM model.

In 1990 the newest survey of the three-dimensional distribution of galaxies in our neighborhood delivered another hard shot. This survey, like the one done by Strauss, Davis, and Yahil, was based on the *IRAS* catalog of infrared sources, but the British team, led by Michael Rowan-Robinson, chose a sample that would probe space less completely but more deeply. With 2,500 new galaxy redshifts they produced the largest contiguous map to date of the nearby universe, and concluded that the galaxy distribution rolled up and down in density like huge ocean swells, in excess of the predictions of biased CDM. The press release that accompanied the paper, whose authors included former CDM zealots George Efstathiou, Carlos Frenk, and Nick Kaiser, declared the CDM model dead.

Rumors of CDM's death, like those of Mark Twain's, have been greatly exaggerated. Although most researchers agree that the *strongly biased* CDM model studied by Davis et al. has failed, a weakly biased or even unbiased

CDM model (in which galaxies fairly trace the dark matter) may in fact reasonably reproduce the features of large-scale structure described above. This model was first rejected because it led to peculiar velocities (departures from smooth Hubble expansion) that were much too high, for galaxies relative to their neighbors. But improved N-body simulations, for example those by James Gelb and Ed Bertschinger at MIT, incorporate more sophisticated computer codes that use vastly larger numbers of "mass points" to represent the distribution of dark matter. Early results of these newer simulations show that less-biased CDM models can produce more "great" structures while holding down the peculiar velocities. This weakens the disagreement between prediction and observations and therefore works toward "saving" the CDM model, though it is not at all clear at this point whether it goes far enough.

Short of finding out that the previous N-body simulations were flawed, the only other way to save the CDM model is to alter the spectrum of input fluctuations, the 1/f form that seemed so natural, to include more power on large scales. One might hope that a better understanding of early-universe physics will call for just such an alteration, but until that happens, there is great reluctance to monkey with the input spectrum: the freedom from arbitrary "adjustments" is perhaps the CDM model's best feature.

To the extent that what "ails" CDM is its anemia on large scales, it is not surprising that the latest effort to "cure" it calls for an injection of large-scale power, not by the introduction of arbitrary adjustments in unknown physics, but by the injection of a dose of hot dark matter. In 1992 some "friends of CDM," including Joel Primak and George Blumenthal at Santa Cruz, and Marc Davis at Berkeley, revived the notion of a universe with *both* hot and cold dark matter, an idea first proposed by Avishai Dekel and explored by one of Sandy Faber's students, Jon Holtzman. *Mixed* dark matter (MDM), as it is called, can provide excellent agreement between observations and prediction. Many astrophysicists will call this folly—needing to appeal to *one* unknown particle was bad enough! But others are already impressed by how well the MDM model works. It is very likely that MDM would achieve the status of "standard model" if even one of its components were confirmed by other means, for example, if direct measurements in earthbound laboratories were ever to measure a small mass for neutrinos or detect a cold dark matter candidate whizzing through a cryogenically cooled crystal. It's a long shot, but HDM + CDM = MDM could be the sought-after equation.

There is good news about the fourth *Day*, the era of nucleosynthesis. Here we finally reach familiar ground, the physics we can test in earthbound laboratories, and our model works extremely well. The right types of atoms— hydrogen, helium, deuterium, lithium—in the right proportions, arise naturally in the application of the big bang model. In comparison to the other missing links and contradictions in this discussion, this one hardly lifts the "worry meter" off its peg.

The fifth *Day*, the era of decoupling, is also known physics, and well understood. Of course, the prediction and subsequent detection of the cosmic microwave background (CMB) was itself one of the great triumphs of modern cosmology. The Cosmic Background Explorer *(COBE)* satellite has been able to accomplish a twenty-year quest to measure the *spectrum* of the CMB—precise measurements of its intensity at different radio frequencies. The exquisite match that has been found to a *blackbody spectrum*, the form nature dictates for equilibrium situations, and the one predicted for the simplest big bang model, is compelling evidence that our tall tale of the story of creation is rooted in fact.

Attention has now shifted to another magnificent achievement: the extraordinarily deep "radio picture" that shows an extremely faint pattern, thought to be the first glimpses of structure formation. Theory had predicted tiny variations in the intensity of the CMB due to embryonic concentrations of matter by gravity, but previous experiments lacked the necessary precision. The first detections of these slight undulations, with the *COBE* satellite and by the MIT group, have replaced somewhat nervous anticipation with admiration for the continuing success of the big bang model. However, the faint patterns that have been detected correspond to extremely large-scale structures, so there is now great activity aimed at detecting fluctuations at smaller scales appropriate for clusters and superclusters—patches of sky about the size of the Moon. Experiments have been inconclusive, but with instruments now reaching the needed sensitivity of one part in a million, conclusive results are expected in this decade.

If the amplitude of these early fluctuations in matter density are measured over a wide range and found to correspond to the $1/f$ form, this will be a ringing endorsement of the inflation model and, indirectly, the idea that seeds of weakly interacting dark matter began the formation of structure. On the other hand, if a form very different from $1/f$ is found, this might point to a more exotic alternative for structure formation, for example, one of the wildest speculations—*cosmic strings*. The idea is that the universe is crisscrossed with ropes of ultra-high energy density, actually regions where the hot big bang phase lives on in today's cold space.

A good visual model for what a cosmic string is, and how one might form, starts with imagining an energy field (like the Higgs field) as a forest of upright pencils, spaced in a regular grid, balanced on their points. The pencils are tied together, and thus kept from falling over, by a "mattress" of springs, four from each eraser—north, east, west, and south. The springs are not really strong enough to hold up the pencils all by themselves, but the system is "hot," like the early universe, so the pencils gyrate wildly in all directions, causing the pencils to wobble around their upright positions. This is the symmetrical state, like the early moments of the Higgs field.

If the system is cooled, the pencils begin to lose energy. Soon their own weight begins to drag the pencils down and they fall over. Some fall to the left and drag their neighbors with them; others some distance away start falling to the right. When the two spreading zones of left-falling and right-falling pencils meet, a line of pencils remains upright, tugged equally to the left and right by the zones of "broken symmetry." They cannot fall and "choose a direction," but remain upright, stuck in the symmetrical "hot" state. Like the line of pencils that remains upright, a cosmic string is a trail of primal energy density that will never cool.

Because high energy density is equivalent to great mass, the gravitational force near a cosmic string is enormous; furthermore, these strings whip through the universe at the speed of light. As a cosmic string sweeps through the smooth sea of mass of a "cold" universe, it will leave a wake capable of initiating the gravitational collapse of structures, from galaxies to great walls.

So far, studies of the behavior, evolution, and effects of cosmic strings suggest that they would not easily give rise to the structures we see in the universe today, and that their analogous two-dimensional forms, called "domain walls," fail utterly. However, this kind of exploration has led to a higher-dimensional analog called a "texture" that looks more promising. Other ways of generating structure may be hiding in the physics of the early universe, but, as the great Russian physicist Yacov B. Zel'dovich was fond of emphasizing, these "cosmic defects" can be exploited only at the cost of losing the benefits of the inflation paradigm. After all, inflation was invented to squelch the production of magnetic monopoles by breaking symmetry in the Higgs field uniformly over vast regions of space. Its ability to mow down all the pencils in the same direction would also prevent the formation of cosmic strings or textures that spring from fractures in the broken field.

From cold-dark-matter fluctuations to cosmic strings, there is a wealth of possibilities for seeding the formation of the galaxies, clusters, and super-clusters we see today. Some guidance from observations is clearly necessary, and it appears that studies of faint patterns in the cosmic microwave back-ground will be the best way to learn how these structures began to form.

The sixth *Day*, when galaxies form, is the beginning of the modern era. We have a long way to go before we can completely describe the process of

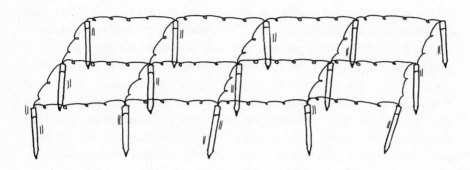

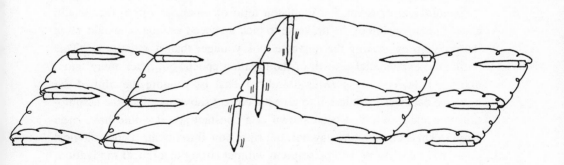

This is a cartoon model for a cosmic string. In the symmetrical "hot" state, the pencils gyrate around their vertical positions, but when the system cools, they fall down. Those pencils forced to remain upright by the tug of other fallen pencils describe a zone where the high-energy state persists.

galaxy formation and how it led to galaxies with a variety of sizes, shapes, and stellar populations. On the other hand, the story, though complicated, does not appear to be mysterious—our knowledge of the physics of matter and energy in this regime should be sufficient to allow us to order the basic steps and fill in the details, eventually.

If there is a major puzzle here, it is whether there is a serious conflict between the expansion age of the universe and the ages of the oldest stars and (radioactive) chemical elements. That these different methods of measuring the age of the universe give similar results is wonderful—the approximate agreement of the time scales is at the very heart of our story. Nevertheless, several recent determinations of the Hubble constant (the rate of expansion) that are judged by many to be the most reliable yet accomplished imply that the universe has been expanding for only about 10 billion years, while the oldest stars are thought to be some 13 to 16 billion years old. For the most part, researchers expect this issue to be cleared up either by an indisputable measurement of the Hubble constant (such as has been planned with the Hubble Space Telescope) that turns out to give an expansion age greater than 15 billion years, or through new insights into stellar evolution that drop the ages determined for the oldest stars to less than 10 billion years.

If this marriage of the "ages" eventually fails, the plan appears to be a reluctant retreat to the idea that gravity and kinetic energy are not the only factors that have ruled during the expansion era. A resurrection of Einstein's "cosmological constant," a repulsion term of unknown origin that would cause the expansion of the universe to pick up speed as it ages, would solve the problem by making the universe look younger than it actually is. (This is like the exponential growth of the inflation epoch, but much more tame, and much later—during *Days* 6 and 7.) That is, running the film of the universe backward would fail to give the correct showtime since the assumption that the movie had been shown at a constant speed would have been wrong. This is a possible way out, but right now introducing a cosmological constant looks like an ad hoc solution without strong theoretical motivation. Rather than embrace this complication, most researchers prefer to wait and hope for a resolution of the "age" problem.

The seventh *Day* is the "present" era, when large-scale structures—superclusters and voids—are just now coming into their own as identifiable structures. Galaxies achieved this status long ago: the density contrast between a galaxy and its surrounding "space" has subsequently grown to be enormous. If the density contrast between superclusters and voids were anywhere near as pronounced, it would have been easy to pick out superclusters in simple maps of galaxy positions on the sky, long before extensive catalogs of redshifts made possible the construction of full three-dimensional maps of the distribution of galaxies. Rather, these giant structures are just beginning to show their architecture; when the universe is twice as old as it is now, in another 10 or 20 billion years, superclusters will rise like the Rocky Mountains, instead of the Appalachians, above the plain.

Measuring the strength of superclustering today is a fairly direct way of investigating the processes in the early universe that initiated the growth of structure. Galaxies grew from early fluctuations too, but the structural evolution of galaxies has been heavily influenced by energy input from star formation. Therefore, much of what could have been learned about the small-scale density fluctuations that gave rise to galaxies has been modified or erased altogether. In contrast to this, the energy released from the birth of stars, though considerable, is far too small to rearrange matter on the larger scales, so observations of clusters, superclusters, and voids are a direct read of primal fluctuations in the density of matter. Together with what we hope to learn from the faint pattern of intensity variations in the cosmic microwave background (which grew into these large-scale structures), observations of the strength of large-scale clustering should show whether weakly interacting particle matter can do the job, or alternatively point to something like cosmic strings or textures as the engines of structure formation.

The Seven Samurai, through their observations of large-scale structure, became involved in elaborating this particular part of our story. The map made by the Samurai has played a special role because only measurements of peculiar velocities—the departures from smooth Hubble expansion—can illuminate the distribution of dark matter and this, rather than the galaxy distribution, is what is predicted by computer simulations of structure formation. Also, comparing a map of the *mass* distribution with a map of the galaxy distribution, for a representative volume of space, is the first step in finding out if galaxies are faithful or "biased" tracers of the dark matter. Eventually, this kind of information will lead to a reliable measurement of matter density in the universe—a direct test of the inflation model and its prediction that the universe has the critical density, $\Omega = 1$, for which gravitational energy exactly balances the energy of expansion.

So how have the large-scale streaming flow, and the Great Attractor (GA) model, fared in the years since the original work of the Seven Samurai? By 1988 there were two additional, independent reports of large peculiar velocities in the Centaurus region. British astronomer John Lucey and Australian David Carter had used the Seven Samurai method for elliptical galaxies to measure the distance of several clusters of galaxies in the GA region, as had a group headed by Jeremy Mould, who applied the Tully-Fisher method of distance determination to spiral galaxies in clusters. Both studies found evidence of peculiar velocities as large as 1,000 km/sec, but both also expressed doubts about the continuity of the flow, where it was centered, and the applicability of the Great Attractor model. Sandy Faber and I followed with our 1989 study of the distances to 136 elliptical galaxies. Our findings worked against this continuing skepticism by substantiating the original Seven Samurai data completely and confirming the prediction of the GA model that the streaming flow would decline to zero peculiar velocity at a distance corresponding to a recession velocity of 5,000 km/sec, about 200 million light-years away.

Sandy and I set out to survey the GA region even more completely, and try our hands at the Tully-Fisher technique. We returned to Las Campanas in April 1989 with finding charts for some 150 spiral galaxies, not just cluster spirals, as the group headed by Mould had chosen, but spiral galaxies in clusters, groups, and the field. Rather than use a radio telescope to measure the rotation of spiral galaxies, we relied on the spectrograph of the du Pont telescope to collect the optical emission of glowing hydrogen gas to map, through the Doppler shift, the rotation speed of each galaxy in the sample. Our effort was complementary to a program by Sandy's student Stephane Courteau to measure peculiar velocities for spirals over the northern sky to a depth comparable to the Seven Samurai sample of elliptical galaxies. Again the weather obliged and we returned to California with spectroscopy and electronic (CCD) images for photometry of 125 spiral galaxies.

My summer project was to teach the computer how to extract a "rotation curve" from the faint trace of the H II (hydrogen) emission line that wandered haltingly, like fossil footprints in a streambed, across the recorded picture of each galaxy's spectrum. With this completed, and the standardization of rotation velocities and photometry that brought the data into accord with other studies, Sandy and I calculated the peculiar velocities and compared them to the previous year's work on ellipticals. We were gratified to find that the spiral galaxies, too, were streaming to the center of the Great Attractor with large velocities; at the putative center of the GA the peculiar velocities fell to zero, as if the center of mass had been reached. Best of all, there were more than a dozen spirals beyond the center of the GA that had negative peculiar velocities—they were headed back our way. Combined with the half-dozen distant ellipticals from our previous sample, these distant spirals provided the first creditable evidence of infall on the backside, toward the GA center. Bob Schommer, from Cerro Tololo (the U.S National Observatory in Chile), and Greg Bothun from the University of Oregon, carrying out their own study of spiral rotation curves with optical techniques, produced results similar to what Sandy and I had found, and these were announced together at the January 1990 meeting of the American Astronomical Society in Washington, D.C., as confirmation of the Great Attractor model. The picture seemed to be in good agreement with what the Seven Samurai had proposed: an enormous region of greater-than-average matter density in the universe that is attracting galaxies from all sides.

Still, there are critics, and the *name* Great Attractor, perhaps more than what it represents, continues to draw fire. British astronomer Michael Rowan-Robinson and his collaborators made a map of the distribution of *IRAS* galaxies, a map that sampled farther into space than the sample used by Strauss, Davis, and Yahil. They used this map to *predict* peculiar velocities throughout the region, by assuming, as Strauss and Co. had done, that *IRAS* galaxies follow the underlying dark-matter distribution exactly. The British team concluded that the distribution of galaxies could indeed account for the *observed* peculiar velocities found by the Samurai and others, and that these departures from smooth Hubble expansion arose from unevenness in the distribution of matter over very large distances. This was exactly the claim, once thought to be radical, made by the Seven Samurai, yet, ironically, a press release that accompanied this paper claimed "the Great Attractor is dead!" The report went on to say that the "dark, mysterious mass" proposed by the Seven Samurai was unnecessary—dark matter associated with the distribution of galaxies would account for the peculiar velocities that had been measured. What was odd about this was that the Seven Samurai had never taken a stand one way or the other as to whether the dark matter of the Great Attractor was accompanied by proportionate increase in galaxies; we simply didn't have the necessary data on the galaxy distribution.

In science, as in Congress, or over dinner, "putting up a straw man"—misunderstanding or even deliberately misrepresenting someone's position and then slaughtering it—has been known to happen. What should have been a satisfying confirmation that there is something to the Great Attractor model—it had been found to match up well to the *IRAS* galaxy distribution—was characterized as a knockdown and paraded before the press. A few heated, private exchanges, some transatlantic, followed. This tussle highlighted the resistance to having the collective name Great Attractor affixed to a region containing several superclusters of galaxies with their own identifications, as if it were superfluous to name a mountain range if each of the peaks had its own name.

So, despite observational confirmations, the Great Attractor model has not been universally embraced. The Seven Samurai never intended the GA model—a perfectly spherical, extended distribution of dark matter centered

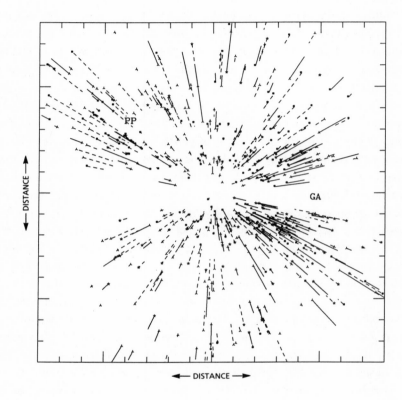

The map of peculiar velocities, updated to 1994. Solid lines show galaxies moving away from the Milky Way (at the diagram's center), dashed lines show motion toward us. As for the original diagram from Lynden-Bell et al., shown earlier, distances are expressed as expansion velocities, in kilometers per second. With much more data the large-scale streaming flow found by the Seven Samurai is now much more clearly defined, particularly in the Great Attractor (GA) area. As predicted, the large outward peculiar velocities disappear beyond the GA center (marked "GA"). The flow of galaxies in the Perseus-Pisces (PP) supercluster toward the Milky Way, found by Willick and Courteau, also shows clearly. *(Figure courtesy of Faber and collaborators.)*

at a particular place—to be taken so literally, but rather to be the simplest representation that would account for our data. And reluctance to embrace such an idealized model makes good sense. Certainly, the universe is not divided neatly into zones: the terrain is varied and it is not clear where one "range of mountains" ends and another begins. Thus, the GA model was undoubtedly an oversimplified way of accounting for the peculiar velocities of galaxies in our corner of the universe.

Hints that the Great Attractor might be too simple a model have come from the work of Jeff Willick, in his Ph.D. thesis at Berkeley, results later confirmed by Stephane Courteau in his thesis at Santa Cruz. Willick made an extensive study of the peculiar velocities of spiral galaxies in the Pisces region and found that galaxies there are, on average, moving in the Great Attractor direction at about 400 km/sec. But the GA model made by the Seven Samurai predicted that these galaxies, so far from the Great Attractor, would be pulled at a speed of only 100 to 200 km/sec. The faster-than-expected flow of the Perseus-Pisces region could be taken as evidence that another concentration of dark matter, one even larger and more distant than the Great Attractor, is tugging our entire region—Perseus-Pisces, Local Supercluster, and the Great Attractor itself—at a speed of around 200 km/sec. Indeed, a candidate for this "giant attractor" has been proposed—Italian astronomer Roberto Scaramella and his collaborators have pointed to a populous association of rich clusters of galaxies, noted long ago by Harlow Shapley, in the same direction as the Great Attractor, but almost three times more distant. The greater distance will make it difficult to confirm if this region contains a surplus of mass large enough to generate a coherent flow pattern over a region more than one billion light-years in diameter. But this grander flow might explain other recent observations that contradict the simple GA model, some further work by Schommer and Bothun, and a study by an Australian team led by Don Mathewson. In mild contradiction with the results Sandy and I found, these groups find little or no evidence for backflow—galaxies on the far side falling back our way—into the Great Attractor. Perhaps the more distant Shapley concentration is preventing these galaxies from falling back.* At this point, however, the limitations of break-

---

*The claim by Mathewson and his colleagues is even more radical. They believe that there is no evidence for the Great Attractor, and that the whole region studied by the Seven Samurai shares in a "bulk flow" of 600 km/sec. One might attribute this kind of flow to the pull of truly giant structures: for example, Hawaii's Brent Tully claims that Abell clusters are roughly arranged in a giant pancake at least one billion light-years across—roughly twice as large (about ten times the volume) as the Great Wall or the Great Attractor. The embryonic form of structures this large would have shown

ing the universe up into discrete structures are crippling: the entirety of this greater volume must be sampled fairly in order to gain a complete picture of how the peculiar velocities arise.

This is precisely why the emphasis has shifted away from pointing a finger at this or that supercluster, and working instead to make a map of the underlying dark matter, with all its hills and valleys. There is much to be learned from a comparison of such a map to the distribution of galaxies. For example, if galaxies could be shown to follow faithfully the swells and troughs in the dark matter, their illumination alone would be sufficient to delineate large-scale structure, while finding the two distributions in only approximate correspondence would throw caution into the effort. One might confirm that the distribution of galaxies is "biased" toward the denser peaks of dark matter, an assumption often made in N-body models of structure growth. If, on the contrary, the galaxy distribution were found to be very poorly correlated with the underlying dark matter, this might be evidence for additional agents in the process of galaxy formation. For example, Jeremiah P. Ostriker at Princeton and Len Cowie at the University of Hawaii have suggested that the first galaxies to form sent out blast waves of energy from the intense initial burst of star births, enough energy to trigger the formation of other galaxies in the "wake," a process that could be largely independent of the location of the dark matter.

The most successful and promising attempt to use measurements of peculiar velocity to produce a true map of the underlying dark matter has come from Ed Bertschinger and Avishai Dekel. The two have been developing a complex computer program (POTENT) that uses peculiar velocities—the departures from a smooth Hubble expansion—to gauge the pull of gravity from swells or troughs in the density of matter. POTENT skirts a difficult problem that has dogged other techniques: it derives the mass distribution from purely *local* measurements so it is unaffected by the gravitational pull of distant structures whose extent and influence is poorly known. How this is done can be visualized by thinking about exploring hilly terrain. Each step that takes you up or downhill is the result of a local change in altitude—the height of distant peaks has no influence. Simply by keeping a careful sum of these steps up and down one can map the *global* profile of the mountain. In the same way, Bertschinger's and Dekel's technique attributes local changes

up, but didn't, as a strong mottling in the cosmic microwave background. So, if these ideas are correct, there is something very wrong with much of the picture painted in this chapter. Science, as you see, can get difficult, confusing, and contentious at one of its frontiers.

in peculiar velocity—whether the galaxies are speeding up or slowing down—to the increase or decrease in the local density of matter, with its attendant gravity. From these purely local measurements, free from the uncertainty of what lies beyond the surveyed region, the program can build up the total profile of the hills and valleys of dark matter.

As more and more measurements of peculiar velocity have been fed into POTENT, a map of the cosmic terrain of dark matter has taken shape that agrees well with the gross features of the galaxy distribution, as mapped by the *IRAS* (infrared-satellite) survey. Most prominent is a mountain of dark matter including and rising up beyond the Centaurus clusters, what the Seven Samurai identified as the Great Attractor. The Local Supercluster also shows as a ridge of higher-than-average mass density that runs toward and merges with the Great Attractor. On the other side of the sky rises another mountain of dark matter, somewhat steeper but much less broad—this matches up with the position of the Perseus-Pisces supercluster of galaxies. In between the two mountains is a valley, an elongated void of lower-than-average matter density.

These four pieces of terrain—the Great Attractor and the Local Supercluster on one side, the Perseus-Pisces Supercluster on the other, and the void in between—are the major features of *both* the dark matter map inferred from the *motions* of galaxies (their peculiar velocities) and the *IRAS* map of the *positions* of galaxies. These maps, though coarse, suggest a good correspondence between the galaxy and dark matter distribution, at least on large scales like superclusters. This is good news for the effort to map large-scale structure by galaxy maps alone, though the data are still too crude to indicate whether there are finer differences between the distribution of galaxies and the underlying dark matter that might tell us more about galaxy formation.

Perhaps the most exciting result to come from the comparison of these maps is the attempt to determine if the universe has the critical density of matter, $\Omega = 1$, that corresponds to a balance between the gravitational energy and the energy of the expansion. Using the *IRAS* map to judge how large is the contrast between hills and valleys, and the peculiar velocities to measure the gravitational force that this degree of contrast generates, Dekel, Bertschinger, and Yahil have derived a value for the mass density that is, within the errors, just the $\Omega = 1$ value that theorists prefer and the model of inflation demands. This is the first time a value close to the critical density has been measured; previous determinations came from much smaller regions, so the hope is that a large enough volume of the universe has been sampled to give the first repre-

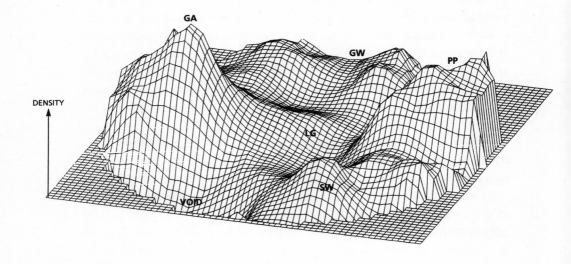

A map of the "terrain" for all matter, dark and visible, in the local universe, made with the POTENT method. The flat region surrounding the landscape shows the limit of the survey, about 300 million light-years, at the level of zero density. Average density is about halfway up. Visible are the "mountains" due to the Great Attractor (GA) and Perseus-Pisces (PP), the ridge of the local supercluster, upon which the Local Group (LG) sits, and a prominent valley—marked "void"—of lower-than-average density. The GA is the largest structure in this volume of the Universe, according to this reconstruction. The GW and SW regions are structures associated with the Great Wall of Galaxies in the north and south found in the CfA surveys. These latter regions lie near the limit of the measurements of peculiar velocity so they are as yet poorly sampled.

sentative measure of the density of the universe. If confirmed by subsequent studies, this will be seen as one of the great accomplishments of the kind of studies the Seven Samurai and others have been doing.

On the whole, the large-scale streaming flow and the Great Attractor model have stood up very well to challenges regarding data and methodology. Now, few doubt that the peculiar velocity of the Milky Way galaxy is shared by neighboring galaxies over a substantial volume of space, and that this flotilla is moving toward a region that contains several superclusters whose combined gravitational effect is at least partly responsible. This is the essence of what the Seven Samurai concluded. Today, criticism of the work of the Seven Samurais seems to be in the sense that the GA is not enough, that the flows are even larger in extent than we imagined. Perhaps the original bulk flow model I disliked was closer to the truth. Time will tell.

My guess is that large-scale flows will be found to be a common feature of the universe—the fact that one has been found "right here" is the strongest indication of that. New techniques are being developed that could yield more accurate measurements of distances to galaxies; this would greatly improve the quality of maps of the mass distribution and our ability to learn to what degree galaxies follow it. Donald Lynden-Bell and Roberto Terlevich are among those trying to determine more accurate distances to rich clusters, and I am collaborating with John Tonry on a method that can make much more accurate distance measurements to individual elliptical galaxies. Many research teams, like the one including Dave Burstein, Gary Wegner, and Roger Davies, are pushing present techniques to larger distances than sampled by the Seven Samurai. No doubt, other "great attractors" will eventually be found, though perhaps we will move beyond the identification of individual structures to a more generic description.

I asked the other Samurai if they would like to add some final words about how they saw our project, in retrospect. Only Sandy and Gary accepted my invitation. Sandy's mood was upbeat:

"Up until about two years ago I was feeling rather uncertain and pessimistic about cosmic [streaming] flows. But since then I have been coming around to think that cosmic flows rank among the fundamental data of cosmology after all, right up there with the cosmic microwave background. Three developments, all moving strongly in the past two years, have been changing my mind. First, there was the smashing discovery by *COBE* of the brightness fluctuations in the CMB. Finding these ripples, which grew in strength and condensed into today's clusters, and superclusters, has really firmed up the basic picture that structure in the universe formed via gravity. Now we can turn the problem on its head and ask, given the observed strength of the CMB ripples, how big should the streaming flows be today? The answer is that typical galaxies should be moving with velocities of a few hundred kilometers per second, exactly what the Seven Samurai reported. In short, if cosmic flows were not already known, *COBE* would have driven us out to find them. I don't think there is much doubt any longer that the streaming flows we found are real.

"The second development is the rapid rise in the number of measurements of peculiar velocities. The Seven Samurai inspired a growth industry— from our sample of only about 400 galaxies there are nearly ten times as many measurements now, enough to paint a realistic picture of the velocity field out to 250 million light-years. Though richer and more detailed, the new picture is basically the same one we saw. The Great Attractor is still there, but drifting slowly at 300 km/sec in addition to drawing galaxies toward it. The other big nearby supercluster, Perseus-Pisces, turns out to be a strong attractor as well, and it is drifting in the same direction as the GA. No other nearby regions compare to these, but then no other regions are as rich in galaxies, either. Clearly, infall, mass, and superclusters go together— the new maps make sense.

"Finally, the third development—key to the whole enterprise—is the invention of clever mathematical tools to analyze cosmic flows. These are big improvements over the analysis we did in Lynden-Bell et al., where we had to shoehorn very complex and highly irregular patterns of peculiar velocities into two idealized, spherical flow patterns [the Local Supercluster and the GA]. Bertschinger and Dekel's POTENT program is one of those flashes of theoretical insight that everyone admires and feels stupid for having missed. With it, we can reconstruct the entire field of peculiar velocities, even the 'sideways' motions we couldn't measure, giving us a full three-dimensional picture out to 250 million light-years. Their mathematical machinery makes use of the fact that flows converge in regions of excess matter density, and diverge in voids, to generate a precise map of the underlying mass distribution. We sensed intuitively that this could be done, but the Samurai use of oversimplified flow models destroyed the very information to do it.

"In addition, Yahil, Dekel and Adi Nusser [Dekel's student] have been working hard with the POTENT machinery to measure the cosmic density parameter; they've been finding values near the critical, $\Omega = 1$. If this holds up it will be very important, and a great accomplishment that cosmic velocity fields were able to do something that all previous cosmic observations have failed to do. Dekel and Nusser have even derived a way to take today's map of mass 'peaks and valleys' back in time to when they were 'born.' These ripples turn out to have the standard pattern generated by Guth's inflation model, and not the weirder shapes predicted by bizarre theories such as cosmic strings or textures. Between the measurement of $\Omega = 1$ and the spectrum of primordial 'ripples,' cosmic streaming flows are providing strong support for the inflationary model of the universe.

"The thought that we might indeed have uncovered one of the fundamental clues to the origin of the universe cheers me considerably."

While Sandy's intense involvement with ongoing analysis of large-scale flows has left her with these strong scientific evaluations, Gary Wegner's look back is more circumspect. He writes: "As time goes on, the ideas and finally the data will be superseded. Still, we must be thankful for the wonderful things that life gives us, and astronomers are more cognizant of this than most other scientists. I like to think of what we found as analogous to William Herschel's early measure of the solar motion from only a handful of stars, for this served as one of the beginnings of stellar dynamics. Nevertheless, we must take care to never take ourselves too seriously.

"In May 1989 at the Rio de Janeiro Meeting on the Large-Scale Structure and Motions in the Universe, much of which was inspired by the work of the Seven Samurai, Donald Lynden-Bell and I were climbing the stairs of the upper level of Sugarloaf, on a beautiful moonlit evening with the myriad lights of Rio twinkling below us, talking about just this subject. Donald became very serious and looking at me asked, 'Tell me what you think, Gary, did we really discover something important?' For a moment we looked at each other in silence, and then bursting out together in laughter, we hurried to regain the throng."

For my own part, I am pleased that what we did has advanced the field, and grateful for the experience, which helped me mature both professionally and personally. No explorer likes to think of his discovery fading in importance or, horror of horrors, being discredited. I am no exception. My confidence in the work of the Seven Samurai remains very high; I know of no study that was done with more care or higher standards. I expect the basic results of our work to endure. However, in time, it may turn out that the identification of a "Great Attractor" was little more than "training wheels" in our attempt to model large-scale structure, just another one of those analogies that allow us to grasp worlds that, at first, seem beyond our comprehension.

And, after all, this is the point. What is truly amazing is the process, that we can use our minds to dare to comprehend the universe and our place in it. Yes, the story is grand, and everyone's first reaction is to feel small and unimportant. But the greatest wonder belongs to the one who wonders.

That we are here, contemplating the universe and trying to tell its tale, rather than the other way around, is a testament to our worth and place in the story of creation. We are possessed with the dream of knowing where we came from, and possessed with the gift—the human mind—to realize that dream.

In the end, whether the Great Attractor survives as seminal nomenclature in the lexicon of astronomy is of no importance—the *idea* of the Great Attractor has played a pivotal role in our understanding of large-scale structure. It has done all that it could have done; it has moved us along our wonderful journey, halfway to creation.

# $12$ • THE CREATURES OF HYPERSPACE

Most scientific papers conclude with a section called "Discussion" where the authors are allowed to muse a bit about the implications of their work. Convention seems to be that they have earned the right to speculate with something less than the full scientific rigor required for the body of the paper. Here I will avail myself of this opportunity to ponder what I see as the implications of this work—not the Great Attractor per se, but the great scientific enterprise that propels us toward what I believe is a future of near unimaginable change. Many of the ideas presented here are not original, but I will try to combine them in a way that is my own.

I have often been asked whether an astronomer, reminded so often about the enormity of the universe, attains a different perspective on the question "What's it all about?" The experiences of my work have influenced me to speculate on the human adventure and its relation to the cosmos; it is a subject that very much interests me. With broad strokes aimed at sketching in the outline and coloring the heart of this ancient question, this discussion is an invitation to think about the evolution of the universe and our own, together.

First, a few words about science, about its worth and role in society. Few if any reading this will need to hear a defense of science as a worthwhile activity. But science is coming under increasing attack, so it is important for all who value it to think about how to answer those who would seek to restrain science or stop it altogether. While the most strident voices come from religious fundamentalists, who clearly see science as a threat to their articles of authority, there is a wider, more troubling opposition that links science to the creation of terrible weapons or the destruction of the environment. This movement mistakes technology for science, and blames the misuse of science on the knowledge itself, rather than recognize that all of us, individually and collectively, bear the responsibility for how and to what purpose that knowledge is used.

Though the spiritual fulfillment of "knowing" has long been used as a *raison d'être* for science (one I strongly believe in), the social contract between science and society is based primarily on science's ability to "provide." One indication of the strain between science and its patron, the public, is the more frequently heard indictment that science has not, after all, improved the quality of life as promised, but that somehow the world is a more gruesome, stressful, unsatisfying place than it ever has been. After all, there are more people starving in the world today than were even alive for most of human history.

Though those who long for the pre-technological past seem blissfully ignorant of the hardship and even barbarism of life ten centuries ago, the question remains: has science failed us? Science through its technological hand gives us the ability to improve the quality of life, at least in a material sense. And while we have all become aware of the limits of materialism in conveying a high quality of life, there is no question, I think, that without adequate food, shelter, and health, a high quality of life is impossible. Science and technology are the only means we know to build this foundation, from which lives of fulfillment and gratification become at least possible. Humans must then define their own paths to these goals, free to choose from a wide range of scientific, cultural, or spiritual resources.

But the promised abundance has not come to most of the world's population. This is not a failing of science, or even technology, but a product of complex social, cultural, and political factors. Some societies on the planet have achieved a remarkable level of success in providing the foundation mentioned above, but many have not. Sometimes this reflects free choice. More frequently it is due to an immiscibility of cultures that, in effect, withholds benefits unless the recipients are willing to adopt the societal precepts and institutions of the donor, that is, there are strings attached. And, too often, it derives from an exploitative arrangement that feeds on a gradient of wealth and power.

How has science increased wealth in what we chauvinistically refer to as the "developed world"? In the simplest terms, the productivity of an individual human is the principal determinant of how much there is to go around—how much food one farmer can grow and how many "man-hours" are required to build a house relate directly to what fraction of the population can live suitably housed and well fed. (This is conceptually correct, but an oversimplification because a combination of capital, resources, and labor is required to accomplish most things in the modern world; therefore, productivity is a more complex measure of the human contribution to all these factors.)

Over the last few centuries, science has had the greatest influence in increasing human productivity. By the application of science in technology, tools of all manner have been developed. Among them are machines (extensions of human muscles) and the power to run them, chemical and biological technology for everything from fabrics to vaccines, and communications devices and computers that augment our sensory and neural abilities. Compared to a few centuries ago, the wealth—the basic necessities and resources for lifestyle choices—that can be generated by each individual with these tools is immensely greater. As a result, more than 10 percent of the world's population lives a life of duration and comfort undreamed of in the past. Millions of people live more comfortable and healthy lives than any Pharaoh ever did. And, contrary to (suddenly outdated) Marxist rhetoric, this abundance has not been gained at the expense of the "downtrodden masses," as it often was in the past. On the contrary, the majority of people on this planet, though they live in poverty compared to the fortunate 10 percent, are materially better off than their ancestors.

The promise of science for a better life is, for most, an unfulfilled promise because almost all that has been gained in productivity has been lost to exploding population, particularly in regions of the world where, for some of the reasons mentioned above, the capacity to produce wealth has never been developed. The problem is not simply one of distribution. It is very unlikely that there is "enough to go around"—the lives of the majority of Earth's citizens would be little improved by evenly dispersing the material wealth, and it is clear that the productive minority would resist violently any attempt to do so. The only promising path is to make more of the world's people productive. The cost of this would probably be a less-rapid increase in wealth of the developed world, but staying the present course of escalating inequity is surely a road to civil strife, suffering, and destruction on a scale never before witnessed.

Of course, in recent decades we have learned that even a full utilization of *human* resources through science and technology will not allow limitless growth of both population and wealth, because we live on a planet of finite *natural* resources and fragile environments. Probably the Earth could support the present 6 billion humans with a relatively affluent lifestyle, given a careful system of resource management and conservation. If it were very tightly managed, our planet might support many times this number, but we might not like such a world. A highly engineered and strictly administered world of 20 or more billion people, one that keeps our few remaining fellow-creatures in zoos and preserves, and maintains a vast park system to retain some glimpse of our native environment, might be a material Eden but a spiritual wasteland.

Science, like nothing else humans have invented, empowers us to grow as a species and improve our collective lot. It may be hard to spread such optimism over a world where selfishness and inequity are fostered by some and accepted by most, but it is important to emphasize our ability to improve the choices and potential for each human life. The message is that we have not run up against any limits. Science can give us the power. The wisdom of command, as always, remains with us.

From science's role in shaping the present and immediate future, let's step back to a broader perspective. Creatures closely related to *Homo sapiens* have been around for a few million years, yet most of us believe we have little in common with people who lived only 20,000 years ago. Not true. For example, in terms of biology, the differences are unimportant—neither the brain nor body has changed substantially in this time. Yet we are different— we have evolved, culturally, through the miraculous invention of complex language, an ability developed in us far in excess of that in any other animal. How language came to drive our "evolution" is not the primary issue here; no doubt it is connected to the adaptation of walking upright, that freed the hand, that developed the brain to manipulate the hand, and so on. The real point is that the vast increase in our ability to communicate led directly to our rapid cultural evolution. We have been able to accumulate experiences not because our brains are capable of storing more memories, but because we can pass them along, by means of language, to other brains. Because of this we have been able to add the knowledge of one generation to another.

Brains were themselves a magnificent invention, one that well predated humans: brains allow the storage of information in addition to that encoded in genes in the sequence of nucleotide building blocks in DNA molecules. This vastly increases the possibility for complex action within an individual's lifetime; as experiences are absorbed and behavior modified, the creature learns how to better survive in a dangerous world. But human evolution cut a new trail in this development of central nervous systems when the development of language allowed the passage of learned experiences to future generations, which led to an accumulation of knowledge *within* brains.

Compound interest, no matter how small the rate, leads inevitably to great wealth. Thus we "learned" how to hunt, how to farm, how to trade— prospering in ever-larger communities and improving our chances for survival. In a biological world that is largely *reactive*, our species was the first to become truly proactive, to extensively reshape the environment to suit our needs. (So "successful" at this are we that it seems we may overshoot our mark and leave the world uninhabitable.) Our first outstanding applications of this ability were sociological in nature. During the last millennium, cultural evolution has been driven by intellectual inventions, including the science and technology that has amplified the power of our bodies, and now, our senses and minds.

Stephen J. Gould brought this into sharp focus in *The Panda's Thumb*:

> Cultural evolution has progressed at rates that Darwinian processes cannot begin to approach. Darwinian evolution continues in Homo sapiens, but at rates so slow that it no longer has much impact on our history. This crux in the earth's history has been reached because Lamarckian processes have finally been unleashed upon it. Human cultural evolution, in strong opposition to our biological history, is Lamarckian in character. What we learn in one generation we transmit directly by teaching and writing. Acquired characters are inherited in technology and culture. Lamarckian evolution is rapid and accumulative. It explains the cardinal difference between our past, purely biological mode of change, and our current, maddening acceleration toward something new and liberating—or toward the abyss.

The dizzying pace of this new kind of evolution, and what the maddening acceleration is leading to, are points worthy of further contemplation. In order to maintain any semblance of species identity and continuity, biological evolution had to build in a speed governor that admitted "a significant change" in the organism no more often than, say, every 10,000 or 100,000 generations.* For many species, particularly the larger life forms, this safety mechanism guaranteed timescales for evolutionary change that were glacial, in every sense of the word. Of course, it makes sense that biology and geology be interconnected at many levels, including timescales.

For humans, those days are gone. Our species is evolving—not biologically, but still in the true sense of the word evolving—on a timeline that is

---

*A "significant change" might be defined as something less than speciation (a new species = breeding incompatibility), but not much less. Whether this change occurs gradually over 10,000 generations, or once suddenly every 10,000 generations or so—a matter of current debate in evolutionary biology—has little bearing on the point under discussion.

closer to *exponential* (the way cells divide when they double in number every fifteen minutes) rather than linear. How can such "maddening acceleration" continue? Where will it end? Consider the search for scientific knowledge, a strong driver of this rapid evolution. Will this enterprise continue to grow without limit? In his book *The Endless Frontier*, which established the guidelines for the National Science Foundation, Carnegie Institution president Vannevar Bush implied that it would.

I do not think so. It is natural to extrapolate from one's own time to the indefinite future, as if the "modern era" has finally been reached and will continue forever. Not surprisingly, then, most scientists see the process of scientific exploration as unbounded—like the breaking of the atom into ever more ultimate particles, a box, within a box, within a box. But our new view of cosmology should help us consider a quite different perspective. The universe, the volume of space-time that conforms to the rules we understand, appears to be *finite*. In all likelihood, *our* universe contains all that we will ever know; the *universe* of other realms of existence that lies beyond is inaccessible, probably forever. If this is true, then it is not outrageous to believe that we can know "all there is to know," though one must be careful here to distinguish the rules of the game from the game itself. In the sense I mean it here, *knowing* means decoding nature's vocabulary and grammar, from which we can read and understand all that she has written. It does not mean complete comprehension of all the literature; the great complexity of the language will guarantee that there will always be new volumes to read, with new dialects and new expressions combining familiar concepts in enlightening ways, but we would have "cracked the code."

Many scientists regard physics as the basic rulebook. Here on Earth, in a small energy range, nature has written many plays describing the interaction of the different elements (we call it chemistry) and volumes of involved histories where the characters are organic molecules (we call it molecular biology or biochemistry). Though we are just beginning the immense task of reading and understanding these texts, there is no fundamental limit to our doing so—the root language is known and the dialect is becoming comprehensible. Only in physics itself is there a domain—at big bang energies—for which the basic rules are not at all well known. Already, however, we have educated guesses that succeed in small but important ways in characterizing the physics of the big bang, and in the next century we are very likely to achieve a model that accounts for the evolution of our universe. Were that to happen, we would have described the foundation of nature *inasmuch as is possible*, over the entire range of energies and densities our universe has ever, or will ever, have. What will be left is the *expression* of these rules in a myriad of physical situations, which, although it could require enormous human endeavor, is, in principle, tractable and achievable.*

It may be, then, that humans will have to come to grips with the idea that there are limits to what they can know. What came before the big bang, and what lies beyond the bubble of our universe, are likely to be fundamentally unanswerable questions, at least within the framework of science, because they deal with a reality so different that we will be unable even to pose questions for which answers can be sought. We should not be disheartened, however, because the vast range of expression of nature will keep scientists occupied for a long time to come, and all of us entertained and amazed, *forever*.

I suggest, then, that the present era is a phase that germinated some 2,500 years ago, blossomed within the last 400 years, and will come to fruition, rather than continue into the indefinite future. Specifically, I believe that the accelerating pace of the exploration of nature will lead to a maturation of scientific endeavor over the next few centuries. When a scientific field becomes mature, as some already have, investigation settles into a sober, meticulous effort to understand the finer points; the surge of activity and discovery that characterized the growth period ends. Even the vast areas of science that are almost untouched, for example, understanding the biochemistry and processes of brain function, could be mastered in the next century. The universe is vast beyond our comprehension, but it is repetitive in its forms; that is its hallmark. To understand all its *examples* of the expression of nature's rules is to "know" it, in a fundamental sense, entirely—this is the purpose that we are about. Even if we encounter other life forms, we should not be surprised to discover that they have much in common, in a broad sense, with life on Earth.

If we are willing to admit the possibility that an accelerating pace of scientific exploration will succeed in its goal of modeling a finite world, it is fair to ask what will come next. An inflationary epoch must continue until there is *fundamental* change. This revolution, I believe, will be the end of a

---

*One must not draw from this the idea of determinism, that with a complete knowledge of the rules we can predict all that can and will happen. Many of the systems of rules, like quantum mechanics, have fundamental limits on specificity, and even the evolution of complex *macroscopic* systems involves random and chaotic behavior. Knowledge of "all the rules" doesn't lead to determinism, just a description of the range of possible outcomes and the probabilities associated with each. Whether we wish it or not, life and all its experiences will remain fresh and unpredictable.

single humanity and the emergence of new, descendent life forms—the same course biological evolution would have followed, in spirit if not in character, had we not usurped its role.

We can already see three types of human activity that will pressure this kind of radical change. One involves an emerging power to rework our biology directly, by genetic engineering. The prospect is more than frightening to most of us—nightmares of mutant monsters are etched into our minds from ancestral memories of the primal forest, and are reinforced by ever-more-effective horror movies. In all probability, the first steps in genetic engineering in humans will involve felicitous, subtle changes that improve health and lengthen life—most of us will find it hard to resist these "cures" when they are offered (though some may). In time, these experiments will almost certainly lead to genetically engineered animals (new plants are already a reality), and though we humans will probably vigorously oppose the idea of tinkering with our own kind, sooner or later an "animal" that looks different enough from us to be an acceptable subject for experimentation may be endowed with mental abilities that rival or even exceed those of present-day people. Regardless of whether or not this creature resembles a human, or whether we eventually break the rules "just a little" in order to produce superathletes, people who function better in zero gravity, or creative super-geniuses, it is likely that humans will "evolve" into, or create, independent forms of sentient life that will be, in reality, our descendants.

A corollary to this hyperspeculation concerns the role of computers in our evolution. For most people, computers are still little more than fancy adding machines or typewriters, or, at best, electronic file cabinets or book-keeping robots. This is, in fact, where one starts in the attempt to build something that functions like a brain, but, again, it is inevitable that these efforts will mushroom in complexity and subtlety until such "machines" challenge and surpass the hardware of the human brain. Even if we find it difficult to decipher and replicate the kind of software in human brains that makes us "conscious," we could at least augment our brains with additional hardware—this alone is likely to lead to startling increases in human brain capabilities. More than likely, however, we *will* figure out how to endow future computers with intelligence like our own, and the union with human hosts is certain to have explosive consequences. Our descendants, if part-human, part-machine (including body parts?), might well feel they have less in common with us than we have with Neanderthals.

It may be difficult to comprehend the circumstances that would make this kind of change acceptable, let alone desirable. Humans have developed with a strong sense of xenophobia, a trait that must have been crucial for the early survival of our species. Our inability to eliminate this psychological baggage has, in the modern world, led to continuous strife between racial and ethnic groups. No wonder that, if we are unable to overcome fear and distrust based on such a trivial difference as skin color, the prospect of far more substantial differences among humans evokes in most of us a terrifying image of a chaotic, perilous world. On the other hand, while we can scarcely imagine all of humanity uniting to direct and control such evolution, isn't it even harder to imagine controls so airtight that all such adventures could be stopped absolutely? And there is likely to be at least one strong driver for such change: sooner or later, barring a biological or environmental disaster that sends us irreversibly back to the Stone Age, some humans are going to leave this planet. For those who attempt this ultimate adventure, the pressure to change the very nature of humans will be intense.

The imperialist mentality has, for centuries, lived by the myth that colonization leads to outposts of dutiful, homeward-looking citizens who faithfully feed and serve the mother country. In fact, never in human history has this expectation been long fulfilled. Resistance to this kind of servitude, for a variety of commonsense reasons, always has led settlers to recoil from what James Joyce called "the dead hand of the past," to find their own way. It is particularly amusing, therefore, that the popular paradigm for space colonization repeats this same old myth, envisaging spaceships ferrying a lively commerce between loyal colonies, happily joined in a "federation" whose benevolent leadership resides (where else?) on Earth.

Some such attempt may be made, but it is more likely that, sometime in the next millennium, people will begin to leave Earth (or its vicinity) in the twenty-eighth-century equivalent of wagon trains or immigrant ships. This would become possible, of course, only if science and technology have evolved the inexpensive construction and powering of spacecraft. There is some reason to think that this will happen: over the last two centuries the cost of fabrication and energy has plummeted. As crazy as it may sound to us, continuation of this trend for any significant length of time should make commercial or even "amateur" efforts feasible.

People embarking on thousand-year journeys to the planets of other star systems are likely to choose hibernation or genetic reconstruction in preference to raising tens of generations en route.* One way or another, such

---

* Space exploration by robot ships opens up enormous possibilities, including smaller, faster, self-regenerating probes that could spread throughout the galaxy in less than a million years—less than the present "age" of our species. Humans might prefer this as a more efficient, safer way to explore the Milky Way, but it would also add a new dimension for spreading earth-life throughout the galaxy. With such remote

journeys seem possible. Their dislocation from Earth, in time as well as space, will free these pioneers to be sole masters of their own destinies. History tells us that, if any humans are likely to be bold enough to alter the biology of mind and body, these will be they: the inducement to adapt into the niches offered by new worlds will be irresistible. In this way, relatively small groups of humans with a compelling incentive for change and the possibility of maintaining some kind of homogeneity within their own number could choose to be the architects of their new forms.

Thus, either here on Earth, or out among the stars, humans may be destined to evolve a variety of sentient creatures who will carry on the exploration of the hyperspatial domain. Of course, even our survival as a species, or at least its continuation in anything like the present direction, is still threatened by global war or biological/environmental holocaust. (Perhaps the strongest argument for a few extraterrestrial settlements is to secure the survival of humanity's hard-won accomplishments in art, culture, and science against the insanity of "autogenocide.") But, if humanity does survive its adolescent phase, it will face strong pressures for *speciation* that could be defeated only by deliberate, absolute controls designed to prevent it. This runs directly counter to the major trend in cultural development in the last few centuries toward empowering individuals and groups with greater self-determination. Even with a consensus to "keep the world as it is," it seems highly improbable that the necessary authoritarian structure could be implemented and enforced, and that the other implications of such controls would long be tolerated.

It is my conclusion, therefore, that we are most likely near the end of what we have known as humanity. Nature's gifts to us have led to the secret keys of evolution, and we are not likely, I think, to long refrain from unlocking this box of treasures and troubles. Coping with the coming cataclysm may be too much for us to contemplate, but our descendants may be better able, the way we have grown accustomed to a world that our great-great grandparents would have judged to be on the verge of anarchy and mass insanity. If such revolutionary change comes, then these "descendants" will probably look back upon ours as an age of awakening—the three-millennia journey from Athens to the Mars colony will appear a short one, and those who took this journey may be regarded with considerable reverence and

outposts in place, it would be possible to transport humans and other biological creatures in "coded" form, for example, by carrying along real cells with DNA or just the stored information of DNA sequences (which could even be "beamed" by radio waves to a new site), and then constructing biological organisms *in situ*. There seems to be no fundamental impediment to this alternative form of space exploration.

admiration. Left from this age will be a million voices of those who recorded the glory and torment of search and discovery, of the world, of the universe, of *humanness* itself—in art, in words, in music. Many will think us lucky to have been a part of the greatest human adventure.

●○

From this perspective, that the human adventure could be radically redirected in the not-too-distant future, I want to step back once more for a still-broader, cosmic perspective. Where does this story "fit" into the evolution of the universe itself?

Life invented brains and sentience, but the universe invented life. The essence of life is complexity, on a scale unknown in "the physical universe." Stars with their nuclear chemistry may seem complicated. Galaxy ecosystems, with their stellar births and deaths enriching and recycling the chemical reservoir, display an intricate mechanism, to be sure. The early phases of the big bang, shadowing complicated relationships of particles and energy fields, remain a puzzle. Yet these pale in complexity compared to what has happened here on Earth in the evolution of the simplest biological systems.

It had to be so. If the defining characteristic of life is replication, its essential ingredient is fantastic variation—an architecture built on complex chemistry. However, chemical complexity is not supported in the vast majority of the universe's environments: stars are too hot to allow combination of atoms into molecules, rocks are too cold to induce chemical reactions within their icy clumps of organic material, and gas clouds are so thin that atoms rarely meet. But complexity is the essential agent that life must have to explore its surroundings and create, through experiment, new combinations that fuel replication and proliferation. The Earth environment slices across just that infinitesimal range of temperature and density where atoms embrace to form molecules with a true penchant for chemical promiscuity—the number of combinations with just carbon, nitrogen, oxygen, and hydrogen alone is staggering.

Young Earth was a kettle of primal soup, a stock of nature's best solvent, liquid water, and chunky bits that included hydrocarbons, organic acids, and salts. The simmering broth was further stirred by lightning, copious ultraviolet rays from the Sun, and geothermal activity. Cooks whose recipes simply call for emptying the contents of refrigerator and cabinets into a pot can count more failures than successes, but once in a great, great while, the results can be spectacular. A billion years after the Earth formed, nature stumbled upon the dream of every working parent—a recipe that cooks itself.

Just as the inflation epoch of exponential expansion cut a path-of-no-return for the physical evolution of the universe, the invention of complex molecular chains able to replicate themselves built a one-way road out of the world of "natural" chemistry. The processes of pre-life chemistry could now be concentrated, amplified, and channeled to build a hierarchy of astounding complexity. Life began—the rest, as they say, is history. The law of this new world order was natural selection, the simple rule that those best adapted would thrive and best succeed in reproduction. To succeed was to survive, and to survive was to succeed. This would be the "way of life" until the next invention, brains, led to another breakaway, in the form of a creature that controlled its own evolution—us.

The chemistry of young Earth had cast about until it "invented" life— something totally new. In a way, the universe itself, by "experimenting" with such a wide range of environments, had done the same thing when it "hit upon" subtle, benign environments like Earth, where complexity could be nurtured. These were the steps—the universe had cooled and produced stable hydrogen atoms, compressed these into stars that assembled heavier atoms, and, in the process, warmed planetary wombs where complexity could be nurtured—that led to our emergence and our attempt to retrace those very steps. The universe had in effect discovered a way to fight the seemingly inevitable triumph of disorder, the so-called "heat death." With so many more disordered states to choose from than ordered states, the natural evolution of any system is toward increasing disorder—scientists call it increasing *entropy*. The universe found a tiny fraction of environments where life— statistical impossibilities representing the ultimate in "order"—could be delivered in phenomenal amounts. The march toward the abyss could not be stopped on a large scale, perhaps, but locally the battle could be fought and won convincingly. In the end, these pockets of increasing complexity and order will make the heat death of the universe irrelevant.

Some see the unlikeliness of all this happening as evidence for a grand *design*, proof of a hand that guided this incredible sequence of events—the traditional role of God the Creator. Others simply say that some sort of chain of events like this had to ensue if creatures like us were to evolve to think about how unlikely it all seems. Regardless, the point is that the "evolution of the universe" has led to sentient creatures—at least one species we know of. The universe tends myriads of "cradles" like Earth; surely some others may have fostered marvels like those here.

Whether we are the first or the millionth civilization to become aware of its place in the universe, the knowledge should fundamentally change us—for the better. Since Copernicus, humanity has suffered the shocks of continual displacement from a central shrine of prominence and importance to a remote crag of insignificance and inconsequence. The road to alienation has led all the way to the hysterical description of human evolution as a senseless drama played on the stage of a hostile universe—the nightmare of existential nihilism.

We continue to take the wrong lesson from what we are now learning. An astronaut who had taken a tethered space walk while on a Gemini flight in the 1960s was recently asked whether the experience had changed him. He had been struck, he recalled, by how small and insignificant were the Earth and the human adventure, "like an ant crawling across the Sahara desert." Exactly. The ant, astronomically outnumbered by the grains of sand, overwhelmed by the size of the inhospitable desert, is nevertheless the greater marvel, by far.

Thirty years ago, one man orbiting the Earth saw human insignificance in comparison with the cosmos. Another man, though stuck on the ground, saw farther and more clearly. Martin Luther King, Jr., said: "As marvelous as the stars is the mind of the person who studies them." With this remarkable thought we can find our way to a new perception of our place in the universe. It is time to take full stock of the discovery that life is the most complex thing we know of in the universe, and, as such, most worthy of our admiration. Yes, the universe dwarfs our world in size and immense power. But the universe of stars, galaxies, and vast gulfs of space is so very, very simple compared to us and our brethren life forms. If we could but learn to look at the universe with eyes that are blind to power and size, but keen for subtlety and complexity, then our world would outshine a galaxy of stars. Indeed, we should marvel at the universe for its majesty, but we must truly be in awe of its greatest achievement—life. The universe has invented a way to know itself, a way to explore itself, to propagate subtle and intricate design throughout itself. The process of creation has been coming in our direction for more than 10 billion years, and now, with us and what others there may be like us, the flow is turned back to the universe, from whence it came. We are ready to explore the universe; whether we launch only our senses and our minds or send our bodies as well, we are bound now to take our gift out into the universe, perhaps to change it forever. We should feel at the center of our universe, for in a very real sense we are its point.

The process of discovering all this is the most amazing accomplishment in human history. Only once in the evolution of a sentient species is this corner turned, and we who are now alive are that most fortunate generation, the people of the awakening. Do the other universes outside our "bubble" have such stories to tell? We will probably never know. Perhaps it required a virtual infinity of universes until one arose with the proper physics to cast the bricks, to build the house, to set the fire and warm the cradle of life. We know these images—they are etched into our lives as humans. Now we must expand our mind's eye to carry them to the larger world from which we came.

Astronomy, with its searches like the voyage to the Great Attractor, is just one tiny part of the knowledge we need to do great things. We leave the twentieth century with burgeoning, godlike powers, though less-than-godlike wisdom of how to use them. We should be eager to seize the opportunity to enter the Promised Land our myths have so long told us about, a land of unlimited possibilities and abundance. And yet, at this critical juncture, there is a falling back from the brink, a reluctance to claim the reward of a thousand generations. Superstition abounds. Many, afraid to take charge of their own destiny, are willing slaves to astrology—a more popular and better-known subject than astronomy—and charlatans of every description cash in on our desire to tame the unknown. Many search the skies for aliens who will deliver us from our own wanton acts of self-destruction and cruelty. Others retreat to fundamentalist religions that promise to remove doubt and uncertainty at the cost of giving up the journey to new wisdom and a greater future.

While some may choose to live in a fantasy world, ignoring or rejecting outright the world of science and rationality, even more disturbing is the self-doubt of some scientists. A trade magazine for physicists carried a cartoon in which a "preacher" delivered the gospel of inflation and particle physics, while the congregation "testified" with such cries as "Verily, it was over $10^{32}$ degrees!" "Amen! Quarks," and "Hallelujah—They annihilated each other." Funny? Sure. But the sad truth is, many scientists have become reluctant to profess any real difference between their attempt to find out how we got here and the religious myths that preceded the age of science. "Relativism" has infiltrated our scientific endeavors as well as our cultural studies. As a society we have become increasingly afraid to claim any greater insight, any greater power, any real responsibility and opportunity for making the lives of all humans fuller and richer.

Jacob Bronowski, in his heroic book *The Ascent of Man*, bore in on this problem, citing what he called Western civilization's "loss of nerve," its falling back from the brink of its long-cherished dreams. The temptation to withdraw from our responsibilities of emerging adulthood, to retreat to a past when we were wards of a dark mystical world, threatens to paralyze us as we poise to take our most important steps. We must find our courage, regain our nerve, and press on in this the highest adventure.

Like a child, humanity awoke to find itself in a strange land, not knowing who we were or where we came from, and puzzled as to what we were to do. As parents often do with children, we created myths to satisfy the incessant questioning from this gift of ours, the mind, and then got along in the business of learning how to survive and prosper. But this long childhood is finally coming to an end; we are growing up and are ready to cope with the real answers and take responsibility for our own welfare and destiny. From cosmology to biology, what we are learning is that the universe is evolving, and we are evolving with it. The universe is about *process*, and we ourselves are part of that process of *becoming*. This is what it is all about. And, in the end, this realization will change us forever. We will become aware and in control of our powers and see the role we might play in the cosmos. And we will be ennobled in the attempt to join consciously, as we have unwittingly, for ten thousand generations, in the evolution of the universe. This is the challenge; these are the possibilities.

We are here, then, a bit less than steady on the deck of our tall ship, its sails filled with a near gale. The rising winds dare us to hold on and steel our resolve in the hope of reaching calmer seas and fresh horizons. As they have from ancient times, the stars guide our voyage, and they urge us on. They bid us every one to look hard into the night sky, at the beckoning Milky Way galaxy, and say: tell me what you are, that I might better know what am I.

# GLOSSARY

*absorption line*—a drop in intensity at a specific color in the spectrum of a star or galaxy, caused by the gobbling up of light by a specific type of atom or molecule in the atmosphere of a star.

*arcsecond*—a unit of size on the sky. Once around the sky amounts to 360°, each degree is divided into 60 minutes of arc, and each of these is divided into 60 seconds of arc. The diameter of the moon is about 30 arcminutes or 1,800 arcseconds. The typical size of galaxies in the elliptical survey was 1 to 2 arcminutes; the most distant galaxies have sizes of only 1 to 2 arcseconds, only a few times bigger than the blurring of optical images, the "seeing," caused by turbulent motions in the Earth's atmosphere.

*baryonic matter*—protons and neutrons, the components of "ordinary" matter.

*black hole*—a region of space with such a high density of matter that space is completely "wrapped around"; because of the intense gravitational field, even light cannot escape, hence the hole is "black." Not to be confused with the more generic term, *dark matter*.

*blackbody spectrum*—the distribution of intensity with wavelength (color or energy) for matter radiating light in (temperature) equilibrium with its surroundings. The temperature determines the wavelength at which the emission reaches peak intensity, and the exact relation of the intensity of light as a function of color—the spectral "shape"—is fixed to a certain form.

*cold dark matter (CDM)*—a model for invisible matter in the universe based on a "gas" of exotic, weakly interacting particles that were moving slowly compared to the speed of light when the embryonic fluctuations in matter density, which grew into galaxies, galaxy clusters, and galaxy superclusters, began to grow.

*cosmic microwave background*—the diffuse glow of light, in the microwave (radio) part of the spectrum, left over from the cooling big bang at an age of a few hundred thousand years.

*critical density*—the density of matter that would, through its self-gravity, exactly counter the dynamical expansion of the universe, bringing it eventually to a stop.

*delta-delta relation*—a correlation of properties from two other relationships comparing properties of elliptical galaxies. The delta-delta relation indicates that, for a sample of elliptical galaxies, an unusually high *velocity dispersion* (see below) for a given luminosity corresponds to an unusually high magnesium line-strength, and vice versa.

*density fluctuation*—a region where the amount of matter per unit volume is slightly higher or lower than average. Somewhat analogous to a high- or low-pressure zone in Earth's atmosphere.

*Doppler shift*—the motion of a source of light (or any kind of wave) will result in a shift to longer wavelengths if the source is receding, and shorter wavelengths if the source is approaching.

*elliptical galaxy*—a system of 1 to 100 billion stars bound together by gravity. Ellipticals have basically round shapes, but they typically have a long and a short axis that may differ by a factor of two; their true shapes are either like onions or footballs, or some of both. Unlike spiral galaxies, they have no thin disks of stars, gas, and dust and appear to have formed few new stars in the recent past. The stars in an elliptical have a wide variety of orbits in various directions, leading to an overall appearance of random motion.

*emission line*—a rise in the intensity of light at a specific color in the spectrum of a star or galaxy, caused by the preferred emission of photons of certain energies by a specific type of atom or molecule in a heated cloud of gas.

*expansion* (or *recession*) *velocity*—the speed at which a galaxy is receding (from our vantage point) as part of the overall expansion of the universe (Hubble expansion).

*hot dark matter (HDM)*—a model for invisible matter in the universe based on exotic, weakly interacting particles, for example, neutrinos that have a small mass, that were moving at near-light speed when the first galaxy and cluster-sized mass fluctuations began to grow.

*Hubble expansion*—the observed characteristic of the universe that galaxies are all receding from each other, with speeds that are in direct proportion to their separations. Thus, a galaxy twice as far away will have a recession velocity twice as fast. Galaxies do not actually move *through* space, but rather it is space itself that is expanding, carrying galaxies with it like sequins pasted on an expanding balloon.

*light-year*—an astronomical measure of *distance*, that which light travels in a year.

*Local Group*—a loose grouping of three large galaxies, the Milky Way, Andromeda, and Messier 33, the Large and Small Magellanic Clouds, and a swarm of dwarfish galaxies.

*Local Supercluster*—the nearest supercluster, a relatively small one, that contains the Virgo and Ursa Major clusters and a diffuse halo of surrounding galaxies. The Local Supercluster has a distinctively flattened shape, resembling almost a giant galaxy of galaxies. The Local Group, and hence, the Milky Way, is a probable member of the Local Supercluster, near the perimeter of this giant system.

*luminosity*—the true, absolute brightness of a star or galaxy, as distinguished from its *apparent* brightness, which depends on distance.

*magnesium line-strength*—a measure of the fractional abundance of heavy elements in the stars responsible for a galaxy's light. Absorption lines in the spectrum grow stronger for greater abundance of elements heavier than hydrogen and helium, of which magnesium is a representative, so the strength of magnesium lines in the spectrum serves as a measure of the overall abundance of heavy elements.

*magnitude*—the astronomers' measure of brightness. In this logarithmic scale each factor of 100 drop in brightness corresponds to an additional five magnitudes. The brightest naked-eye stars are first magnitude (there are a couple a bit brighter), and the faintest naked-eye stars are sixth magnitude. The faintest galaxies yet recorded are 100 million times fainter, about twenty-sixth magnitude.

*Malmquist bias*—the effect of systematically underestimating the distance to a galaxy that arises when distance predictions are not precise. Because the sampled volume of space increases rapidly as one looks to greater distance, brighter, more distant objects incorrectly included as within the survey volume outnumber the fainter, closer objects that are incorrectly excluded as lying beyond the boundary. Thus, the sample is "biased" to include more distant galaxies than a fair sample, so the distance to the average galaxy is underestimated.

*N-body (computer) simulation*—following the growth of structure through a computer program that calculates the gravitational force between N bodies (where N is a large number) representing the total mass.

*peculiar velocity*—motion that a galaxy has in addition to the Hubble expansion of the universe, as can be induced by gravitational acceleration.

*photon*—packet of electromagnetic energy, synonymous with *light*.

*recession velocity*—see *expansion velocity*.

*redshift*—the shift of all the light in a spectrum toward longer wavelengths (redder colors) caused by the Doppler shift.

*rotation velocity*—the orbital speed of stars in a spiral galaxy. A spiral galaxy does not rotate as a solid body (it does not turn as a unit), because a star's orbital speed is fixed by the gravity: the amount of mass enclosed by the orbit and the distance to the center of the galaxy. A "rotation curve" plots orbital speeds of stars as a function of distance from the center of the galaxy. Such rotation curves usually reach a maximum (the stars at this point are circling with higher speed than those on closer-in orbits) and level out for as far as can be measured. This maximum speed is often called, simply, the rotation velocity.

*spectrograph*—an instrument that splits light into its component colors. Abundant information on the physical conditions at the source of the light can be deduced from a detailed analysis of the amount of light at each color, particularly from the presence of *absorption* or *emission* lines (see above).

*spiral galaxy*—a galaxy, like our own Milky Way, made up of some 1 to 100 billion stars bound together by gravity. The spiral pattern, found in the thin disks of stars, gas, and dust that surround a spheroidal "bulge" of stars, is highlighted by sites of continuing star birth. The stars in the bulge swarm in many directions, as in the closely related elliptical galaxies, while those in the disk trace near-circular orbits around the center of the galaxy.

*stellar kinematics* (or *dynamics*)—a description of the motions of the stars in a cluster or galaxy, including their speeds of motion, and shapes and distribution of their orbits.

*stellar population*—the mix of stars of different masses, with their various temperatures, metal abundances, and ages, that makes up a large system like a star cluster or galaxy.

*supercluster*—a collection of thousands of galaxies over a region with a diameter on the order 100 million light-years. The overall density is low, only a few times the average density of the universe, but the contrast with nearby voids, which are several times less dense than average, can be quite striking. Often, superclusters contain one or more rich clusters of galaxies, higher-density knots of several hundred galaxies, some 10 to 20 million light-years across.

*Supergalactic Plane*—a collection of many prominent local superclusters into a flat distribution that stretches for at least several hundred million light-years. The Supergalactic Plane includes the Local Supercluster, the Hydra-Centaurus and Pavo-Indus superclusters, and the Perseus-Pisces supercluster, among others.

*surface brightness*—the brightness per unit area of an image, for example, the magnitude per square arcsecond of a galaxy within a certain diameter.

*velocity dispersion*—a measure of the random motions of stars in a bound system. In a spheroidal distribution of stars, like an elliptical galaxy, stars move with a range of speeds on a variety of orbits in many different directions. The velocity dispersion is a measure of the typical speed of these millions of stars; it can be measured from a spectrum of their combined light. Small Doppler shifts in the colors of each absorption line, due to the motions of the individual stars, broaden the absorption line in proportion to the velocity dispersion.

## Symbols

*1/f* noise—the particular distribution of fluctuations where the total "power" is directly proportional to the scale size. Seen commonly in nature, as in the case where the number of peaks in a mountain range decreases in a predictable way for higher and higher peaks.

*omega* ($\Omega$)—the density of the universe compared to the critical density at which the energy of the expansion exactly balances the gravitational energy. $\Omega > 1$ is the case where there is more gravitation energy than expansion (kinetic) energy: the universe will eventually recollapse. $\Omega < 1$ is the case where expansion energy is greater than gravitational energy: the universe will continue to expand forever.

# ACKNOWLEDGMENTS

The author gratefully acknowledges the many people who contributed to the writing of this book. First and foremost, of course, are my six colleagues, Dave Burstein, Roger Davies, Sandy Faber, Donald Lynden-Bell, Gary Wegner, and Roberto Terlevich, whose work forms the heart of this book, and who revealed something about their personal lives for the biographies included here.

Much of the manuscript was written at the University of California at San Diego. I wish to express my great appreciation to the astronomy and physics departments, and the Center for Astronomy and Astrophysics, for their generous support, and in particular to Arthur Wolfe, Geoffrey Burbidge, David Tytler, and Ken Lanzetta for helping me think through parts of the text. The photographic work on the illustrations was carried out with artistry by Steve Padilla. And of course, this book could not have been undertaken without the unfaltering support I have received from the Carnegie Institution—Carnegie's encouragement and commitment to my scientific work has made all the difference.

Several people gave a great deal of their time to reading the manuscript and working with me to improve it: Sandy Faber, Dave Burstein, Ann Shipley, George Lake, Russell Galen, David Dressler, and Wendy Boren—I am deeply appreciative of all their efforts. But no one contributed more than my editor at Knopf, Jonathan Segal, who read and reread, argued and criticized, and challenged me to achieve what I set out to do, to talk to people about science.

# INDEX

Note: Page numbers in *italics* refer to illustrations

Aaronson, Marc, 153, 154, 165, 167, 189, 198, 203, 216, 258; death of, 260; and expansion of universe, 209

Aarseth, Sverre, 228

Abell, George, 62, 63, 140, 234, 235

Abell cluster, 144

absolute zero, 93 *n.*

absorption line, 96, *97*, 98, 101–4, *105*, 156, 176; definition of, 339

acceleration model, 225, 238–54. *see also* Great Attractor model

Alpher, Ralph, 267–70

Ambartsumyan, V.A., 213

American Astronomical Society, 157, 209, 243, 313

American Physical Society, 253

Andromeda galaxy, 12, 14, 19, 21; distance from Earth, 27, 48–9; Great Spiral galaxy, 13, *15;* rotation curve, 214 *n.*

Anglo-Australian telescope, Siding Spring Observatory, 18, 55, 106, 154, 157, 190

angular momentum, 70

Antares (star), 103

anthropic principle, 132

antimatter, 288, 299

Antlia cluster, *250*

Apollo space missions, 8–11, 44,54

Aristarchus of Samos, 112, 113

Aristotle (philosopher), 9

Armstrong, Neil, 52

Ascent of Man, The (Bronowski), 336

Astronomical Society of the Pacific, 110

*Astrophysical Journal,* 167, 240, 251

AT&T Bell Laboratories, 269

atoms, 217, 222, 223, 255, 271, 289, 299, 306

Attracting mass model. *see* Great Attractor model

Babcock, Horace, 85, 214 *n.*

backfalling, 189, 315

baryonic matter, 217–24, 255, 280, 299, 339

baryons, 225, 231, 291

Bertschinger, Ed, 256, 306, 316–17, 320

beta-decay, 223

Betelgeuse (star), 103

biasing, 255, 304

big bang model, *x, xi,* 5, 33, 48, 60, 62, 114, 122–3, 139, 157, 220–5, 254, 267–73, 280, 286, 292–5, 298, 307; glow of, 267; vs. steady state theory, 265, 266, 271, 272

Big Dipper, 10, *11,* 12, 20

blackbody, 97, 270, 274

blackbody spectrum, 270, 271, 274, 275, *276,* 307; definition of, 339

black hole, 17, 121, 127, 218, 300; definition of, 339

blueshift, 29–30, 104, 202 and *n.,* 214

Blumenthal, George, 228, 236, 255, 306

A NOTE ABOUT THE AUTHOR

Alan Michael Dressler was born in Cincinnati, Ohio, on March 23, 1948.
He received a B.A. in physics from the University of California, Berkeley,
in 1970, and a Ph.D. from the University of California, Santa Cruz, in 1976.
From there he went to the Hale Observatories (formerly Mount Wilson and
now Carnegie Observatories) in Pasadena, California, on a postdoctoral
Carnegie Fellowship. Dressler became a member of the scientific staff of the
Observatories in 1981, and has lived in nearby Altadena since that time. He
received the Newton Lacey Pierce Prize of the American Astronomical
Society in 1983 and was elected to the American Academy
of Arts and Sciences in 1993.

A NOTE ON THE TYPE

The text of this book was set in a face called Times Roman,
designed by Stanley Morison for *The Times* (London),
and first introduced by that newspaper in 1932.
Among typographers and designers of the twentieth century,
Stanley Morison has been a strong forming influence, as typo-
graphical advisor to the English Monotype Corporation, as a
director of two distinguished English publishing houses, and
as a writer of sensibility, erudition, and keen practical sense.

Composed by Crane Typesetting Service, Inc.
West Barnstable, Massachusetts
Printed and bound by Quebecor Printing Martinsburg,
Martinsburg, West Virginia
Designed by Iris Weinstein